W0235263

Practices, Perceptions and Prospects for Climate Change Education in Africa

Marcellus Forh Mbah · Petra Molthan-Hill ·
Ernest L. Molua
Editors

Practices, Perceptions and Prospects for Climate Change Education in Africa

Editors
Marcellus Forh Mbah
University of Manchester
Manchester, UK

Petra Molthan-Hill
Nottingham Trent University
Nottingham, UK

Ernest L. Molua
University of Buea
Buea, Cameroon

ISBN 978-3-031-84080-7 ISBN 978-3-031-84081-4 (eBook)
https://doi.org/10.1007/978-3-031-84081-4

This work was supported by University of Manchester.

This Springer imprint is published by the registered company Springer Nature Switzerland AG
The registered company address is: Gewerbestrasse 11, 6330 Cham, Switzerland

Foreword

The science is unequivocal that climate change continues to worsen, as reported by key flagship global climate assessments, including by the Intergovernmental Panel on Climate Change (IPCC). One of its peculiar consequences is its disproportionate affliction on the continent. Africa hosts up to 17% of the global population but accounts for the least emissions at only 3%–4%. However, it stands out for being disproportionately vulnerable to the changing climate because of highly variable weather and a low socio-economic base. For example, Africa already heating up twice as fast as the rest of the globe, sees progressive temperature increases that take a far worse toll. In the past 50 years, drought-related disasters have claimed over half a million lives and triggered up to $70 billion in economic losses. No other region of the globe has experienced such loss within this period. Agricultural productivity growth has declined by 34% since 1961—the highest globally. Sea level rise in Africa is projected to be higher than the global average, especially along the Red Sea and along the western Indian Ocean. Cumulatively, while climate change is projected to shrink global economies by about 23%, incomes in the poorest 40% of countries, most of which are in Africa, are projected to fall by a drastic 75%, further widening inequalities. While climate change effects are global, the poor are disproportionately vulnerable to its effects because they lack the resources they need to afford the goods and services to buffer against the worst of the changing climate effects. This means that effective responses will centre on the simultaneous realisation of socioeconomic imperatives.

Education stands out as a core building block to this end. From enhanced pro-climate beliefs to behavioural changes and pro-climate policy preferences, the impact of education in catalysing structural shifts towards a climate-positive development is undeniable. One study has found that having up to 16% of secondary school students across the globe study climate change would result in cutting almost 19 gigatons of CO_2 by 2050, especially in high-emitting countries. In this book, the authors have provided deep insights into different aspects based on the foundational premise that education is a cornerstone to equip individuals with climate literacy and a deeper understanding of climate change, which is essential to harness their potential and capability for Africa's climate response. It makes a case for mainstreaming ecocentric perspectives in sustainability in business and management education in universities. It presents innovative and structured frameworks designed to systematically address climate change education across diverse educational levels in Africa. It makes a case for mainstreaming climate change mitigation and adaptation across all courses at African universities. It explores the reality of climate change impacting different genders differently and the need to actualise gender-transformative outcomes in climate change responses. In this tech-driven era, it makes a case for leveraging ICT as a tool for enhancing climate change education in Africa. It gets deep into the theory of education and examines educational pedagogies and approaches needed to foster an effective climate change education, to help develop sustainability and adaptability mindsets in Africa. Considering the importance of early warnings, the book further explores the importance of embracing climate change education in Africa's Regional Platforms for disaster risk reduction (DRR). This book is a worthy read that applies both case studies and theoretical explorations to make a solid case for increased investment and prioritisation of climate education as a critical pillar for strengthening Africa's climate change response. This is a worthy, timely, and interesting read.

Nairobi, Kenya

Dr. Richard Munang
United Nations Environment
Programme (UNEP) Head of Global
Environment Monitoring Systems

The original version of the book has been revised. A correction to this book can be found at https://doi.org/10.1007/978-3-031-84081-4_18

Acknowledgements

This book is part of a project funded by Research England via Nottingham Trent University (NTU) on the Indigenisation of Climate Change Education in Africa. While the empirical work that led to this output was supported by NTU, open access was made available by the University of Manchester in England. We are grateful to both universities for the valuable support.

The realisation of this book has been a team endeavour and profound gratitude goes to the editors and chapter authors who responded to the call. Thank you for sharing your in-depth insights and analysis of your work. We are particularly elated for the diversity of contributions from authors from a range of countries and disciplinary backgrounds focusing on different facets of climate change education in the continent of Africa.

Every chapter manuscript went through rigorous peer review and to the many reviewers who took time out of their busy schedule to ensure accepted contributions were of high standard, we say "Thank You!".

Contents

Editors and Contributors

About the Editors

Marcellus Forh Mbah is an academic in Manchester Institutre of Education and the Race and Equality Lead in the School of Education, Environment, and Development (SEED) at the University of Manchester. Before joining the Institute, he held teaching and research positions in Bournemouth and Nottingham Trent. He obtained a Master's degree in Educational Leadership and Management from the University of Nottingham in 2011 and a bachelor's degree in Management from the University of Buea in 2005. He obtained a Ph.D. from Canterbury Christ Church University in the broader subject of Education and sub-area of Development Education in 2015. He is particularly interested in researching the role of Indigenous Knowledge Systems within the nexus between Education (broadly defined) and Sustainable Development, with a key focus on SDG 13 (Climate Action) and other SDGs.

Prof. Petra Molthan-Hill Ph.D., is the co-chair of the UN PRME Working Group on Climate Change, a professor at NTU (UK), and an author of the 'Handbook of Carbon Management.' She is an international multi-award-winning expert for Climate Change Mitigation Tools and leads the 'Climate Literacy Training-CLT-ECOS' distributed worldwide with QS Impact and UN PRME, which won the QS Reimagine Education Award in Sustainability (Gold) in 2022.

Ernest L. Molua is a Professor of Agricultural Economics in the Faculty of Agriculture and Veterinary Medicine, University of Buea, Cameroon. He obtained academic degrees in Agricultural Economics and Policy Analysis from the Georg-August University, Göttingen, Germany and the Royal Veterinary and Agricultural University, Copenhagen, Denmark. He is a member of the College of Research Associates of the United Nations University Institute for Natural Resources in Africa, a continent-wide network of senior research scientists for natural resources management. A Fulbright Research Fellow at Yale University, USA, he also serves as a Visiting Professor to the United Nations Institute for Economic Development and Planning in Dakar, Senegal and the Africa Program of the United Nations University for Peace in Addis Ababa, Ethiopia. He is a member of the International Association of Agricultural Economists, the African Association of Agricultural Economists, as well as the Cameroon Association of Agricultural Economists.

Contributors

Mufti Nadimul Quamar Ahmed Department of Sociology, Anthropology and Criminal Justice, Utah State University, Logan, Utah, United States

Kemi Funlayo Akeju Economics Department, Ekiti State University, Ado Ekiti, Nigeria

Edidiong Samuel Akpabio Department of Political Science, Trinity University, Yaba, Nigeria

Nyong Princely Awazi Department of Forestry and Wildlife Technology, The University of Bamenda, Bamenda, Cameroon;
FOKABS, Ottawa, Canada;
FOKABS, Yaounde, Cameroon

Henry Ngenyam Bang Department of Disaster and Emergency Management, School of the Environment, Coventry University, Coventry, UK

Mariusz Baranowski Faculty of Sociology, Department of Sociology of Social Stratification, Adam Mickiewicz University, Poznań, Poland

Bruno Borsari Department of Biology, Winona State University, Winona, MN, USA

Bhagwan Das Centre for Artificial Intelligence Research and Optimization (AIRO), Design and Creative Technology, Torrens University, New South Wales, Australia

Moses Metumara Duruji Political Science Department, Covenant University, Ota, Nigeria

Osuji Emeka School of Management and Social Sciences, Pan Atlantic University, Lagos, Nigeria

Sophie E. Etomes Department of Educational Foundation and Administration, Faculty of Education, University of Buea, Buea, Cameroon

Chidi Ezegwu Manchester Institute of Education, School of Environment, Education and Development, University of Manchester, Manchester, UK

Zaina Gadema Newcastle Business School, Northumbria University, Newcastle, UK

Jennifer E. Givens Department of Sociology, Anthropology and Criminal Justice, Utah State University, Logan, Utah, United States

Brian Harman Nottingham Business School, Nottingham Trent University, Nottingham, England

Tony Jan Centre for Artificial Intelligence Research and Optimization (AIRO), Design and Creative Technology, Torrens University, New South Wales, Australia

Lionel P. Kemeni Kambiet Mohamed VI Polytechnic University, Ben Guerir, Morocco

Perez L. Kemeni Kambiet International Water Research Institute (IWRI), Mohammed VI Polytechnic University (UM6P), Ben-Guerir, Morocco

Helen Kopnina Newcastle Business School, Northumbria University, Newcastle, UK

Christopher Liberty Ashinaga Uganda, Kampala, Uganda

Thato Majola University of Botswana, Environmental Education Unit, Gaborone, Botswana

Wilson Muyinda Mande Nkumba University, Entebbe Nkumba, Entebbe, Uganda

Marcellus Forh Mbah Manchester Institute of Education, School of Environment, Education and Development, University of Manchester, Manchester, UK

Tracey J. M. McKay Olympus Online Education, Gauteng, South Africa

Petra Molthan-Hill Nottingham Business School, Nottingham Trent University, Nottingham, England

Ernest L. Molua Department of Agricultural Economics and Agribusiness, University of Buea, Buea, Cameroon;
Centre for Independent Development Research, Buea, Cameroon

Marco Alberto Nanfouet Faculty of Agriculture, Institute for Food and Resource Economics, University of Bonn, Bonn, Germany

Liz Nantunda Newcastle Business School, Northumbria University, Newcastle, UK

Francis E. Ndip Centre for Development Research (ZEF), Bonn, Germany

Sihle Ndlovu GARD Division, Higher Colleges of Technology, Abu Dhabi, UAE

Tabani Ndlovu Faculty of Business, Higher Colleges of Technology, Abu Dhabi, UAE

Neo Scholarstic Ntirase University of Botswana, Environmental Education Unit, Gaborone, Botswana

Evans Olaniyi School of Management and Social Sciences, Pan Atlantic University, Lagos, Nigeria

Eregha Perekunah School of Management and Social Sciences, Pan Atlantic University, Lagos, Nigeria

Tatjana Radovanovic Engineering Faculty, Seoul National University, Seoul, South Korea

Henry Bikwibili Tantoh Department of Environmental Sciences, College of Agriculture and Environmental Sciences, University of South Africa, Pretoria, South Africa;
Department of Geography and Planning, Faculty of Arts, The University of Bamenda, Bamenda, Cameroon

Suiven John Paul Tume Department of Geography and Planning, Faculty of Arts, The University of Bamenda, Bamenda, Cameroon

Kgosietsile Velempini Department of Environmental Sciences, University of North Carolina, Wilmington, USA

Part I
Introduction, Conceptual Premises and Frameworks

Chapter 1
Climate Change Education in Africa: Setting the Scene

Marcellus Forh Mbah, Petra Molthan-Hill, and Ernest L. Molua

Abstract The continent of Africa has enormous resources and potential to meet its long-term development needs. However, the challenges posed by climate change in the continent are frequently worsened by underlying vulnerabilities, intensifying their influence on the progress of the United Nations Sustainable Development Goals (SDGs). Africa remains disproportionately affected by the consequences of climate change despite having the smallest share of global greenhouse gas emissions. In this chapter, we advance opportunities and solutions that could be maximised to strengthen the continent's resilience and sustainable future. In particular, we draw attention to climate change education and policy drives across the continent to promote climate literacy as demonstrated in the African Union's Climate Change and Resilient Development Strategy and Action Plan (2022–2032). While it is not out of place to assert that there are some policy drives across Africa for climate change literacy and education, the gap between policy and practice needs scrutiny. An insight into existing policies and practices in Africa as they relate to climate change can bring to light the gaps in educational frameworks intended to support a climate-resilient future. A framework for an impactful climate change education in Africa is advanced. The framework recommends that Indigenous knowledge holders can play a crucial role by working with other holders of knowledge such as academics with Western insights to accentuate a CCE in the continent that is critical of imposed systems, is relatable to places, promotes collective participation and is holistic in nature. The latter part of the chapter introduces the different chapters.

Keywords Africa · Climate change · Climate change education · Indigenous knowledge holders

---

M. F. Mbah (✉)
School of Environment, Education and Development, University of Manchester, Manchester, England
e-mail: marcellus.mbah@manchester.ac.uk

P. Molthan-Hill
Nottingham Business School, Nottingham Trent University, Nottingham, England
e-mail: petra.molthan-hill@ntu.ac.uk

E. L. Molua
Department of Agricultural Economics and Agribusiness, University of Buea, Buea, Cameroon

M. F. Mbah et al. (eds.), *Practices, Perceptions and Prospects for Climate Change Education in Africa*, https://doi.org/10.1007/978-3-031-84081-4_1

Introduction

The continent of Africa is endowed with an abundance of natural resources (Lebert, 2015; Tumushabe, 2018), with an estimated 30% of the world's recognised mineral reserves (Sharaky, 2014). On the other side, African countries continue to encounter common global challenges such as climate change, the COVID-19 pandemic, fuel prices, and poverty. Africa remains disproportionately affected by the consequences of climate change. Czechowski (2020) asserts that Africa remains one of the most vulnerable continents to the impacts of climate variability and climate change, despite having the smallest share of global greenhouse gas emissions (Ogwu, 2019; Niang & Ruppel, 2014). The challenges posed by climate change in the continent are frequently worsened by underlying vulnerabilities, intensifying their influence on the progress of the United Nations Sustainable Development Goals (SDGs). Africa is especially susceptible to several social, economic, political, and geographic factors. One significant issue is the limited ability to adjust to climate change. Poverty leads to fewer options for households, while governance often neglects to prioritize and act on climate change. However, development can still be achieved, especially via investments in individuals, particularly women and youth, especially through education.

With regards to climate change adaptation and mitigation, African countries are especially suited to many of the climate solutions. As shown in Fig. 1.1, Suri et al. (2020: x) point out that African countries could fulfil all their energy demands through solar power, for example "Ethiopia would need only 0.003% of its land area to be covered in solar PV to meet its annual energy needs in recent years." Many of the most impactful climate solutions proposed by 'Project Drawdown' (2024) could also be implemented and upscaled in African countries, an example would be 'Clean Cooking'. Project Drawdown's (2024) Clean Cooking solution involves the use of solar-powered or fuel-burning household stoves that reduce greenhouse gas emissions by increasing thermal efficiency or ventilation. Another example calculated and recommended by Hawken (2018) would be 'Indigenous Peoples' Forest Tenure', which includes agroforestry and swidden cultivation among others. Often seen as damaging to the environment, swidden cultivation encompasses the burning and clearing of forestland for annual cultivation and regeneration of the same land in the following years, and—as Hawken (2018:125) highlights—leads to "more carbon (…) being sequestered under shifting cultivation than under annual cropping or plantations."

The resilience plan for the continent is anchored on the African Union Agenda 2063, which serves as a comprehensive plan to transform Africa (African Union Commission, 2015). The agenda aims to achieve inclusive and sustainable development and is a tangible expression of the pan-African pursuit of unity, self-determination, freedom, progress, and collective prosperity under the principles of Pan-Africanism and African Renaissance (Nwozor et al., 2021). Agenda 2063 encompasses Africa's future aspirations and outlines important flagship programs that might enhance economic growth and development, ultimately leading to the

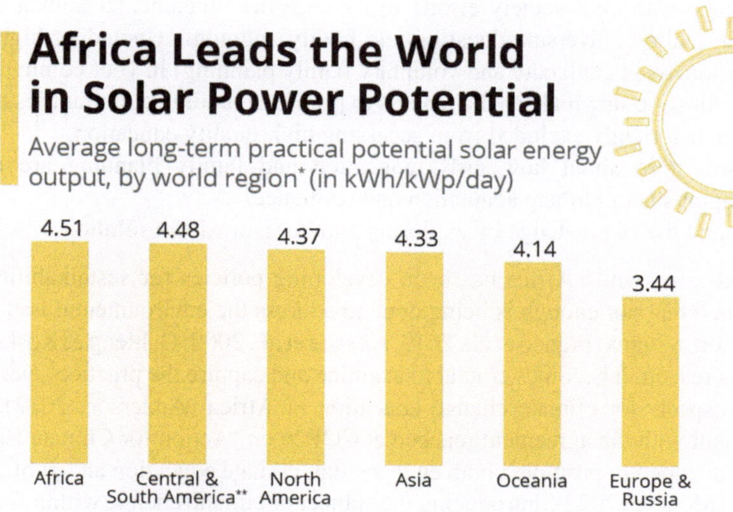
Africa Leads the World in Solar Power Potential
Average long-term practical potential solar energy output, by world region* (in kWh/kWp/day)
4.51
4.48
4.37
4.33
4.14
3.44
Africa
Central & South America**
North America
Asia
Oceania
Europe & Russia
* Based on national averages from a total 209 countries. Calculated for utility-scale installations of monofacial modules at optimum tilt. Excl. land with identifiable physical obstacles but ignoring possible restraints due to land use regulations.
** Including the Caribbean

Fig. 1.1 Africa leads the world in solar potential. *Source* Global Solar Atlas/The World Bank via Statista

fast transformation of the continent (Royo et al., 2022). Public education investment continues to be a fundamental aspect of this new development agenda. Education that incorporates understanding of environmental change would contribute to revitalized economic growth and social advancement, promoting development that prioritizes people, gender equality, and youth empowerment (Addaney, 2018).

Irrespective of its wealth and emerging prospects for a sustainable future, the adverse effects of climate change threaten some of the gains that are being made toward the realisation of the sustainable development goals in the African continent (Ogwu, 2019; Tumushabe, 2018). As the climate crisis worsens, studies suggest education as a key strategy for changing people's attitudes and promoting positive behavioural practices to adapt to climate change impacts as well as mitigate global warming (Anderson, 2012; Cordero et al., 2020; Lehtonen et al., 2019; Simpson et al., 2021). Consequently, various governments, especially in the developed world, are incorporating education as part of their commitment toward a long-term solution (Lehtonen et al., 2019).

The aforementioned 'Project Drawdown' (2024) includes 'education and family planning' among its 100 most impactful climate solutions. In this context they suggest that everyone could:

- Engage with civil society efforts to advance the fundamental human rights of high-quality, universal education and bodily autonomy (including high-quality reproductive health care and voluntary family planning) in your community.
- Encourage policymakers and leaders to prioritize funding communities that have been historically excluded from accessing high-quality education.
- Learn more about how girls' education and family planning are essential components of climate adaptation and resilience.
- Expand their knowledge by exploring another Drawdown solution.

Contrarily, while Africa has been developing policies for sustainability, it can be argued that not enough is being done to address the environmental issues via its education systems (Franco et al., 2019; Læssøe et al., 2009; Ochieng & Koske, 2013). For this reason, it becomes crucial to examine and capture the practices, perceptions and prospects for climate change education in Africa (Anderson, 2012). This is consistent with the agreement reached at COP26 on "Action for Climate Empowerment" intended to promote youth engagement, climate education and public participation (McGhie, 2022). Introducing the subject of climate change within formal and informal sectors of education is key to building the capacity of current and future generations to address what has been conceived as the most pressing global issue (Molthan-Hill et al., 2021; Centeno and Todd, 2020; Lehtonen, Salonen and Cantell, 2019).

Some progress has been made in African states with regards to tackling climate change, and education provides an immersed opportunity to create awareness for behavioural change, capacity building to implement effective solutions. The significance of having policy drives across the continent to promote climate literacy and climate change education (CCE) is demonstrated in the African Union's Climate Change and Resilient Development Strategy and Action Plan (2022–2032). Within the context of its strategic intervention axes, the need to increase regional climate change literacy across all levels of formal and informal education curricula was highlighted by the AU (2022). Three suggested actions are advanced. Namely: (a) Develop and include climate change literacy curricula for formal education (primary, secondary and tertiary levels), extending skills and knowledge for responses to climate change; (b) school girls need to be the focus of gender-sensitive approaches to education, emphasising attendance and completion of their schooling, and (c) Develop regional climate change literacy programmes for informal education (e.g., civil society and other partnering actors), extending skills and knowledge for responses to climate change. While it is not out of place to assert that there are some policy drives across Africa for climate change literacy and education (Apollo & Mbah, 2021), the gap between policy and practice needs scrutiny. An insight into existing policies and practices in Africa as they relate to climate change can bring to light the gaps in educational frameworks intended to support a climate-resilient future (Simpson et al., 2021). From this, opportunities and future possibilities for climate change education can be discussed. For policymakers, governments, and other stakeholders, this would mean creating pathways and implementing effective strategies to address SDG-13 and its target 3, which underscores the need to improve education, awareness-raising and human and institutional capacity on climate change mitigation, adaptation, impact reduction and early warning.

Despite efforts to achieve the SDGs, most developing countries especially in Africa have economies that are tied to nature and driven by natural resource exploitation. Climate-induced environmental changes have a significant impact on various natural systems, including the ecology, agricultural production, water availability, public health, and energy systems (Loucks, 2021; McMichael, 2001). These impacts exacerbate the process of desertification and the decline in biodiversity, resulting in more frequent occurrences of food insecurity. Primary stakeholders such as farmers closely monitor variations in both temperature and precipitation, in addition to noting other consequences of climate change. Rural women, who are among the most impoverished, are the most disadvantaged and vulnerable populations when it comes to education regarding climate change (Benevolenza & DeRigne, 2019; Denton, 2002). Both rural and urban entities utilize diverse agro-ecological and socioeconomic adaptation mechanisms in response.

Undoubtedly, it is imperative to ensure the implementation of climate change adaptation and mitigation policies. This will be facilitated by a strong foundation of climate literacy and climate change education (CCE) across Africa. Nevertheless, although numerous educational institutions throughout Africa currently provide opportunities for climate change literacy, the notion suffers from a lack of practicality. In order to effectively address climate change education across various educational levels, it is necessary to develop innovative and well-structured frameworks that focus on climate literacy and bridge the gaps between different generations. Empirical data indicates that the implementation of climate change education in African Higher Education Institutions (HEIs) is currently limited. Only a limited number of HEIs provide specialist courses on the subject.

Nevertheless, CCE could also open new opportunities and experiences. The UN Principles for Responsible Management Education (PRME) working group on climate change for example was inspired by its African scholars to set up a COIL (Collaborative Online International Experience) for the next academic year: Students are travelling the world virtually, ‘landing’ in different countries. The host countries showcase the climate solutions in their country, but they are also setting one of their unsolved climate problems for the international students to solve. Working in small groups of four or five students from diverse countries, these experience international collaboration and different cultures without causing massive emissions. In addition, COIL is inclusive offering students from lower socio-economic backgrounds to participate, including those who might not be able to leave their country due to financial and/or political constraints. Similarly, the UN PRME working group is offering free climate literacy training to academics; scholars from many different African countries have chosen to participate, please visit https://www.unprme.org/events/ to register for the next training. Another free online training that can be completed within three weeks is FutureLearn’s ‘Climate Literacy and Action for all’, please visit https://www.futurelearn.com/courses/climate-literacy-and-action-for-all.

Leveraging African Indigenous Knowledges for CCE

The incorporation of Indigenous knowledge into education will have significant implications for climate management. However, the existing educational paradigms in the continent, which are influenced by colonial legacies and oppression of Indigenous knowledge and local realities, are insufficient in providing effective solutions to the climate-related challenges in the continent. Climate change education is now considered vital in policy-making circles, since it is acknowledged as an essential instrument for developing skills and motivating necessary actions (Narksompong & Limjirakan, 2015; O'Brien et al., 2013).

African CCE strategies must be customised to conform to the lived experiences of those within the community (Shava & Nkopodi, 2020). To achieve this goal, CCE strategies must leverage the Indigenous knowledge (IK) of the local community. Indigenous Knowledge Systems (IKS) encapsulate the local skills, knowledge, cultural components and inter-generational traditions that allow indigenes to work in concert with nature (Ubisi et al., 2019). IKS's are especially relevant to farming communities within developing countries where the intergenerational transmission of knowledge is commonplace (Greenwood & Lindsay, 2019). IKS provide "ground truth" (Praskievicz, 2022) and may be considered integral to securing sustainable development in developing countries (Thaman, 2002). A recent systematic review finds that IKS can shape influential climate change adaptation strategies that are transferable across regions (Schlingmann et al., 2021). However, scientists need to exercise restraint and ensure these IKS are not exploited, monetised, or misused (Latulippe & Klenk, 2020). The exploitation of IKS by non-indigenous scientists simply reinstates a new brand of colonialism (Chavez & Gavin, 2018). Nevertheless, embedding IKS within CCE strategies presents exciting opportunities for impactful knowledge co-creation (Mbah, 2019). In addition, African states have an important and expanding role in progressing sustainability goals through co-creation activities (Stein, 2023) that brings to the fore contextual realities.

Since places have distinctive cultures, it follows that CCE strategies should preserve and protect both place and culture (Ajaps and Mbah, 2022). In the past, "colonially induced environmental changes altered the ecological conditions that supported Indigenous peoples' cultures, health, economies, and political self-determination" (Whyte, 2017: 154). The "epistemic violence" perpetrated by colonial forces has been institutionalised through Eurocentric education systems that inculcate and indoctrinate Western values and cultures within the developing world. To remedy this situation, there is a growing movement to decolonise curricula that ultimately serve as vehicles for driving societal change. To this end, the decolonisation of curricula can be viewed as an evolving process that shrugs off past colonial influence through the restoration of IKS (Adebisi, 2016). "The foundational intent of decolonisation is to equip students with "diverse academic learning environments, curricula and approaches to research within which Indigenous cultures, histories, and knowledge are embedded" (Waghid & Hibbert, 2018 as cited by Lumadi, 2021: 2).

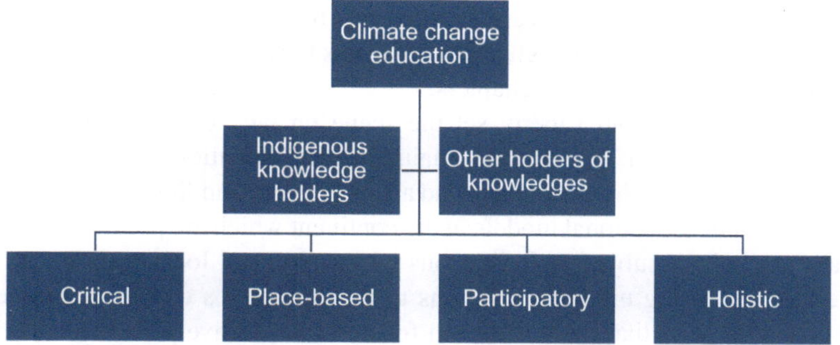

Fig. 1.2 A framework for an impactful climate change education in Africa. *Source* Adapted from Mbah et al., 2021: 17

Mbah et al. (2021) made the case for an impactful framework of IKS-based climate change education in contexts such as in Africa. The impact of such a framework can be embedded in critical, place-based, participatory and holistic approaches as illustrated in Fig. 1.2.

At the centre of this framework is the realisation that the vast number Indigenous knowledge holders in the continent can contribute to framing the content and context of an impactful climate change education, in the spirit of Ubuntu, an ideology of shared solidarity (Mbah, 2016). The need to actively create spaces for Indigenous knowledge holders is fitting with their vulnerability to climate change as a result of their dependence upon, and close relationship, with the environment and its resources (UN, 2008). Furthermore, Indigenous peoples and their communities have some valid and tested contributions to make to climate change education, as they have been able to preserve biodiversity and create some sophisticated food systems over generations (FAO, 2021). Therefore, in the context of knowledge cocreation, Indigenous knowledge holders can work together with other holders of knowledge such as academics with Western insights to accentuate a CCE in the continent that is critical of imposed systems, is relatable to places, promotes collective participation and is holistic.

Structure of the Book

In the next section, we give an overview of the different parts and chapters included in the book and describe how the authors of this book have expounded on the theme of CCE in Africa, providing deep insights into different aspects ranging from case studies to regional and continental premises. The book is divided into four interconnected parts. Part one, consisting of this chapter and Chaps. 2, 3, and 4, introduces the book and captures some conceptual premises and frameworks relevant to climate change education in Africa. Part two, covering Chaps. 5, 6, 7, 8, 9, and 10, touches on pedagogical insights, approaches and capabilities for an impactful climate change education in the continent. Part three, comprising Chaps. 11, 12, 13, 14, and 15, advances regional and country case studies. Part four, consisting of Chaps. 16 and 17 concludes the book with insights into prospects for the future. We will now delve into a glimpse of the remaining chapters.

In Chap. 2, Mbah and Liberty set the scene on why education matters for a climate-resilient Africa. The authors maintain that education is a cornerstone to equip individuals with climate literacy and a deeper understanding of climate change. However, current educational models in the continent which are products of colonial legacies, and which subjugate Indigenous knowledge and local realities are inadequate in contributing effective solutions to the continent's climate-related crises. Their chapter asserts that there is a need for a transformative pedagogical shift for CCE, underpinned by an appropriate framework. The chapter draws on the New Green Learning Agenda Framework and an education continuum to underscore the place for learning content and delivery for climate justice to advance place-based, decolonised, experiential and holistic pedagogies for climate change education in the continent. It underscores that it is only through the operationalisation of these approaches for climate justice via education that a climate-resilient future can be guaranteed in the continent.

Pursuant to the previous chapter, Das and Jan argue in Chap. 3 for the need to bridge generations using climate change education in early years, primary, secondary, and tertiary levels in the continent. The authors present innovative and structured frameworks designed to systematically address climate change education across diverse educational levels in Africa. Starting with the pre-primary level, the framework incorporates elements such as childhood awareness, nature activities, and strategic utilization of social media platforms to teach environmental consciousness. Moving through primary and secondary education, emphasis is placed on integrative teaching methodologies, obstacle identification, and the facilitation of hands-on experiential learning. The Technical and Vocational Education (TVET) framework integrates experiential learning and collaborative initiatives with industry stakeholders. Secondary education underscores interdisciplinary integration and the introduction of specialized electives. At the post-secondary level, the focus shifts to academic research, industry partnerships, and active engagement with SDGs. The chapter concludes by discussing the role of climate change education, advocating for policy adjustments, increased research initiatives, and sustained efforts. The envisioned outcome is the cultivation of an ecologically conscious generation adept at navigating the complex challenges raised by climate change in the African context.

Central to climate change education in Africa is the role of Higher Education. In Chap. 4 on conceptual premises for climate change adaptation education in African universities, Mbah, Harman and Molthan-Hill maintain that there is a pressing need to ensure that climate change mitigation strategies (i.e. strategies to reduce carbon emissions) and climate change adaptation strategies (i.e. strategies to circumvent the deleterious effects of climate change) are rolled out across all courses at African universities. Universities serve as both knowledge hubs and vehicles for societal change. However, African universities have traditionally adopted a Eurocentric approach to education that delegitimizes Indigenous knowledge and reinforces colonial narratives. To overcome these historical shortcomings, the chapter maintains that African universities must engage with local populations and leverage Indigenous knowledge systems to co-create place-based climate solutions that provide transformative change for all. In the chapter, the authors call for African universities to reposition their orientation by reconsidering their conceptualisation of climate change education. It elucidates on co-creation of climate change adaptation education via transformative social learning and the enabling of epistemic plurality and polycentricity.

In the ensuing Chap. 5, Evans, Eregha and Osuji examine the context of ICT-enabled climate change education and adaptation in Africa. Despite the importance of ICT, the authors note that there is limited information available in the literature about how it is presently and will be used in Africa's efforts to adapt to climate change. The chapter highlights that ICT is crucial in addressing the significant challenges posed by climate change in Africa and can be employed to facilitate the dissemination of knowledge required for climate change adaptation at the community level. The authors posit this can be accomplished by raising awareness, providing access to critical information, and promoting learning and sharing of experiences.

In Chap. 6, exploring further the subject of knowledge dissemination on climate change, Mbah and Ezegwu provide in-depth insight into educational pedagogies and approaches needed to foster an effective climate change education (CCE) and help people develop sustainability and adaptability mindsets in the continent. The authors argue that this is very crucial, considering the continent is vulnerable to the adverse effects of climate change and education is recognised as an important tool for capacity building and mobilising requisite actions against. The chapter contends climate change education pedagogies in the continent need to be decolonised and decentred to address context-driven capacity needs of the continent. It also argues that effective pedagogy for climate change education in Africa should integrate Indigenous approaches for dealing with localised environmental realities and challenges. Different perspectives on pedagogies are discussed in the chapter, namely: behavioural, cognitive, sociocultural, critical, post-structuralist and connectivism perspectives.

In recognition of international frameworks endorsing climate change (CC) and disaster risk reduction (DRR), Bang maintains in Chap. 7 that education is pivotal in enhancing resilience to climate/disaster risks. The chapter explores the depth and scope to which CC education has been inculcated in Africa's high-level discourse for enhanced resilience to climate and disaster risks. By adopting a qualitative research methodology, the study scrutinised the depth to which the determinants of CC/DRR education have been inculcated into Africa's regional platforms (ARP) for DRR. By applying a Likert scale analysis, the findings fall within the range of "satisfactory" and "mediocre-poor". Unarguably, the findings paint a bleak picture of the importance of CC/DRR education during the Platforms. This evidence provides a compelling argument for mainstreaming CC/DRR education into high-level discussions to expedite uptake by African countries for enhanced resilience to climate-related disaster risks.

Tume, Awazi and McKay ascertain in Chap. 8 that gender and CCE in Sub-Saharan Africa are the missing components in climate change adaptation for an effective management of natural resources. The authors highlight that rural women, for example, who are amongst the poorest, are one of the most marginalized groups and vulnerable in terms of education about climate change. Given the fast pace at which the world is changing, rural women must contend with the challenges faced in the management of natural resources, which are exacerbated by the rising population and their limited level of education. A major question is how to effectively improve the education of rural women so that they may be able to adapt to the effects of climate change in natural resource management? The chapter, therefore, focuses on the significant role of gender education in climate change adaptation and effective natural resources management. The authors argue that a gendered approach that values the capacities, limits and vulnerabilities of rural women is required for effective management of climate change and accompanying education.

Similar to the subject of gender, Ahmed and Givens delve in Chap. 9 into African smallholders' perceptions of climate change, their adaptation techniques, and various underlying factors that have implications for environmental awareness and education. Findings from their research show farmers observe fluctuations in both temperature and rainfall as well as observing other impacts of climate change. In response, farmers employed various adaptation strategies such as diversification of crops, crops rotations, planting drought-resistant crops, incorporating livestock into crop production, shifting the time of agricultural operations, homestead gardening, increasing irrigation, engaging in mixed farming, and migration. Various household-related, farm-related, institutional, and other factors drive local-level adaptation strategies among farmers. The authors maintain that in order to contribute to the progress on the SDGs, farmers' insights and experiences should be included in the dominant policies and plans, including coordination of education about climate change and adaptation strategies, to achieve effective adaptation.

Although there is growing recognition for African states to engage their citizenry in climate change education, there is need to evaluate their capacities to undertake such a venture. In this regard, Ezegwu and Mbah in Chap. 10 employed Almond and Powell's spheres of state capability to discuss factors affecting African countries' capacity to respond to the demands of youths—particularly with respect to climate change. Based on the observable state of many African states' extractive, regulative and distributive capacities, the authors argue that they have limited capacity to provide effective climate change education. The chapter recommends the need to explore and build efficient and capable institutions through the strengthening of democracy and governance, as well as dealing with corruption and the root causes of diverse conflicts plaguing the continent. Also, the authors highlight the need to buttress civil service capacity, empowerment of the citizenry, enacting measures to address the immediate and long-term engagement of youth for a sustainable future.

In Chap. 11, Ndlovu examines the antecedents of climate change literacy in Africa. The author asserts that there is a missing link that maps the calamity of climate change with human activities in the continent on one hand, and remedies to mitigate the effects on the other hand. The chapter maintains that this can be traced back to climate change literacy and conceptualisation of this phenomenon, leading to compromised climate change mitigation practices in Africa, especially among the vulnerable communities that bear the bulk of the brunt of this phenomenon.

Despite many education institutions across Africa now offering climate change literacy education, the concept remains somewhat abstract, lacking practicability as many graduates face the harsh realities of securing jobs and earning a living as opposed to climate change mitigation endeavours. In sum, the chapter focuses on Africa's bottom of the pyramid populations and how their views of, and engagement with climate change literacy may be shaped by their socio-economic realities.

The next set of chapters draw on case studies within the continent. In Chap. 12, Kopnina, Baranowski, Natunda, Mande, Radovanovic and Gadema argue for an urgent need to reorient anthropocentric normative framings of sustainability in business and management education to ecocentric ontologies and epistemologies within pedagogical praxes of design and delivery. The chapter draws upon two examples from university business schools' sustainability programmes in Africa and The United Kingdom as comparative cases to illuminate the commonalities and differences in ontological and epistemological characteristics. Uganda and England are contextually different, whether it be geography, economy(ies), or how these contexts, in turn, shape/influence pedagogical approaches and practices in each of those business schools. Findings in the chapter show that Education for Sustainable Development Goals (ESDGs) increasingly feature within courses across 'sustainable' management-centric pedagogies such as 'sustainable marketing' and 'sustainable supply chain management'. However, these normative framings of sustainability pedagogies were found to potentially negate the inclusion of 'deep green' rooted concepts, ontologies and epistemologies in business and management education. In contrast to these normative approaches, the chapter presents and analyses opportunities for critically evaluating sustainability education that centres on biodiversity through eco-pedagogy and eco-literacy.

In Chap. 13, Borsari highlights global climate change as the major anthropogenic factor that continues to impact the livelihoods of millions of people, forcing human displacement and mass migrations. Climate-driven environmental changes amplify desertification, and biodiversity loss, leading to frequent crop failures and food shortages. Industrial agriculture is heavily implicated for exacerbating climate change on a planetary scale, with expanding effects on vast, world regions. Yet, farming remains a primary economic activity and in Africa this is accomplished for providing food for local communities primarily, through peasant agriculture. Thus, curricula in agroecology can be transformative and have potential for realizing an effective implementation of climate change education. The assessment study reports about the benefits that agroecology has for Sierra Leone and from it, a model that could serve local stakeholders and more in African countries was produced to employ agroecology as a mean to foster climate change education in all curricula.

In a different country context, Molua, Ndip, Nanfouet, Kambiet, and Etomes examine in Chap. 14 the indigenization of education for climate management in Cameroon. The authors underscore the significance of education to farmers as a crucial measure beyond food security. Farmers need education to remain abreast of technological innovations that affect agricultural operations. However, farmers encounter unique obstacles and must receive education and training to be successful. Indigenizing education and providing information about climate change are emerging as means of securing livelihoods, particularly in vulnerable sectors such as tropical African agriculture. The field study the chapter draws on shows positive perceptions about a changing climate, and that farming households are employing local context-specific coping and adaptive measures. The quantitative analysis of the study demonstrate that education is significantly positive in influencing farm outcomes. The generated results are crucial for influencing climate change policy related to awareness-building, education, and training for optimal adaptation efforts.

The case of Botswana was also brought to the fore in Chap. 15, where Majola, Ntirase and Velempini uncover challenges and opportunities for climate change literacy in the country. The depicted findings show that even though there are opportunities for climate change literacy in Botswana, the challenges play a critical role in the failures of opportunities for climate change literacy. The paper recommends that local knowledge can be used for sustaining climate change literacy.

Similarly, in Chap. 16 on prospects and problems that underpin climate change education in West Africa, Akpabio, Akeju and Duruji address concerns associated with the uptake of climate change education in the sub-region, drawing on a synthesis of secondary literature that engage a cross-national comparisons of climate change education.

The book closes with Chap. 17 on current practices and future directions for climate change education in African higher education institutions. The authors, Kambiet and Mbah maintain that CCE can be proposed as a suitable strategy in Africa partly because of its mitigation and adaptation role. However, evidence suggests that the adoption of CCE in African Higher Education Institutions (HEIs) remains low. Few HEIs have specialized courses on the subject, while others have an integrated approach within related subjects like geography and environmental sciences. This underscores the necessity to identify the factors behind the slow uptake of CCE across the continent. Given this premise, the authors explore the nature of the barriers as well as existing and future opportunities for scaling up climate change education in the continent's HEIs. Among other factors, the authors identify resource limitations, institutional barriers, and socio-political challenges as the primary constraints to CCE uptake in Africa. On the other hand, the increasing climate commitment and the development of regional climate institutions equally provides a unique opportunity for capacity building, knowledge exchange and the spread of CCE across the continent. Moreover, the rapid development of affordable digital communication and computing has the potential to increase networking and collaborative efforts between climate change learners and scientists across the region. These prospects signal a need for regional governments to multiply their efforts towards providing an environment for the development and uptake of suitable CCE programs.

To put it briefly, the conclusions of this book emphasize the need to operationalize various strategies for climate literacy and education to achieve climate justice and a future for the continent that is robust to climate change. Effective pedagogies for teaching about climate change should incorporate Indigenous and innovative ways of addressing specific environmental issues and realities. This will necessitate the development of an environmentally aware generation that can successfully navigate the many issues brought forth by climate change in Africa. The many chapters urge African education systems to reorient by rethinking how they conceptualize climate change education to realize ambitious goals that contribute to a realization of the SDGs. Through the revamped CCE framework important information will be accessible, awareness will be increased, and learning and experience sharing will be encouraged. More crucially, successful management of climate change and related education necessitates a gendered approach that appreciates the skills, limitations, and vulnerabilities of rural women. In general, in order to accomplish effective adaptation, experiences and insights of local stakeholders should be incorporated into the prevailing policies and programs, including the coordination of education regarding climate change and adaptation measures. In sum, increasing the capacity of the civil service, empowering the populace, while enacting policies to address youth engagement both now and in the future will all be necessary to achieve sustainable development for the continent via CCE.

References

Addaney, M. (2018). The African union's agenda 2063: Education and its realization. *Education Law, Strategic Policy and Sustainable Development in Africa: Agenda, 2063*, 181–197.

Adebisi, F. I. (2016). Decolonising education in Africa: Implementing the right to education by re-appropriating culture and indigeneity. *Northern Ireland Legal Quarterly, 67*, 433–451.

African Union Commission. (2015). Agenda 2063: Background Note, Agenda 2063 Document No. 01. Addis Ababa: African Union. Retrieved 03 July 2024, from https://au.int/sites/default/files/documents/33126-doc-06_the_vision.pdf.

African Union (AU). (2022). African Union Climate Change and Resilient Development Strategy and Action Plan (2022–2032). Retrieved 03 July 2024, from https://au.int/en/documents/20220628/african-union-climate-change-and-resilient-development-strategy-and-action-plan.

Ajaps, S., & Mbah, M. F. (2022). Towards a critical pedagogy of place for environmental conservation. *Environmental Education Research, 28*(4), 508–523.

Anderson, A. (2012). Climate change education for mitigation and adaptation. *Journal of Education for Sustainable Development, 6*(2), 191–206.

Apollo, A., & Mbah, M. F. (2021). Challenges and opportunities for climate change education (CCE) in East Africa: A critical review. *Climate, 9*(6), 93.

Benevolenza, M. A., & DeRigne, L. (2019). The impact of climate change and natural disasters on vulnerable populations: A systematic review of literature. *Journal of Human Behavior in the Social Environment, 29*(2), 266–281.

Chavez, D. M., & Gavin, M. C. (2018). A global assessment of Indigenous community engagement in climate research. *Environmental Research Letters, 13*(12), 123005.

Cordero, E. C., Centeno, D., & Todd, A. M. (2020). The role of climate change ducation on individual lifetime carbon emissions. *PLoS ONE, 15*(2), e0206266.

Czechowski, A.S. (2020). Benchmarking progress towards climate safe cities, states and regions. *CDP Africa Report.*

Denton, F. (2002). Climate change vulnerability, impacts, and adaptation: Why does gender matter? *Gender & Development, 10*(2), 10–20.

FAO. (2021). Indigenous Peoples' food systems: Insights on sustainability and resilience in the front line of climate change. *Food and Agriculture Organization of the United Nations and Alliance of Bioversity International and CIAT Rome.* https://doi.org/10.4060/cb5131en

Franco, I., Saito, O., Vaughter, P., Whereat, J., Kanie, N., & Takemoto, K. (2019). Higher education for sustainable development: Actioning the global goals in policy, curriculum and practice. *Sustainability Science, 14*(6), 1621–1642.

Hawken, P. (Ed.). (2018). The most comprehensive plan ever proposed to reverse global warming. *Drawdown* (pp. 124–127).

Læssøe, J., Schnack, K., Breiting, S., Rolls, S., Feinstein, N., & Goh, K. C. (2009). Climate change and sustainable development: The response from education. In *A cross-national report from international alliance of leading education institutes*. The Danish School of Education, Aarhus University.

Latulippe, N., & Klenk, N. (2020). Making room and moving over: Knowledge co-production, Indigenous knowledge sovereignty and the politics of global environmental change decision-making. *Current Opinion in Environmental Sustainability, 42*, 7–14.

Lebert, T. (2015). Africa: a continent of wealth, a continent of poverty. *New Internationalist, 24*.

Lehtonen, A., Salonen, A. O., & Cantell, H. (2019). Climate change education: A new approach for a world of wicked problems. In *Sustainability, human well-being, and the future of education* (pp. 339–374). Palgrave Macmillan.

Loucks, D. P. (2021). Impacts of climate change on economies, ecosystems, energy, environments, and human equity: A systems perspective. In *The impacts of climate change* (pp. 19–50). Elsevier.

Lumadi, M. W. (2021). The pursuit of decolonising and transforming curriculum in higher education. *South African Journal of Higher Education, 35*(1), 1–3.

Mbah, M. F. (2016). Towards the idea of the interconnected university for sustainable community development. *Higher Education Research & Development, 35*(6), 1228–1241.

Mbah, M. (2019). Can local knowledge make the difference? Rethinking universities' community engagement and prospect for sustainable community development. *The Journal of Environmental Education, 50*(1), 11–22.

Mbah, M. F., Shingruf, A., & Molthan-Hill, P. (2022). Policies and practices of climate change education in South Asia: Towards a support framework for an impactful climate change adaptation. *Climate Action, 1*(1), 1–18.

McGhie, H. A. (2022). Action for Climate Empowerment, a guide for galleries, libraries, archives and museums. Curating Tomorrow, UK. Retrieved 03 July 2024, from https://www.curatingtomorrow.co.uk/wp-content/uploads/2022/04/action-for-climate-empowerment_2022.pdf.

McMichael, A. J. (2001). Impact of climatic and other environmental changes on food production and population health in the coming decades. *Proceedings of the Nutrition Society, 60*(2), 195–201.

Molthan-Hill, P., Blaj-Ward, L., Mbah, M. F., & Ledley, T. S. (2021). Climate change education at universities: relevance and strategies for every discipline. In: M. Lackner, B. Sajjadi, W.-Y.Chen (Eds.), *Handbook of climate change mitigation and adaptation* (pp. 1–64). Springer.

Narksompong, J., & Limjirakan, S. (2015). Youth participation in climate change for sustainable engagement. *Review of European, Comparative & International Environmental Law, 24*(2), 171–181.

Niang, I., Ruppel, O. C., Abdrabo, M. A., Essel, A., Lennard, C., Padgham, J., & Urquhart, P. (2014). Africa. In: V. R. Barros, C. B. Field, D. J. Dokken, M. D. Mastrandrea, K. J. Mach, T. E. Bilir, M. Chatterjee, K. L. Ebi, Y. O. Estrada, R. C. Genova, B. Girma, E. S. Kissel, A. N. Levy, S. MacCracken, P. R. Mastrandrea, & L. L. White (Eds.) *Climate Change 2014: Impacts, Adaptation, and Vulnerability. Part B: Regional Aspects*. Contribution of Working Group II to the Fifth Assessment Report of the Intergovernmental Panel on Climate Change (pp. 1199–1265). Cambridge University Press, Cambridge, United Kingdom and New York, NY, USA.

Nwozor, A., Okidu, O., & Adedire, S. (2021). Agenda 2063 and the feasibility of sustainable development in Africa: Any silver bullet? *Journal of Black Studies, 52*(7), 688–715.

O'Brien, K., Reams, J., Caspari, A., Dugmore, A., Faghihimani, M., Fazey, I., Hackmann, H., Manuel-Navarrete, D., Marks, J., Miller, R., & Raivio, K. (2013). You say you want a revolution? Transforming education and capacity building in response to global change. *Environmental Science & Policy, 28*, 48–59.

Ochieng, M. A., & Koske, J. (2013). The level of climate change awareness and perception among primary school teachers in Kisumu municipality, Kenya. *International Journal of Humanities and Social Science, 3*(21), 174–179.

Ogwu, M. C. (2019). Towards sustainable development in Africa: The challenge of urbanization and climate change adaptation. In *The geography of climate change adaptation in Urban Africa* (pp. 29–55). Palgrave Macmillan.

Praskievicz, S. (2022). Ground truth: Finding a "place" for climate change. *Progress in Environmental Geography, 1*(1–4), 137–162.

Project Drawdown. (2024). Table of solutions. Retrieved 03 July 2024, fromhttps://drawdown.org/solutions/table-of-solutions.

Royo, M. G., Diep, L., Mulligan, J., Mukanga, P., & Parikh, P. (2022). Linking the UN sustainable development goals and African agenda 2063: Understanding overlaps and gaps between the global goals and continental priorities for Africa. *World Development Sustainability, 1*, 100010.

Schlingmann, A., Graham, S., Benyei, P., Corbera, E., Sanesteban, I. M., Marelle, A., Soleymani-Fard, R., & Reyes-García, V. (2021). Global patterns of adaptation to climate change by indigenous peoples and local communities: A systematic review. *Current Opinion in Environmental Sustainability, 51*, 55–64.

Sharaky, A. M. (2014). *Mineral resources and exploration in Africa.* Institute of African Research and Studies, Cairo University, Egypt.

Shava, S., & Nkopodi, N. (2020). Indigenising the university curriculum in southern Africa. In *Indigenous studies: Breakthroughs in research and practice* (pp. 243–254). IGI Global.

Simpson, N. P., Andrews, T. M., Krönke, M., Lennard, C., Odoulami, R. C., Ouweneel, B., Steynor, A., & Trisos, C. H. (2021). Climate change literacy in Africa. *Nature Climate Change, 11*(11), 937–944.

Stein, S. (2023). Universities confronting climate change: beyond sustainable development and solutionism. *Higher Education*, 1–19. https://doi.org/10.1007/s10734-023-00999-w

Suri, M., Betak, J., Rosina, K., Chrkavy, D., Suriova, N., Cebecauer, T., Caltik, M., & Erdelyi, B. (2020). *Global Photovoltaic Power Potential by Country (English).* Energy Sector Management Assistance Program (ESMAP) Washington, D.C.: World Bank Group. http://documents.worldbank.org/curated/en/466331592817725242/Global-Photovoltaic-Power-Potential-by-Country. Accessed 03 July 2024.

Thaman, K. H. (2002). Shifting sights: The cultural challenge of sustainability. *Higher Education Policy, 15*(2), 133–142.

Tumushabe, J. T., (2018). Climate change, food security and sustainable development in Africa. In *The Palgrave handbook of African politics, governance and development* (pp. 853–868). Palgrave Macmillan.

Ubisi, N. R., Kolanisi, U., & Jiri, O. (2019). Comparative review of indigenous knowledge systems and modern climate science. *Ubuntu: Journal of Conflict and Social Transformation, 8*(2), 53–73.

UN (2008). Inter-agency support group on indigenous peoples' issues collated paper on indigenous peoples and climate change. Permanent Forum on Indigenous Issues Seventh session. https://www.un.org/esa/socdev/unpfii/documents/2016/egm/IASG-Collated-Paper-on-Indigenous-Peoples-and-Climate-Change.pdf. Accessed 03 July 2024.

Waghid, Z., & Hibbert, L. (2018). Advancing border thinking through defamiliarisation in uncovering the darker side of coloniality and modernity in South African higher education. *South African Journal of Higher Education, 32*(4), 263–283.

Whyte, K. (2017). Indigenous climate change studies: Indigenizing futures, decolonizing the Anthropocene. *English Language Notes, 55*(1), 153–162.

Chapter 2
Why Education Matters for a Climate-Resilient Africa

Marcellus Forh Mbah and Christopher Liberty

Abstract Given the quest for a climate-resilient Africa, education emerges as a cornerstone in equipping individuals with climate literacy and a deeper understanding of climate change. However, current educational models on the continent, which are products of colonial legacies and subjugate Indigenous knowledge and local realities are inadequate for contributing effective solutions to the continent's climate-related crises. There is therefore a need for a transformative pedagogical shift for climate change education, underpinned by an appropriate framework. We draw on the New Green Learning Agenda Framework to underscore the urgent need for a transformational model of education that fosters climate justice in Africa. Central to this are the crucial roles of place-based, decolonised, experiential, and holistic pedagogies. It is only through the implementation of these approaches for climate justice through education that a climate-resilient future can be guaranteed for the continent.

Keywords Education · Africa · Climate change · Resilience

Introduction

Climate change is one of the most crucial issues facing humanity in the twenty-first century, impacting all facets of human and natural systems, including ecosystems, biodiversity, and food security (IPCC, 2023). Africa contributes only 4% of the world's greenhouse gas emissions but it bears a disproportionate share of the costs and hazards associated with climate change, endangering its development opportunities and aspirations (IPCC, 2023; World Bank, 2020). The continent is especially vulnerable to the effects of climate change due to its high exposure to extreme weather

M. F. Mbah (✉)
Manchester Institute of Education, University of Manchester, Manchester, United Kingdom
e-mail: marcellus.mbah@manchester.ac.uk

C. Liberty
Ashinaga Uganda, Kampala, Uganda
e-mail: liberty.christopher@alumni.lse.ac.uk

M. F. Mbah et al. (eds.), *Practices, Perceptions and Prospects for Climate Change Education in Africa*, https://doi.org/10.1007/978-3-031-84081-4_2

events, low capacity for adaptation, and several stressors such as poverty, inequality, violence, and disease (IPCC, 2023). The 2015 Paris Agreement reiterated a commitment to recognising the crucial role that education can play (Gandhi et al., 2024; Zheng et al., 2024).

Consensus-building initiatives focused on prevention, reducing carbon emissions, adaptation, and mitigation measures could be among the significant contributions that Africa can makes to global climate change action efforts, given the continent's low greenhouse gas emissions and high vulnerability to climate change. These contributions can be amplified by climate change education (Lin & Agyeman, 2020; Ssekamatte, 2022). Education can plays a pivotal role in enhancing climate literacy and fostering climate resilience in Africa. According to a study conducted by Afrobarometer, individuals with higher levels of education are significantly more aware of climate change and its underlying causes. Notably, completing high school increases the likelihood of climate change literacy by 19% compared to those without formal schooling (Afrobarometer, 2023). Ssekamatte, (2022) contends that climate change and sustainability education have profound potential for fostering behavioural shifts toward pro-sustainability practices. These shifts, he asserts, significantly advance our collective journey toward a more sustainable global environment. Therefore, formal, non-formal, and informal education sectors are essential, and investing in comprehensive climate change education is crucial for building and ensuring climate resilience in Africa.

Africa and Climate Change

The Intergovernmental Panel on Climate Change has identified Africa as the continent most vulnerable to climate change globally. This is due to complex problems such as extreme poverty, large segments of the population lacking access to essential services like water and sanitation, wealth disparities, gender inequality, and difficulties with governance (Birkmann et al., 2022; Calvin et al., 2023). Climate change has far-reaching implications for agricultural productivity, particularly in African countries that heavily rely on agriculture, especially in sub-Saharan Africa (Bedasa & Deksisa, 2024; Kangama, 2024). African economies are disproportionately dependent on climate-exposed sectors, with 55–62% of the sub-Saharan workforce employed in agriculture (Adelekan et al., 2022).

The Notre Dame Global Adaptation Initiative Country Index (ND-GAIN, 2023) (see Fig. 2.1) illustrates the comparative resilience of countries to climate change and shows how Africa remains the most vulnerable continent and the least prepared to tackle climate-related challenges as depicted by Fig. 2.1. The five most vulnerable countries include Guinea-Bissau and Chad. Among the metrics used in the indicator is social structures, including the quality of education systems (Serrano-Candela et al., 2024). This emphasises the importance of education as a key marker

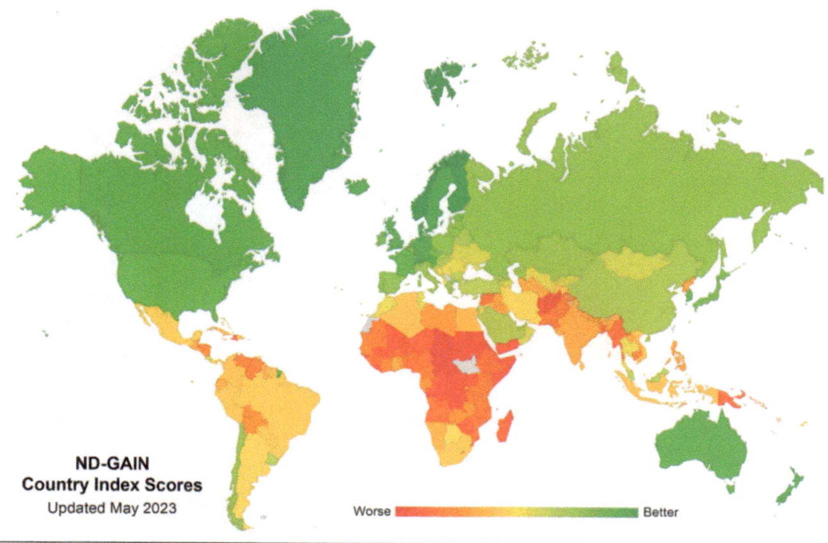

Fig. 2.1 ND-GAIN (2023) country index. *Source* Country Index // Notre Dame Global Adaptation Initiative // University of Notre Dame

for climate resilience in Africa. The fact that the index considers tertiary education completion rates as an indicator of climate resilience further underscores the value of education in ensuring climate resilience in Africa.

Moreover, the economic growth of countries in sub-Saharan Africa is at risk. Changes in temperature and carbon dioxide emissions have significant negative implications with a modest increase of 1 °C in temperature leading to a notable decline of 1.8 % points in economic growth, as highlighted by Rafique et al. (2024). This sobering reality underscores the urgent need for climate change education to support sustainable climate action, protect African livelihoods and foster both economic and climate change resilience on the continent.

Climate change also influences precipitation patterns and variability across Africa, leading to more extreme rainfall events, floods, and landslides, as well as longer and more intense dry spells. These changes have significant ramifications for the hydrological cycle, water resource management, agriculture, food security, and disaster risk reduction (Kusangaya et al., 2021; Ofori et al., 2021). For example, the 2019–2020 rainy season in East Africa was one of the wettest on record, causing widespread floods, landslides, crop losses, and outbreaks of diseases such as cholera and malaria (Gebrechorkos et al., 2023; Wainwright et al., 2021). Conversely, the 2019–2020 rainy season in Southern Africa was one of the driest on record, resulting in severe drought, water shortages, food insecurity, and power outages (Olivier et al., 2020). The impacts and risks of climate change are not uniformly distributed across Africa, as different regions, sectors, and groups face different levels of exposure, sensitivity, and adaptive capacity (Calvin et al., 2023). Furthermore, Africa's carbon emissions

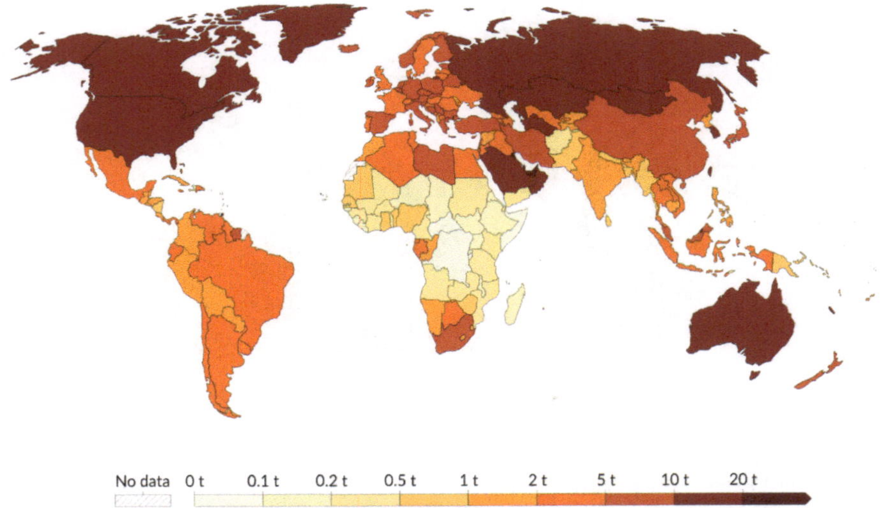

Fig. 2.2 Annual CO_2 emissions including land-use change, 2022. *Source* Our world in data (OurWorldInData.org/co2-and-greenhouse-gas-emissions)

are comparatively low, despite accounting for nearly one-fifth of the world's population (Lin & Agyeman, 2020; Zhu & Wang, 2024). This is due in part to the continent's unequal distribution of energy use, as well as lower consumption of goods and services compared to more developed parts of the globe. Additionally, Africa's energy systems have not been industrialised to the extent seen on other continents, which typically results in higher emissions. Figure 2.2 presents annual CO_2 emissions including land-use change, highlighting Africa's significantly lower emissions compared to those of other continents. This disparity underscores the continent's minimal contribution to global carbon output and sets a precedent for Africa to serve as a global example of sustainable development. Through enterprises such as education and training that emphasise climate change resilience, Africa has a unique opportunity to maintain low-to-zero carbon emissions.

Why is Climate Change Education Important?

Climate change education is crucial for fostering resilience and empowering communities to adapt to climatic changes. It enhances understanding of sustainable practices, promotes the adoption of renewable energy, and fosters innovation in low-emission technologies (*UNESCO*, n.d.). An important point for climate change education lies in integrating climate change into educational curricula. Education at all levels in Africa—pre-primary, primary, secondary, and higher education should be reoriented not only towards achieving climate literacy but also towards climate action and climate justice. This transformation extends beyond curricula reform or the mere inclusion of climate change concepts in the existing curriculum, even though that is an important starting point (Nusche et al., 2024). The process raises ethical considerations about how we disseminate climate change knowledge to students in various cultural contexts and what learning objectives we set for them. Aligning the curriculum, the body of climate change knowledge, and the local cultural context within a single framework requires a complex integration of diverse evidence-based theories and cultural practices. Such collective efforts can result in the development of robust curricula (Eilam, 2024; Khupe et al., 2024). The ultimate shift calls for a comprehensive transformation of education towards the appropriate implementation of climate change education.

Effective climate action driven by education requires that citizens possess sufficient critical knowledge and skills to both contribute to climate change management and advocate for their governments' adoption of suitable and just climate change mitigation measures. This competence plays a pivotal role in promoting informed decision-making and impactful action towards a climate-resilient Africa (Apollo & Mbah, 2021; Svarstad, 2021). Furthermore, education can be considered a key factor in enhancing the continent's resilience and adaptation to climate change, as it fosters understanding and encourages actionable measures in communities. Education can also contribute to climate change mitigation, by promoting low-carbon lifestyles, technologies, and innovations (Gebrekidan & Gebremedhin, 2024; Léna & Lescarmontier, 2023).

However, the current forms of education, including those related to climate change and the environment, are largely influenced by Western perspectives and do not reflect the continent's diverse Indigenous knowledge and realities. These forms of education often fail to address the complex and context-specific challenges and opportunities that climate change poses for Africa, as well as fall short of empowering learners to take informed and effective action in their local and global contexts (Apollo & Mbah, 2021; Kwauk & Wyss, 2023). Therefore, while education is indeed crucial for fostering a climate-resilient continent, it is not just any type of education—It should be contextually relevant.

Climate Change Literacy as a Stating Point

The first contribution of education to climate resilience is through climate literacy, which provides learners with accurate and relevant information, enhances their critical thinking and problem-solving abilities, and fostering their awareness and motivation to address climate issues (Okada & Gray, 2023). Climate literacy, as a critical instrument in enabling citizens to comprehend and make informed choices regarding climate change, has the potential to facilitate a trajectory from individual awareness to collective engagement for a climate-resilient Africa (Hoydis et al., 2023; Kumar et al., 2023; Pan et al., 2023).

Charbonnier (2023) posits that certain graduate skills including critical and analytical thinking, are essential in the learning process for adults regarding climate change.

These skills are necessary to assess the reliability of learning resources. Graduates can also exercise pro-environmental tendency and educational institutions serve as ideal spaces for fostering this behaviour (Charbonnier, 2023). Climate literacy also encompasses an awareness of and concern about the environment and its associated problems, as well as the knowledge, skills, and motivation to work towards solutions for current problems and the prevention of new ones (Molthan-Hill & Winfield, 2023).

However, knowledge alone is not enough. Becken and Coghlan (2024) argue for regenerative literacy, which incorporates a wide range of ideas and practices related to sustainability, education, and social change. It is based on the belief that we need to move beyond simply learning about environmental problems to taking action to create a more just and sustainable world (Becken & Coghlan, 2024). Regenerative literacy also embraces cultural and contextual knowledge and practices which are vital for fostering a climate-resilient continent. According to Afrobarometer (2023), an average of 58% of respondents across 34 African countries declared they were familiar with the concept of climate change as shown in Fig. 2.3. Although climate literacy is a starting point for a climate-resilient Africa, it alone is insufficient. Efforts to go beyond literacy and a basic understanding of the concept of climate change should focus on education that ensures a more environmentally engaged citizenry to strengthen resilience across the continent.

Afrobarometer (2023) survey further demonstrates that groups with lower familiarity with the concept of climate change include rural residents, women, the poor, the less-educated, people who work in agriculture and those with limited exposure to news media. Fundamentally, access to education is the strongest predictor of the rate of climate change literacy. This assertion is highlighted in the African Union's (AU) Climate Change and Resilient Development Strategy and Action Plan (2022–2032). Specifically, the AU (2022) maintains that Africans with post-secondary education are more likely to have some awareness of climate change.

Making Education Work for Climate Resilience in Africa

Education is widely recognised as a key driver in building climate resilience, as it enhances awareness, knowledge, skills, and values that enable people to cope with and adapt to climate change, as well as contribute to mitigation and transformation efforts (UNESCO, 2017). However, not all forms of education are equally effective or relevant for addressing the complex and multidimensional nature of the climate crisis. Literature on how to improve Africa's climate resilience through education suggests that educational approaches should be culturally relevant and context-specific, as climate change poses complex challenges for the continent (Apollo & Mbah, 2021; Gyimah et al., 2023; Sherpa, 2018).

Mbah and Ezegwu (2024) advance a decolonised approach to climate change and environmental education that challenges the assumptions and practices of Western academic research and acknowledges Indigenous epistemologies, ontologies, and

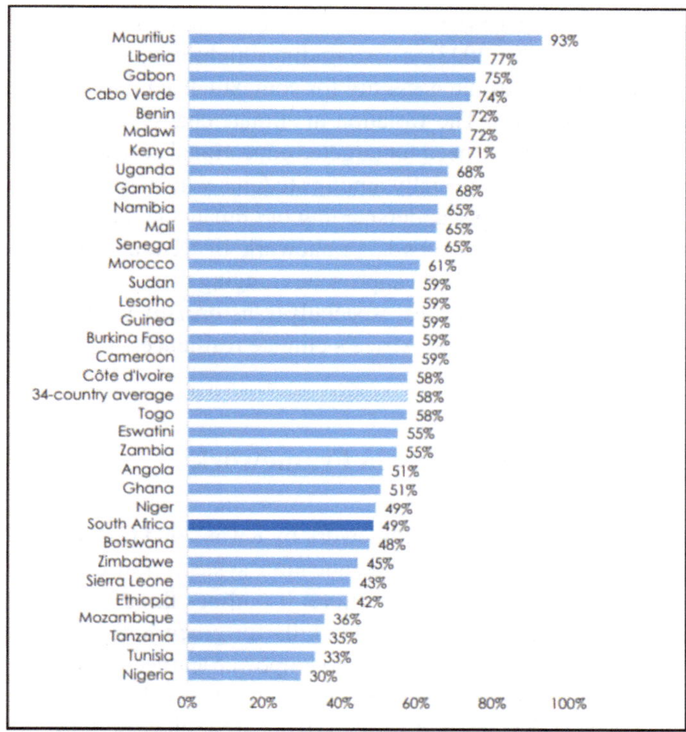

Fig. 2.3 Climate change awareness / 34 countries / 2019/2021. *Source* Afrobarometer, Dispatch No. 615, 2023, p.3

methodologies. They provide examples of how Indigenous researchers and communities have engaged in decolonising research projects that address their own needs and aspirations. These include the practices of knowledge co-creation, participatory action research and international partnership. Drawing on the authors' insights, education needs to be decolonised and indigenised to enhance climate change awareness. Additionally, education should enable and empower learners and educators to act in pursuit of climate justice and climate action in ways that are relevant to local contexts.

In 2023, 70% of 17,000 young people across 166 countries surveyed by UNESCO expressed concerns about the quality of climate change education (UNESCO, 2023). The Global Education Monitoring Report indicated that, although 9 in 10 students in Africa reported learning about climate change in school, 27% could not explain it, and 41% could only describe broad principles (UNESCO, 2023). From the same report, 91% of students wanted more comprehensive climate education, particularly emphasising practical solutions and local relevance. Many expressed concerns that climate change is often taught in a generic manner, without meaningful connections to real-world solutions (UNESCO, 2023). These findings highlight gaps within the education system and underscores the necessity for contextualised interventions, particularly at the curriculum level. The variation in climate change understanding across African countries suggests the need for targeted educational policies. Moreover, tools that enhance peer learning, especially among students who already comprehend the concepts, would foster a deeper understanding of climate change.

The New Green Learning Agenda Framework by Kwauk and Casey (2021) provides a continuum of educational approaches ranging from conformative to transformative insights. These approaches, especially transformative ones, underpin ideals of education that can contribute to Africa's climate resilience.

A New Green Learning Agenda Framework

The New Green Learning Agenda Framework by Kwauk and Casey (2021) presents a heuristic perspective for reorienting education systems towards climate action and achieving a green and sustainable future. It outlines three key approaches: (1) Technical education and training to build skills for green jobs and industries, (2) Transformative climate change education to empower learners to become agents of change, and (3) Eco-social-emotional learning to nurture environmentally conscious and resilient mindsets. Kwauk and Casey (2021) argue that the feasibility of each approach varies by context; however, all three should be pursued simultaneously to achieve climate resilience.

In essence, the New Green Learning Agenda Framework underscores the critical role of education in climate resilience. By equipping individuals with the knowledge, skills, and mindsets necessary to navigate a changing climate, education becomes a powerful tool for fostering resilience and driving sustainable development. The framework provides a continuum of education that aligns with the learning agenda (Kwauk & Casey, 2021; Lough, 2023). The continuum encompasses conformative, reformative and transformative practices, where conformative education represents 'Education as we know it' at one end of the spectrum, associated with climate injustice while 'Education for climate justice' sits at the aspirational end, as illustrated in Fig. 2.4 below:

Education as We Know It & About Climate Change

Starting with **Education as we know it**, which fails to address climate change and, as a result is complicit in perpetuating its drivers and unequal impacts (Mallow, n.d.; Wawersik, 2023), there is need to consider alternative educational models. This form of education sustains the status quo of a socio-economic system based on exploitation, extraction, consumption, and competition, which are factors that generate and

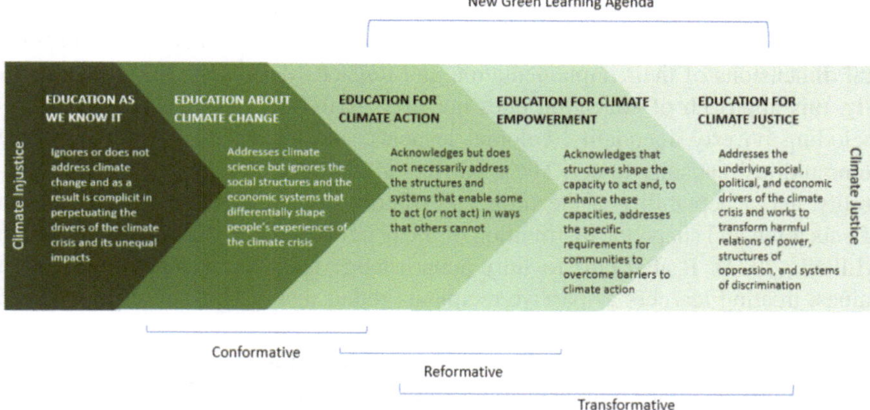

Fig. 2.4 Education continuum from education as we know it to education for climate justice. *Source* Illustration by Kwauk, 2022, p. 26

exacerbate social and environmental injustices. It also disregards Africa's diversity and complexity by imposing a standardised and homogenised curriculum that does not reflect the realities, needs, and aspirations of the learners and communities (Eilam, 2024).

The second category, **Education about climate change**, addresses climate science but overlooks the social structures that shape people's experiences of the climate crisis (Kwauk & Wyss, 2023; Xu & Iyengar, 2023). This approach provides learners with factual insights into climate change—covering greenhouse gas emissions, global warming, rising sea levels, and biodiversity loss (Calvin et al., 2023). However, it fails to account for the socioeconomic factors influencing people's contributions to and experiences of climate change, such as poverty, inequality, colonialism, and capitalism (Kwauk & Casey, 2021; Xu & Iyengar, 2023). This educational model also neglects Africa's cultural diversity and complexity, instead enforcing a universal and standardised curriculum that does not align with specific needs and aspirations of the learners and their communities.

Education for Climate Action

The third category in the continuum is **Education for Climate Action**, which acknowledges the climate crisis but does not sufficiently address the structures and systems that enable some individuals or communities to act while others are constrained (Kwauk & Casey, 2021; Xu & Iyengar, 2023). This form of education aims to motivate learners to engage in climate change mitigation and adaptation by equipping them with practical skills such as renewable energy solutions, water conservation, disaster risk reduction, and low-carbon lifestyles. However, these solutions are often framed as purely technical and neutral, ignoring their social and political dimensions of their implementation and impact (McCowan, 2023). Additionally, this approach of education does not necessarily address systemic barriers—including poverty, inequality, colonial legacies, and capitalism—that limit some communities' capacity to act. These barriers affect the access, availability, affordability, and acceptability of the skills and solutions that are proposed by this form of education and shape the distribution of the benefits and burdens of climate action (Hillock, 2024). It also fails to fully acknowledge local knowledge, culture, and values, treating learners as passive recipients rather than active participants in the design, delivery, and evaluation of the curricula.

Although **Education about climate change** and **Education for climate action** represent progress, they remain insufficient for achieving climate justice. While they acknowledge the climate crisis and promote action, they often overlook the deep-rooted social structures that shape people's experiences of climate change and their capacity to respond.

Education for Climate Empowerment

The fourth category, **Education for climate empowerment**, recognises that structural barriers impact individuals' capacity to act. To enhance agency, this approach identifies and seeks to eliminate the barriers that limit communities' ability to engage in climate change. It encompasses programmes that empower marginalised or underserved communities through the development of life skills, critical thinking, and agency (Kwauk & Casey, 2021; McGuire, 2023). Education for empowerment has been implemented in diverse global settings, including rural communities in Nigeria, secondary schools addressing climate change, and Muslim youth groups in South Florida. The core objective is to nurture youth leadership and foster skills and agency necessary for addressing social, economic, and environmental challenges (Blanchet-Cohen & Brunson, 2014). For learners, this approach enhances cognitive, socio-emotional, and behavioural outcomes, including increased climate literacy, critical thinking, creativity, collaboration, communication, and problem-solving skills (Fields et al., 2023). For educators, it facilitates professional opportunities, curriculum innovation, and pedagogical transformation. At a societal level, it contributes to reducing greenhouse gas emissions, promoting green jobs, and fostering civic engagement and democracy (Hügel & Davies, 2024; Okada & Gray, 2023).

Education for Climate Justice

The fifth and final category in the continuum is **Education for climate justice**, which addresses the underlying social, political, and economic causes of the climate crisis and works to transform harmful power structures, systems of oppression and patterns of discrimination (Ryder, 2024). This is a transformative approach to education focusing on the root of climate injustice and working to dismantle oppressive systems that sustain environmental inequalities (Kwauk & Casey, 2021; Rolleston et al., 2023). Sakellari (2023) asserts that a strong commitment to climate activism is intrinsically tied to social justice, suggesting that incorporating climate justice as a focal point in climate change education can inspire greater engagement and action.

Unlike other approaches, **Education for climate justice** goes beyond simply raising awareness, fostering action or empowerment—it actively challenges the socio-political order and reimagines alternative, sustainable futures. It scrutinises the root causes of climate vulnerability, including the fossil fuel industry, neoliberal economic policies, colonial legacies, patriarchal norms, and racial inequalities (Adebisi, 2016; Kayira, 2015; Trott et al., 2023).

Rather than treating learners as passive recipients of knowledge, this approach empowers them to critically engage with climate injustices and participate in collective action and policy advocacy for climate justice. It fosters skills, attitudes, and values that support climate activism, community engagement, and the co-creation of equitable solutions (Harris et al., 2023; Roderick, 2023; Trott et al., 2023). Kwauk and Casey (2021) argue that this revolutionary and transformative educational approach is essential for addressing the urgent and multifaceted nature of the climate crisis. At the highest level of the continuum, **Education for climate justice** fully embodies climate justice principles. It not only seeks to mitigate climate change but also aims to rectify historical injustices and redistribute power, resources, and decision-making authority to marginalised communities. This approach is especially crucial for Africa, where communities face disproportionate climate impacts due to colonial legacies, socio-economic disparities, and a limited access to climate finance. In light of the urgency to establish climate change education frameworks that promote climate justice in Africa, we advance a number of pedagogical approaches that align with transformative climate change education.

Pedagogical Approaches for a Climate Resilient Africa

The New Green Learning Agenda framework, proposed by Kwauk and Casey (2021), serves as the foundation for five pedagogical strategies that are integral to fostering climate resilience in Africa. These strategies encompass place-based, decolonised, experiential, and holistic education approaches. Rather than being isolated, these pedagogical strategies are interconnected, with each reinforcing and being reinforced by the others. They are context-specific and designed to address the unique environmental challenges and socio-economic conditions of each region in Africa. By recognising cultural nuances, values, and local realities, these strategies can enrich the learning experiences of African students, thereby nurturing a resilient and sustainable future for the continent.

Place-Based Education (PBE)

This is a pedagogical approach that emphasises the connection between the learning process and the geographical place in which students are located. PBE underscores the importance of localised learning, which is deeply rooted in the unique history, environment, socio-cultural context, economy, literature, art, experiences, and meanings of a given locality—extending beyond the formal school setting. The approach gained renewed attention during the outbreak of the COVID-19 pandemic, which caused large-scale school closures and the adoption of alternative learning environments, including teaching outdoors and from home (Yemini et al., 2023). It underscores the value of educational practices that prioritise contextual, experiential, and community-based learning to foster connections with local cultures, environments, and challenges (Nichols et al., 2016; Smith & Sobel, 2014).

The PBE model engages the local community and environment as the starting point for curriculum design. Smith and Sobel (2014) maintain that PBE can help students develop a deeper connection to their locality and culture, a critical understanding of the global issues and their local implications, and a sense of responsibility and agency in promoting environmental and social justice. Some examples of PBE activities for Climate Change Education in Africa are: (1) Exploring and documenting the local flora and fauna and their adaptations to climate change, (2) conducting community surveys and interviews to identify local environmental problems and solutions, and (3) organising cultural events and exchanges to celebrate and share the Indigenous knowledge and practices for environmental stewardship (Hügel & Davies, 2024; Kassim et al., 2024).

Concerning climate resilience in Africa, place-based education anchors learning in local realities and experiences, equipping learners with the knowledge and skills necessary to assess and respond to the impacts of climate change in their communities (Hügel & Davies, 2024). This approach, underpinned by The New Green Learning Agenda Framework (Kwauk & Casey, 2021) empowers learners to actively participate in climate resilience-building, fostering a sense of ownership and agency in tacking environmental challenges. When effectively implemented, PBE can serve as a powerful tool for promoting climate resilience in Africa. By fostering a comprehensive understanding of local environments and global issues, and by encouraging active engagement and responsibility, it empowers learners to contribute meaningfully to climate resilience efforts in their communities.

Decolonised Education

This approach aims to empower learners to critically analyse and dismantle colonial and neocolonial legacies that have contributed to environmental degradation and the marginalisation of Indigenous African epistemologies and cultural practices (Fanon, 1967a, 1967b). Decolonised education integrates Indigenous ways of knowing that respects and incorporates the cultural backgrounds of students into the learning process, with a particular focus on Indigenous knowledge systems and practices. It equips learners with the tools to assess and challenge prevailing narratives and practices that have historically undermined African knowledge systems and exacerbated ecological decline. The process goes beyond analysing curriculum materials and developing a decolonisation model for education (Adebisi, 2016; Khupe et al., 2024; Olstead & Chattopadhyay, 2024; Stein, 2019). It also fosters the reclamation and affirmation of African identities, agency, and creativity, nurturing a positive self-concept and resilience against internalised subjugation. For instance, educational activities within this framework include: (1) Critically analysing and deconstructing media and curriculum content that portray Africa in a biased manner, (2) documentating African contributions to environmental and social justice, and (3) expressing independent narratives and visions for a sustainable future through artistic and media channels (Neag et al., 2021). Moreover, the decolonisation of education offers a potent counter-narrative to the dominant knowledge paradigms that are deeply-rooted in Euro-American educational traditions (Smith, 2021). By engaging with Indigenous communities and their deep-rooted connection to the land and natural world, learners can gain invaluable insights and practices that promote ecological balance and sustainability. This approach not only facilitates the preservation of Indigenous knowledge pertinent to climate change mitigation and adaptation but also advocates for the rights and aspirations of Indigenous peoples, their languages, cultures, and traditions.

In essence, decolonised education embodies dynamic, experiential, holistic, and relational forms of knowledge that function as tools of resistance. These forms of knowledge are crucial in the decolonisation of climate change education and the empowerment of African communities. By fostering an educational ethos that values and integrates these diverse epistemologies, we can cultivate a generation of learners equipped to navigate and address the complexities of climate change, thereby contributing to the creation of a more resilient and equitable society. This vision aligns with the transformative goals highlighted in The New Green Learning Agenda Framework, which advocates for pedagogical models that drive both technical and social transformation (Eilam, 2024; Fricker, 2007; Kwauk & Casey, 2021).

Experiential Education

This is a pedagogical approach that promotes learning through first-hand, direct engagement and reflection to facilitate a deeper understanding of the subject matter (Adusu et al., 2023; Marx, 2023). This approach enables learners to acquire and apply practical knowledge and skills that are essential for coping with and contributing to climate change mitigation and adaptation. It also fosters attitudes and values such as empathy, solidarity, and citizenship.

Some examples of experiential education activities for Climate Change Education in Africa include: (1) Conducting experiments and observations to understand the scientific principles behind climate change, (2) engaging in service-learning and community-based projects to address local environmental challenges, and (3) developing and implementing action plans and campaigns to raise awareness and advocate for policy and behavioural changes (Kolb, 1984). Experiential learning is based on the principle that people learn best through experience. Kolb's (1984) experiential learning model outlines four key stages in the learning cycle: concrete experience, reflective observation, abstract conceptualisation, and active experimentation. The approach presents a typology of individual learning styles, which are based on the preferences and strengths of learners in different stages of the learning cycle.

A study by Adusu et al. (2023) on climate perceptions in Sunyani, Ghana, highlights the gap between scientific knowledge and local experiences of climate change. The residents, despite lacking scientific expertise, recognise the signs of climate change through their lived experiences. The study reinforces the need for experiential learning models that integrate local knowledge with scientific evidence to enhance climate change adaptation and mitigation strategies.

The integration of experiential education within The New Green Learning Agenda Framework is essential for building climate resilience. By combining hands-on learning experiences with a deep understanding of environmental challenges, education can empower individuals and communities to drive sustainable and equitable solutions. This approach not only equips learners with the necessary technical skills but also instils the values and attitudes needed to navigate the complexities of climate change, making education a cornerstone of a climate-resilient African society.

Holistic Education

As conceptualised by Miller (2000), holistic education represents a transformative, postmodern, ecological, and spiritual pedagogical approach that addresses global environmental challenges. It is grounded in the principles of balance, inclusion, and connectedness and challenges the fragmented, reductionist assumptions of mainstream education (Miller, 2019). This educational philosophy seeks to reshape human relationships with one another and the natural world, fostering a collective consciousness essential for tackling climate change and environmental challenges (Miseliunaite et al., 2022; Rudge, 2021). Holistic education is vital for fostering a climate-resilient Africa, as it stresses the interconnectedness between humans and their environment, both living and non-living. It aims to cultivate a sense of collective responsibility that is essential for addressing climate change's complex challenges (Hare, 2006; UNESCO, 2016). This approach also advocates for the integration of all effective solutions, be they scientific advancements or Indigenous knowledge systems, to facilitate climate change adaptation or mitigation (Jiri et al., 2015; Mbah et al., 2021). Holistic education is not merely an academic concept but a practical framework for societal transformation. Through this perspective, education becomes a crucial tool for climate action empowering individuals and communities to adopt proactive environmental practices.

Conclusion

Evidence indicates climate change is having a devasting impact across Africa. The critical role of education in fostering a climate-resilient continent cannot be overstated. It is through education that climate literacy is cultivated, enabling individuals to become more informed about climate change and its far-reaching consequences. The knowledge and skills imparted through education are fundamental for meaningful climate action and essential for advocating for fair and effective climate change adaptation and mitigation policies. However, current educational models, including those associated with climate change education in Africa are largely shaped by Western narratives and fail to incorporate the continent's diverse Indigenous knowledge systems and local realities. Western-centric education models are inadequate for addressing the unique challenges and opportunities that climate change presents in Africa. Moreover, they fail to equip learners with the necessary tools to drive sustainable solutions within their own communities. To secure Africa's future in light of climate change, education must demonstrate relevance by aligning with local contexts and needs. Specifically, climate change education must be context-sensitive, decolonised, experiential, and holistic. This necessitates a shift from climate change education as we currently know it towards a transformational model that actively promotes climate justice.

References

Adebisi, F. I. (2016). Decolonising education in Africa: Implementing the right to education by re-appropriating culture and indigeneity. *Northern Ireland Legal Quarterly, 67*, 433.

Adelekan, I. O., Simpson, N. P., Totin, E., & Trisos, C. H. (2022). IPCC Sixth Assessment Report (AR6): Climate change 2022-Impacts, adaptation and vulnerability: Regional factsheet Africa.

Adusu, D., Anaafo, D., Abugre, S., & Addaney, M. (2023). Experiential knowledge of urbanites on climatic changes in the Sunyani municipality, Ghana. *Journal of Urban Affairs, 45*(3), 488–504.

Afrobarometer. (2023). Despite growing evidence, climate change is still unknown to many South Africans. Dispatch No. 615. Retrieved 05 July 2024, from https://www.afrobarometer.org/wp-content/uploads/2023/03/AD615-Climate-change-still-unknown-to-many-South-Africans-Afrobarometer-8mar22.pdf.

Anderson, A. (2010). Combating climate change through quality education. Brookings Global Economy and Development Washington, DC.

Apollo, A., & Mbah, M. F. (2021). Challenges and opportunities for climate change education (CCE) in East Africa: A critical review. *Climate, 9*(6), 93.

AU (2022). African Union Climate Change and Resilient Development Strategy and Action Plan. Retrieved 05 July 2024, from https://au.int/sites/default/files/documents/41959-doc-CC_Strategy_and_Action_Plan_2022-2032_08_02_23_Single_Print_Ready.pdf.

Battiste, M. (2013). *Decolonizing education: Nourishing the learning spirit*. Purich Publishing.

Becken, S., & Coghlan, A. (2024). Knowledge alone won't "fix it": Building regenerative literacy. *Journal of Sustainable Tourism, 32*(2), 385–401.

Bedasa, Y., & Deksisa, K. (2024). Food insecurity in East Africa: An integrated strategy to address climate change impact and violence conflict. *Journal of Agriculture and Food Research*, 100978.

Birkmann, J., Liwenga, E., Pandey, R., Boyd, E., Djalante, R., Gemenne, F., Leal Filho, W., Pinho, P., Stringer, L., & Wrathall, D. (2022). Poverty, livelihoods and sustainable development.

Blanchet-Cohen, N., & Brunson, L. (2014). Creating settings for youth empowerment and leadership: An ecological perspective. *Child & Youth Services, 35*(3), 216–236.

Calvin, K., Dasgupta, D., Krinner, G., Mukherji, A., Thorne, P., Trisos, C., Romero, J., Aldunce, P., Barrett, K., & Blanco, G. (2023). IPCC, 2023: Climate change 2023: Synthesis report. In *Contribution of Working Groups I, II and III to the Sixth Assessment Report of the Intergovernmental Panel on Climate Change*. IPCC, Geneva.

Case Studies on Adaptation and Climate Resilience in Schools and Educational Settings. (n.d.). Global Center on Adaptation. Retrieved 24 January 2024, from https://gca.org/reports/case-studies-on-adaptation-and-climate-resilience-in-schools-and-educational-settings/.

Charbonnier, J. (2023). Exploring climate literacy: Unveiling the motivation and skills of climate-literate higher education graduates. https://gupea.ub.gu.se/handle/2077/77341.

Chirisa, I., & Matamanda, A. R. (2022). Science communication for climate change disaster risk management and environmental education in Africa. In *Research anthology on environmental and societal impacts of climate change* (pp. 636–652). IGI Global.

Country Index // Notre Dame Global Adaptation Initiative // University of Notre Dame. (n.d.). Retrieved March 14, 2024, from https://gain.nd.edu/our-work/country-index/.

Daniels, E., Bharwani, S., Swartling, Å. G., Vulturius, G., & Brandon, K. (2020). Refocusing the climate services lens: Introducing a framework for co-designing "transdisciplinary knowledge integration processes" to build climate resilience. *Climate Services, 19*, 100181.

Dei, G. J. S. (2000). Rethinking the role of indigenous knowledges in the academy. *International Journal of Inclusive Education, 4*(2), 111–132.

Digital, S. (n.d.). Programs for men who have used violence against women: Recommendations for Action and Caution. Save the Children's Resource Centre. Retrieved 14 August 2022, from https://resourcecentre.savethechildren.net/document/programs-men-who-have-used-violence-against-women-recommendations-action-and-caution/.

Eilam, E. (2024). Interrogating climate change education epistemology: Identifying hindrances to curriculum development. *ECNU Review of Education*, 20965311241240491.

Fanon, F. (1967a). *Black skin, white masks*. Grove Press.

Fanon, F. (1967b). *The Wretched of the Earth*. Penguin.

Fields, L., Moroney, T., Perkiss, S., & Dean, B. A. (2023). Enlightening and empowering students to take action: Embedding sustainability into nursing curriculum. *Journal of Professional Nursing, 49*, 57–63.

Fricker, M. (2007). Epistemic injustice: Power and the ethics of knowing. Oxford University Press.

Gandhi, K. A., Seba, R. O. C., & Hadiwijoyo, S. S. (2024). Implementation of the Paris agreement in handling climate change due to forest fires in Indonesia 2015–2019. *Jurnal Impresi Indonesia, 3*(2), 145–154.

Gansser, O. A., & Reich, C. S. (2023). Influence of the new ecological paradigm (NEP) and environmental concerns on pro-environmental behavioral intention based on the theory of planned behavior (TPB). *Journal of Cleaner Production, 382*, 134629.

Gay, G. (2010). *Culturally responsive teaching: Theory, research, and practice*. Teachers College Press.

Gebrechorkos, S. H., Taye, M. T., Birhanu, B., Solomon, D., & Demissie, T. (2023). Future changes in climate and hydroclimate extremes in East Africa. *Earth's Future, 11*(2), e2022EF003011.

Gebrekidan, T. K., & Gebremedhin, G. G. (2024). Integration and effectiveness of formal environmental education in Africa and India: Review. *European Journal of Sustainable Development Research, 8*(2).

Gyimah, A. B. K., Antwi-Agyei, P., & Adom-Asamoah, G. (2023). Educating the rural woman farmer for climate resilience in the global south: enablers and barriers. In *University Initiatives on Climate Change Education and Research* (pp. 1–23). Springer.

Hare, J. (2006). Towards an understanding of holistic education in the middle years of education. *Journal of Research in International Education, 5*(3), 301–322.

Harnessing Education for Effective Climate Action. (n.d.). World Bank Live. Retrieved 24 January 2024, from https://live.worldbank.org/en/event/2022/cop27-harnessing-education-effective-climate-action.

Harnessing education to build climate resilience—World | ReliefWeb. (2024). https://reliefweb.int/report/world/harnessing-education-build-climate-resilience.

Haro, A., Mendoza-Ponce, A., Calderón-Bustamante, Ó., Velasco, J. A., & Estrada, F. (2021). Evaluating risk and possible adaptations to climate change under a socio-ecological system approach. *Frontiers in Climate, 3*, 674693.

Harris, E., Kang, P., Manango, J., & Melwani, N. (2023). Fueling change: Rethinking education for climate justice in BC schools.

Hillock, S. (2024). *Greening social work education*. University of Toronto Press.

Hoydis, J., Bartosch, R., & Gurr, J. M. (2023). *Climate change literacy*. Cambridge University Press.

Hu, X., & Zhao, Y. (2024). Decoding the green supply chain: Education as the key to economic growth and sustainability. *Environmental Science and Pollution Research, 31*(6), 9317–9332.

Hügel, S., & Davies, A. R. (2024). Expanding adaptive capacity: Innovations in education for place-based climate change adaptation planning. *Geoforum, 150*, 103978.

Impact Report 2020–2022. (n.d.). Retrieved 24 January 2024, from https://www.rockefellerfoundation.org/reports/impact-report-2023/?utm_source=facebook&utm_medium=organic_social&utm_campaign=impactreport&utm_content=granteetoolkitoverview.

IPCC. (2023). Climate change 2023: Synthesis report. In Core Writing Team, H. Lee & J. Romero (Eds.) *Contribution of Working Groups I, II and III to the Sixth Assessment Report of the Intergovernmental Panel on Climate Change*. IPCC. https://www.ipcc.ch/report/ar6/syr/downloads/report/IPCC_AR6_SYR_LongerReport.pdf.

Jiri, O., Mafongoya, P. L., & Chivenge, P. (2015). Indigenous knowledge systems, seasonal "quality" and climate change adaptation in Zimbabwe. *Climate Research, 66*(2), 103–111.

Kangama, C. O. (2024). African agriculture facing climate change. *Journal of Agricultural Earth & Environmental Sciences, 3*(1), 1–2.

Kassim, N. F., Saidon, M. K., Bahador, Z., & Zulkifli, M. I. (2024). A scoping review of place-based education in a context of environmental education, outdoor education, and history education. *Proceeding of International Conference on Multidisciplinary Research, 6*(2), 369–378.

Kayira, J. (2015). (Re) creating spaces for uMunthu: Postcolonial theory and environmental education in southern Africa. *Environmental Education Research, 21*(1), 106–128.

Kayira, J., Lobdell, S., Gagnon, N., Healy, J., Hertz, S., McHone, E., & Schuttenberg, E. (2022). Responsibilities to decolonize environmental education: A co-learning journey for graduate students and instructors. *Societies, 12*(4). https://doi.org/10.3390/soc12040096

Khupe, C., Seehawer, M., & Keane, M. (2024). Decolonising curriculum policy research through community-centredness. In *Analysing Education Policy* (pp. 201–214). Routledge.

Kolb, D. A. (1984). *Experiential learning: Experience as the source of learning and development.* Prentice Hall.

Kumar, P., Sahani, J., Rawat, N., Debele, S., Tiwari, A., Emygdio, A. P. M., Abhijith, K. V., Kukadia, V., Holmes, K., & Pfautsch, S. (2023). Using empirical science education in schools to improve climate change literacy. *Renewable and Sustainable Energy Reviews, 178*, 113232.

Kusangaya, S., Mazvimavi, D., Shekede, M. D., Masunga, B., Kunedzimwe, F., & Manatsa, D. (2021). Climate change impact on hydrological regimes and extreme events in southern Africa. *Climate Change and Water Resources in Africa: Perspectives and Solutions towards an Imminent Water Crisis* (87–129).

Kwauk, C. T., & Wyss, N. (2023). Gender equality and climate justice programming for youth in low-and middle-income countries: An analysis of gaps and opportunities. *Environmental Education Research, 29*(11), 1573–1596.

Kwauk, C. T. (2022). Towards climate justice: Lessons from girls' education. In *"Education in Times of Climate Change" NORRAG SPECIAL ISSUE 07*. Reteieved 04 July 2024, from https://resources.norrag.org/resource/view/732/424.

Kwauk, C., & Casey, O. (2021). A new green learning agenda: Approaches to quality education for climate action. Center for Universal Education at The Brookings Institution.

Le Grange, L. (2016). Decolonising and Africanising curriculum studies in South Africa: Tensions and possibilities. *South African Journal of Higher Education, 30*(2), 69–83. https://doi.org/10.20853/30-2-549

Léna, P., & Lescarmontier, L. (2023). Climate education. In *Handbook of the Anthropocene: Humans between Heritage and Future* (pp. 1401–1407). Springer.

Lin, B., & Agyeman, S. D. (2020). Assessing Sub-Saharan Africa's low carbon development through the dynamics of energy-related carbon dioxide emissions. *Journal of Cleaner Production, 274*, 122676.

Lotz-Sisitka, H., Wals, A. E., Kronlid, D., & McGarry, D. (2016). Transformative, transgressive social learning: Rethinking higher education pedagogy in times of systemic global dysfunction. *Current Opinion in Environmental Sustainability, 20*, 73–80. https://doi.org/10.1016/j.cosust.2016.04.017

Lough, D. M. (2023). Curriculum and learning for climate action: Toward an SDG 4.7 roadmap for systems change. In R. Iyengar, & C. T. Kwauk (Eds.) Boston: Brill, 2021, 368 pp. 52(paperback), 132 (hardback), ISBN 978-90-04-47181-8 (eBook); ISBN 978-90-04-47180-1 (hardcover); ISBN 978-90-04-47179-5 (paperback). Taylor & Francis.

Mallow, M. S. (n.d.). Flexible working arrangement and flexible education: How flexibility can transform education as we know it.

Marx, K.-A. H. (2023). Experiential education and climate change.

Mbah, M. F., & Ezegwu, C. (2024). The decolonisation of climate change and environmental education in Africa. *Sustainability, 16*(9), 3744.

Mbah, M., Ajaps, S., & Molthan-Hill, P. (2021). A systematic review of the deployment of indigenous knowledge systems towards climate change adaptation in developing world contexts: Implications for climate change education. *Sustainability, 13*(9), 4811.

McCowan, T. (2023). The climate crisis as a driver for pedagogical renewal in higher education. *Teaching in Higher Education, 28*(5), 933–952.

McGuire, C. (2023). Culture Greening Communities Brief.

Miller, J. P. (2007). *The holistic curriculum.* University of Toronto Press.

Miller, J. P. (2019). The holistic curriculum. University of Toronto press.

Miller, R. (2000). Beyond reductionism: The emerging holistic paradigm in education. *The Humanistic Psychologist, 28*(1–3), 382–393.

Miseliunaite, B., Kliziene, I., & Cibulskas, G. (2022). Can holistic education solve the world's problems: A systematic literature review. Sustainability. 2022; 14: 9737. Note: MDPI stays neutral with regard to jurisdictional claims in published.

Molthan-Hill, P., & Winfield, F. (2023). Climate literacy training for all. In *The Handbook of Carbon Management* (pp. 33–55). Routledge.

Mullen, N. (2022). Integrating green education into Ghana's basic education curriculum. *World Education*. https://worlded.org/integrating-green-education-into-ghanas-basic-education-curriculum/.

ND-GAIN. (2023). Country index scores. Reteieved 03 July 2024, from Country Index // Notre Dame Global Adaptation Initiative // University of Notre Dame.

Neag, A., Bozdag, C., & Leurs, K. (2021). Media literacy education for diverse societies.

Nelson-Barber, S., Rechebei, E. D., Limes, J. T., & Johnson, Z. (2023). Developing a framework for integrating systems of local indigenous knowledge with climate education in the Mariana Islands. In *Indigenous STEM Education: Perspectives from the Pacific Islands, the Americas and Asia* (Vol. 1, pp. 79–91). Springer.

Nichols, J. B., Howson, P. H., Mulrey, B. C., Ackerman, A., & Gately, S. E. (2016). The promise of place: Using place-based education principles to enhance learning. *The International Journal of Pedagogy and Curriculum, 23*(2), 27–41. https://doi.org/10.18848/2327-7963/CGP/v23i02/27-41

Nusche, D., Rabella, M. F., & Lauterbach, S. (2024). Rethinking education in the context of climate change: Leverage points for transformative change.

Ofori, S. A., Cobbina, S. J., & Obiri, S. (2021). Climate change, land, water, and food security: Perspectives from Sub-Saharan Africa. *Frontiers in Sustainable Food Systems, 5*, 680924.

Okada, A., & Gray, P. (2023). A climate change and sustainability education movement: Networks, open schooling, and the 'CARE-KNOW-DO'Framework. *Sustainability, 15*(3), 2356.

Olivier, C., Engelbrecht, C., Bopape, M.-J., & Botai, J. (2020). Status of climate in South Africa and predictions for the 2020–2021 summer season.

Olstead, R., & Chattopadhyay, S. (2024). Circles and lines: Indigenous ontologies and decolonising climate change education. *Settler Colonial Studies, 14*(1), 41–58.

Ostrom, E. (2009). A general framework for analyzing sustainability of social-ecological systems. *Science, 325*(5939), 419–422. https://doi.org/10.1126/science.1172133

Pan, W.-L., Fan, R., Pan, W., Ma, X., Hu, C., Fu, P., & Su, J. (2023). The role of climate literacy in individual response to climate change: Evidence from China. *Journal of Cleaner Production, 405*, 136874. https://doi.org/10.1016/j.jclepro.2023.136874

Rafique, K., Abbas, S., Abbas, H., & Ullah, K. (2024). Assaying ramifications of climate change over productivity growth in developing countries. *Gondwana Research, 130*, 278–290. https://doi.org/10.1016/j.gr.2024.01.014

Roderick, T. (2023). Teach for climate justice: A vision for transforming education. Harvard Education Press.

Rolleston, C., Nyerere, J., Brandli, L., Lagi, R., & McCowan, T. (2023). Aiming higher? Implications for higher education of students' views on education for climate justice. *Sustainability, 15*(19), 14473.

Rudge, L. (2021). The growth of independent education alternatives in New Zealand. *International Journal of Progressive Education, 17*(6), 324–354.

Ryder, A. (2024). "We're here, the next generation": Exploring possibilities for children's citizenship in local climate policy in Ōtautahi Christchurch. Open Access Te Herenga Waka-Victoria University of Wellington.

Sakellari, M. (2023). Key opinion shapers' perceptions of climate migration: Why and how to put climate justice at the centre of climate change education. *Open Research Europe, 3*(213), 213.

Scholes, L. (2024). Reading for digital futures: A lens to consider social justice issues in student literacy experiences in the digital age. *Cambridge Journal of Education, 54*(1), 71–88.

Serrano-Candela, F., Estrada, F., Raga, G., & Salazar, C. G. (2024). Development of metrics and indices for assessing national progress on climate risk, adaptation and dissonance based on observations, perceptions, and projections.

Sherpa, P. D. (2018). Interfacing indigenous knowledge and climate change education. *Journal of Education and Research, 7*(1), 52–64. https://doi.org/10.3126/jer.v7i1.21240

Singh, S., & Shah, J. (2022). Case studies on adaptation and climate resilience in schools and educational settings.

Smilan, C. (2023). Visualizing climate change: Here we come to save the day!

Smith, G. A., & Sobel, D. (2014). *Place-and community-based education in schools*. Routledge.

Smith, L. T. (2021). Decolonizing methodologies: Research and indigenous peoples. Bloomsbury Publishing.

Sobel, D. (2004). *Place-based education: Connecting classrooms and communities*. Orion Society.

Ssekamatte, D. (2022). The role of the university and institutional support for climate change education interventions at two African universities. http://umispace.umi.ac.ug:80/xmlui/handle/20.500.12305/1421.

Stein, S. (2019). The ethical and ecological limits of sustainability: A decolonial approach to climate change in higher education. *Australian Journal of Environmental Education, 35*(3), 198–212.

Svarstad, H. (2021). Critical climate education: Studying climate justice in time and space. *International Studies in Sociology of Education, 30*(1–2), 214–232. https://doi.org/10.1080/09620214.2020.1855463.

Teixeira, J. E., & Crawford, E. (2022). Climate change education and curriculum revision.

Thornton, S., Graham, M., & Burgh, G. (2019). Reflecting on place: Environmental education as decolonisation. *Australian Journal of Environmental Education, 35*(3), 239–249.

Trisos, C., Krönke, M., Simpson, N. P., & Andrews, T. M. (2021). Africa's first continent-wide survey of climate change literacy finds education is key. *The Conversation*. http://theconversation.com/africas-first-continent-wide-survey-of-climate-change-literacy-finds-education-is-key-169426.

Trott, C. D., Lam, S., Roncker, J., Gray, E.-S., Courtney, R. H., & Even, T. L. (2023). Justice in climate change education: A systematic review. *Environmental Education Research, 29*(11), 1535–1572.

UNESCO promotes climate change education in Africa | UNESCO. (n.d.). Retrieved 14 March 2024, from https://www.unesco.org/en/articles/unesco-promotes-climate-change-education-africa.

UNESCO. (2016). Education 2030: Incheon declaration and framework for action for the implementation of sustainable development goal 4: Ensure inclusive and equitable quality education and promote lifelong learning opportunities for all; UNESCO: Paris, France, 2016; pp. 24–29. https://unesdoc.unesco.org/ark:/48223/pf0000245656.

UNESCO. (2017). Education for sustainable development goals: learning objectives. UNESCO. https://unesdoc.unesco.org/ark:/48223/pf0000247444.

UNESCO. (2021). Getting every school climate-ready: How countries are integrating climate change issues in education.

UNESCO. (2023). Climate change communication and education country profiles: Approaches to greening education around the world. Unesdoc.unesco.org. https://unesdoc.unesco.org/ark:/48223/pf0000387867.

Valentín, B., Abreu, D., Ramírez, O., Bynoe, P., Ferguson, T., Simmons, D., Gokool-Ramdoo, S., Lotz-Sisitka, H., Mandikonza, C., & Sweeney, D. (2015). Not just hot air: Putting climate change education into practice.

Wainwright, C. M., Finney, D. L., Kilavi, M., Black, E., & Marsham, J. H. (2021). Extreme rainfall in East Africa, October 2019–January 2020 and context under future climate change. *Weather, 76*(1), 26–31.

Wawersik, D. (2023). Plenary: Revolutionizing education through the power of simulation.

World Bank. (2020). CO_2 emissions (metric tons per capita). https://data.worldbank.org/indicator/EN.ATM.CO2E.PC.

Xu, L. C., & Iyengar, R. (2023). Climate change education: An earth institute sustainability primer. Columbia University Press.

Yemini, M., Engel, L., & Simon, A. B. (2023). Place-based education—a systematic review of literature. *Educational Review*, 1–21. https://doi.org/10.1080/00131911.2023.2177260

Yiwen, Z., Maxwell, S., Runting, R., Venter, O., Watson, J., & Carrasco, L. R. (2020). Environmental destruction not avoided with the Sustainable Development Goals.

Zheng, L., Umar, M., Safi, A., & Khaddage-Soboh, N. (2024). The role of higher education and institutional quality for carbon neutrality: Evidence from emerging economies. *Economic Analysis and Policy, 81*, 406–417.

Zhu, L., & Wang, Y. (2024). Entrepreneurship and carbon footprints in Sub-Saharan Africa. *Problemy Ekorozwoju, 19*(1), 221–231.

[illegible]

Chapter 3
Bridging Generations Using Climate Change Education in African Early Years to Tertiary Levels

Bhagwan Das and Tony Jan

Abstract This chapter offers an overview of climate change education strategies across four distinct educational stages in Africa, focusing on the proposed frameworks tailored for each level. Rather than analysing the current state of climate change teaching, it introduces innovative approaches designed to enhance educational practices and outcomes. These frameworks were developed to provide adaptable solutions that can be customized to meet the specific needs of different educational contexts within the continent. The Pre-Primary Education framework encompasses childhood awareness, nature activities, environmental stewardship, social media showcase, behaviour impact, and attitude shifts. Primary schools prioritize integrative methods, cohesive teaching blueprints, hurdle addressing, educator development, and hands-on guidance. At the post-secondary level, the framework emphasizes the importance of academic research, specialized courses, climate research labs, industry project collaborations, and active engagement in Sustainable Development Goals (SDGs) and Technical Societies such as Institute of Electrical and Electronics Engineers (IEEE). The framework within Technical and Vocational Education and Training (TVET) institutions encompasses the integration of syllabus incorporation, experiential education, cross-disciplinary methodology, industrial collaborations, pedagogical enhancement, climate-adaptive competencies, stimulating innovations, assessment and oversight, international collaboration, and vocational counsel. Secondary schools emphasize the integration of different disciplines, specialized elective courses, discussions about the benefits, evaluation of the impact of electives, functionality of extracurricular activities, and outstanding achievements. The conclusion emphasizes the crucial significance of climate change education at all levels, with future directions focusing on policy modifications, research, and continuous endeavours. Suggestions involve integrating visual aids, such as charts and graphs, to demonstrate the progress of climate change education programs and measures such as school acceptance rates, student engagement, and influence on awareness. This framework offers customized strategies for each educational level with the

B. Das (✉) · T. Jan
Centre for Artificial Intelligence Research and Optimization (AIRO), Design and Creative Technology, Torrens University, New South Wales, Australia
e-mail: bhagwan.das@torrens.edu.au

M. F. Mbah et al. (eds.), *Practices, Perceptions and Prospects for Climate Change Education in Africa*, https://doi.org/10.1007/978-3-031-84081-4_3

goal of developing an ecologically aware generation capable of tackling the intricate challenges of climate change in Africa.

Keywords Climate education · Cross-disciplinary climate curricula · Eco-conscious learning models · Innovative pedagogical practices · Sustainable development engagement

Introduction

The devastating effects of climate change on our planet's ecosystems, communities, and economy make it one of the most critical environmental challenges that we face today. The global community is hindered in its ability to address and mitigate the effects of climate change due to a serious lack of climate education at all levels, despite the undeniable importance of this issue.

The global climate change vulnerability dilemma requires a comprehensive answer. The global community recognizes the gravity of this issue and offers several climate change mitigation plans (Seddon, 2021). Globally, the adoption of renewable energy sources is crucial. Renewable energy sources such as solar and wind power may dramatically reduce greenhouse gas emissions (Gielen, 2019). Solar and wind energy constitute a large share of the 2800 gigawatts of worldwide renewable energy capacity as of 2023 (Luo, 2022). This shift further reduces climate change, improves energy security, and fosters sustainable development.

Climate change education empowers individuals and communities to actively create resilience (Bell, 2024). We can build a global community that can handle climate change by incorporating climate change education into the curriculum and by raising awareness at all levels. However, educational institutions segregate disciplines, which makes it difficult to appreciate the interconnectedness of climate change. Education regarding climate change involves learning in the face of danger, uncertainty, and rapid change. The current global scenario is unprecedented (Geneva, 2013). We cannot guarantee young people a steady environment in their lifetime, so how do we teach? How can pupils engage in and study in an uncertain future? These questions are relevant to educators worldwide. The good news is that global public awareness of climate change action has grown in recent years. Climate change education can address and develop effective solutions. The program teaches students about climate change, prepares them for their effects, and encourages them to choose sustainable lives (Nepraš, 2022). Climate change and climate change education are global issues that may be included in the curriculum to teach local lessons and broaden climate change mitigation mindset. Learning how to deal with climate change is greater than climate change literacy (Kuthe, 2020). Policymakers, educators, communities, and stakeholders must collaborate to create and implement comprehensive, accessible, and current climate education programs that empower people to understand, confront, and adapt to climate change.

The major focus of this chapter is the complex issue that hinders the effectiveness of climate education, specifically when it comes to teaching climate change at different educational levels ranging from early years to university education. The issue at hand involves insufficient incorporation of the curriculum, lack of teacher readiness, use of methodology that is not suitable for the children's age, and a failure to adopt interdisciplinary methods. These factors impede students' ability to build a complete knowledge of climate change. Climate change education is a frequently overlooked yet potent approach for mitigating vulnerability to climate change (Ngcamu, 2023). Providing education to people at all stages of development, from early childhood to higher education, is crucial for constructing a strong and knowledgeable society. In Africa, where susceptibility to climate change is conspicuous, primary strategies encompass afforestation, reforestation, and climate-smart agriculture (Belay, 2023). Efforts such as Ethiopia's tree-planting efforts effectively address deforestation, save biodiversity, and provide valuable ecosystem services (Beyene, 2023). Climate-smart agriculture, as demonstrated by successful initiatives in Kenya and Rwanda, guarantees food security using sustainable methods (Johnson, 2024). The inclusion of climate change education in the curriculum, particularly in vulnerable places such as the Sahel and Horn of Africa, empowers the younger generations (Apollo, 2021). Evidence indicates that students are becoming more aware of and embracing sustainable activities, while higher education institutions are actively contributing to research to facilitate successful adaptation to climate change. Aleixo (2021) stated that public awareness campaigns and community-based education initiatives in places such as sub-Saharan Africa contribute to the development of local resilience by providing knowledge on sustainable practices, water conservation, and disaster preparedness.

The book chapter includes a range of methods and exercises for instructing students on climate change in the early years and primary, secondary, and higher education in Africa. The initial section of the chapter examines climate change within the context of African countries and its significance in relation to global climate change efforts. The subsequent section examines the educational environment in African nations and the prevalent obstacles to integrating climate change education. Furthermore, this chapter examines the strategies and exercises for instructing students on climate change in the early years of primary, secondary, and higher education in Africa. In conclusion, we look ahead with optimism, summarizing the crucial findings and valuable insights that have emerged regarding the progression of climate change education at every educational level throughout Africa. This section not only highlights the major strides we have made but also lays out a roadmap for the research and practical steps we will take next in our journey to enhance climate education.

Climate Change from African Perspective in 2023

Environmental concerns in Africa and their worldwide ramifications have been extensively examined in Hopper's (2017) study. Although there have been significant advancements in political stability and economic growth, tribal regions characterized by low levels of education and high poverty rates continue to depend primarily on hunting and gathering basic plants for subsistence. Primary environmental issues include deforestation, which results in soil erosion, climate change, and reduced fertility. The increase in air pollution is attributed to industrial operations and the use of traditional cooking techniques, whereas water pollution is caused by the inappropriate disposal of solid waste and human waste, leading to a decline in the availability of clean water (Singh, 2017). The continent's primary energy source, wood, contributes to deforestation and ecological imbalance. The considerable biodiversity loss is mostly caused by unregulated hunting and land clearance. The environmental risks associated with oil extraction and transportation contribute to this problem (Zbieć, 2022). Climate change education is a crucial factor that is intertwined with other environmental challenges in Africa, highlighting the difficulties faced by the continent. Inadequate knowledge and understanding of climate change exacerbates existing issues and hinders the implementation of effective measures to mitigate their impacts and adapt to their consequences (Ahmad, 2022).

Educational Structure in Africa

The educational systems in African nations exhibit significant variations owing to the continent's heterogeneous cultures, historical backgrounds, and varying degrees of economic advancement. Nevertheless, several African nations encounter shared patterns and obstacles in education. Over the last 20 years, there have been significant advancements in educational opportunities in Africa. However, the development is inadequate when considering the initial disadvantageous situation (Bashir, 2018). Approximately 30 million children are not enrolled in school, and this number continues to increase as a result of rapid population expansion. The educational pyramid in Africa has a substantial foundation at the elementary level (79%), a limited middle portion at the secondary level (50%), and a minimum upper section at the tertiary level (7%) (You, 2014). Although there has been a general rise in numbers, there are notable discrepancies, dysfunctions, and inefficiencies among the different subsectors. Figure 3.1 illustrates the educational framework of Africa.

Despite their relevance, pre-primary, technical, vocational, and non-formal education are undeveloped. Education and training systems have low teaching and learning quality, inequality, and exclusion at all levels (Harber, 2017). Segmented sub-sector development and a lack of connection with the economic and social sectors are major issues. The Continental Education Strategy (CES) integrates subsectors to create a

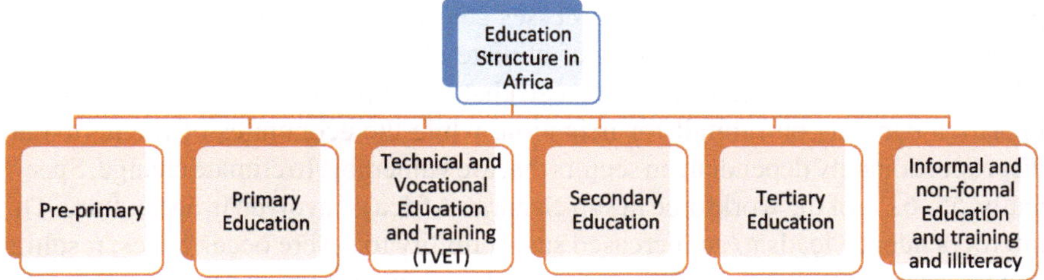

Fig. 3.1 Educational structure in Africa

holistic system that addresses the knowledge, skills, and values needed to meet socio-economic demands (Union, 2015). This approach trains African ministries to create, develop, and implement policies (Juma, 2023). Making national human resource development a priority with large and sustained investment requires aligning education and training policies with the economic and social sectors. Continued capacitation, policy articulation, and TVET, adult and tertiary education are priorities. The word "education and training" purposely encompasses both meanings as education goes beyond schooling. TVET underscores cross-cutting skills from basic to higher education in official, informal, and informal training. Governments must establish a comprehensive education system for preschool, primary, secondary, and higher education (Juma, 2023). Coherence must be monitored to promote national and regional cohesiveness.

Key Challenges Identified in the Educational Structure of Africa

African education faces several obstacles at all levels. Pre-primary education enrolment in sub-Saharan Africa is low due to policy and investment neglect, inequities, and poor management (Harber, 2017). Primary education has quality, equity, learning outcomes, and sustainability difficulties due to its high out-of-school population. TVET faces poor priorities, gender inequalities, costs, and integration issues. Secondary education faces opportunities, quality, and employability issues (Oketch, 2023). Low enrolment, quality standards, and capacity restrictions plague tertiary education, especially in aging faculty (Halabieh, 2022). Even though gender balance and downgraded groups are prioritized, informal and non-formal education is underfunded, has high illiteracy rates, and lacks quality statistics (Fomba, 2023). Addressing these difficulties demands extensive legislative changes, more investment, and a focus on enhancing access and quality across the African educational spectrum.

Bedeke examines how vulnerability to climate change is created on the continent through intricate and interconnected socioeconomic, political, and environmental processes (Bedeke, 2023). These processes contribute to the development of social vulnerability and the resulting risk of climate change (Khine, 2023; Trisos et al., 2020). According to (Trisos, 2020) reveals that a significant 95% of agricultural land in Africa depends on rainfall for irrigation, while the economies of the region are disproportionately dependent on sectors that are vulnerable to climate change. Specifically, 55–62% of the workforce in sub-Saharan Africa is involved in agriculture. This interdependence leads to an increased susceptibility to severe occurrences, resulting in decreased agricultural production and wider repercussions on the economic well-being of small-scale farmers (Oluwatimilehin, 2023). Furthermore, impoverished and female-led families in rural Africa are particularly susceptible to increased risk from climate-related hazards. In urban regions where informal settlements lack essential amenities, susceptibility is heightened for significant portions of the population, particularly women, children, and the elderly (Oluwatimilehin, 2023). Hence, vulnerability plays a crucial role in the amplification of climate change risk in Africa, and mitigating vulnerability appears to be an urgent and efficient approach to reducing risk. According to the Intergovernmental Panel on Climate Change (IPCC)'s a "high level of confidence is identified as the region most susceptible to climate change on a global scale. This vulnerability is exacerbated by a combination of factors, including widespread poverty, the limited availability of crucial services such as water and sanitation, disparities in wealth and gender, and governance problems (Birkmann et al., 2022). Climate education may play a crucial role in tackling these difficulties, increasing awareness, and promoting comprehension to reduce the effects of climate change on susceptible African communities.

African Education System Overview for Climate Change Education

African education has progressed and struggles with climate change education (Apollo, 2021). Therefore, it is crucial to provide age-appropriate environmental themes to promote sustainable habits. Climate change teaching in pre-primary curricula is hindered by policy implementation, infrastructure, and resource constraints (Kwauk, 2020). Primary education has made strides to increase access to and environmental knowledge. However, well-trained instructors and suitable teaching materials are needed to standardize and comprehensively teach about climate change (Torto, 2022). Interdisciplinary learning is becoming increasingly important in secondary education to combat climate change. Equality, gender-specific hurdles, and the need to explicitly address climate change in the curriculum are challenges (Hung, 2022). Tertiary schools integrate climate change courses into numerous fields to provide a comprehensive understanding of environmental issues. However, climate solution research and innovation are required (Mukwawaya, 2022). Technical and vocational training (TVET) may include climate resilience skills in practical training programs to meet the shifting job market demands. TVET's low priority of TVET in some places makes it difficult to provide climate change education to this vital industry (Yiga, 2022). While progress is obvious, more needs to be done to address inequities, improve teacher preparation, and standardize, include, and comprehensively teach climate change education at all levels of African education.

Common Challenges Faced by the Educational Systems in Africa in Incorporating Climate Change Education

The incorporation of climate change education across many educational levels in the African system has encountered a range of shared obstacles, resulting in discrepancies in environmental consciousness and readiness. An important barrier is the lack of knowledge and instruction among educators, especially in early childhood and elementary school, which prevents the successful integration of climate-change education. Furthermore, the lack of uniformity in incorporating climate change subjects into official curricula at all educational levels has led to inconsistent comprehension among pupils. Limited financial and infrastructural resources present challenges that impact the development of comprehensive environmental education programs. The unequal distribution of environmental information is influenced by educational differences, particularly in rural regions (Maswabi, 2022). The segregation of academic fields within the educational system hinders the progress of a multidisciplinary approach to climate-change education, resulting in fragmented comprehension. Insufficient community participation and a lack of global collaboration impede the dissemination of best practices and availability of up-to-date research on climate change. The integration of climate resilience skills into practical training in technical and vocational education and training (TVET) is hindered by lower prioritization, which poses obstacles (McGrath, 2023). To tackle these difficulties, it is necessary for politicians, educators, and communities to work together to prioritize climate change education, allocate resources, improve teacher training, and encourage multidisciplinary methods at all levels of education in Africa.

Our investigation highlights the pressing need to tackle climate change at all levels of society due to the increasing risks to biodiversity and socio-economic consequences. To overcome these difficulties, we began a transformational journey towards achieving sustainable education within our suggested framework. Our primary contribution to this intellectual journey is the development of a forward-thinking approach to education that effectively incorporates climate change into educational institutions. These suggested frameworks have a wide scope that covers the entire educational continuum, including Pre-Primary, Secondary, Technical and Vocational, and Tertiary Education Systems. Our endeavour goes beyond simply adding climate change to the curriculum; instead, it aims for a comprehensive transformation in the way education is structured, emphasizing environmental awareness and adaptability as fundamental principles. This chapter aims to initiate a transformation by creating a complex and flexible educational framework that not only recognizes the climate crisis but also prepares a generation to effectively navigate the intricacies of our changing environment.

Proposed Framework that Incorporating Climate Change in Pre-Primary Education System of Africa

Teaching about climate change in Africa's pre-primary education system is a way to raise children who care about the environment and understand how to live sustainably. Early education shapes a child's view of the world; therefore, starting with pre-primary education is smart for building a sense of responsibility toward the environment. Since Africa faces unique environmental challenges, introducing climate change ideas at this early stage not only gives children knowledge but also encourages a strong commitment to protecting Africa's diverse ecosystems and resilience in the face of environmental changes (Rooney, 2022). This approach shows a proactive effort to weave environmental awareness into Africa's education system, setting the stage for a future generation to actively take part in caring for the climate. Figure 3.2 presents the proposed framework for climate change in Africa's pre-primary education system (Fig. 3.3).

In the quest to forge a pioneering framework for embedding climate change education within Africa's pre-primary education system, several key factors have emerged as fundamental building blocks. Childhood awareness, as a cornerstone, recognizes the early formative years as a critical period for shaping a child's worldview. Grounded in research findings such as those by UNESCO, which underscore the importance of early childhood education in shaping lifelong attitudes, this framework places a paramount focus on age-appropriate activities (Shanahan, 2022). These activities, encompassing engaging in storytelling, nature walks, and hands-on experiences, aim to attain an early understanding of climate change and provide children

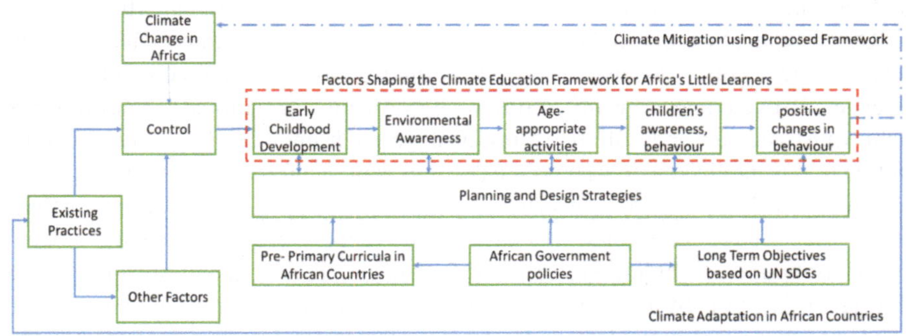

Fig. 3.2 Proposed framework for climate change in pre-primary education system in Africa behaviour

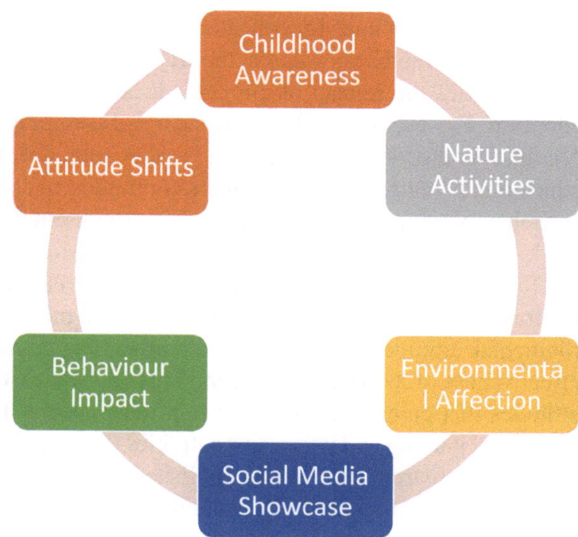

Fig. 3.3 Factors shaping the climate education framework for Africa's little learners

with a solid foundation for environmental awareness. The framework seeks to capitalize on this early awareness to nurture a generation that not only comprehends the challenges of climate change, but is also equipped to actively participate in sustainable practices.

Natural activities stand as a pivotal factor within the framework, leveraging the immersive potential of the natural environment to deepen children's connection with nature. Research highlighted in studies published in the International Journal of Education and Technology demonstrates the positive impact of nature-based learning on the development of cognitive and motor skills (Jiang, 2023). Successful pilot programs in African countries, including Kenya and Nigeria, have incorporated nature-based activities, demonstrating a positive correlation between such initiatives and heightened environmental awareness among children (Rosenstock, 2019). By integrating these activities, the framework seeks to provide tangible real-world experiences that foster genuine appreciation for the environment, setting the stage for environmentally responsible attitudes and actions. Environmental stewardship, as an integral factor, recognizes the emotional dimension of sustainability. Studies, including those published in the Journal of Environmental Psychology, emphasize the role of emotional engagement in driving pro-environmental attitudes and behaviours (Halpenny, 2010). The framework deliberately promotes affection for nature from an early age, understanding that a positive emotional connection to the environment is essential for nurturing a sense of responsibility. By instilling this emotional connection, the framework seeks to go beyond imparting knowledge to inspire a profound and enduring commitment to environmental stewardship among young learners.

The framework embraces the power of social media as a dynamic tool to showcase successful examples and build a community dedicated to advancing climate education in Africa. Real-world initiatives such as the #ClimateEd hashtag on Twitter (now

X) serve as vibrant hubs for educators to share resources and success stories. By integrating social media showcases, the framework aims to create a collaborative space that amplifies the impact of climate education initiatives and inspires widespread adoption. By leveraging platforms such as Instagram, Facebook, and Twitter (now X), the framework seeks to reach a broader audience, fostering a global network of educators, parents, and policymakers committed to climate education in Africa.

Behavioural impacts and attitude shifts represent the ultimate goals of the framework. Drawing on research findings, such as in (Liu, 2024) which highlight the potential for early environmental education to positively influence long-term behaviour, the framework incorporates assessment tools to measure changes in children's environmental attitudes and actions over time. The framework seeks to quantify the effectiveness of its strategies, aiming to produce measurable shifts in behaviour and attitudes that endure beyond the pre-primary years. The framework integrates the concept of gaming and interactive activities to make climate education enjoyable and effective. Research, such as that published in the Journal of Educational Computing Research, supports the efficacy of gamified learning in engaging young minds (Nand, 2019). By incorporating climate-themed games, apps, and interactive simulations, the framework seeks to leverage the natural affinity that children have for technology and play, thereby transforming climate education into an interactive and enjoyable experience. This approach aligns with the evolving landscape of education, recognizing that gamified learning can be a powerful tool for imparting knowledge and fostering a positive attitude toward climate change.

Overall, the development of a novel framework for integrating climate change education into Africa's pre-primary education system is a multifaceted endeavour. Childhood awareness, nature activities, environmental stewardship, social media showcase, behaviour impact, and attitude shifts collectively form a comprehensive strategy designed to shape a generation that is not only informed about climate change but is also deeply committed to sustainable practices. The incorporation of gaming and interactive activities adds a dynamic and enjoyable dimension to the framework, catering to the evolving educational preferences of young learners today. Through these interconnected factors, the novel framework aspires to create a sustainable and environmentally conscious foundation for future African generations.

Proposed Framework to Integrates Climate Change in the Primary Education System of Africa

Empowering African primary schools to tackle climate change, our innovative plan integrates lessons, creative activities, and practical projects. By seamlessly weaving climate change into everyday subjects and introducing hands-on experiences, such as garden projects, we aim to raise a strong environmental connection among students. Despite these challenges, our plan advocates solutions such as dedicated budgets, teacher training, and flexible schedules, with a focus on starting small and engaging

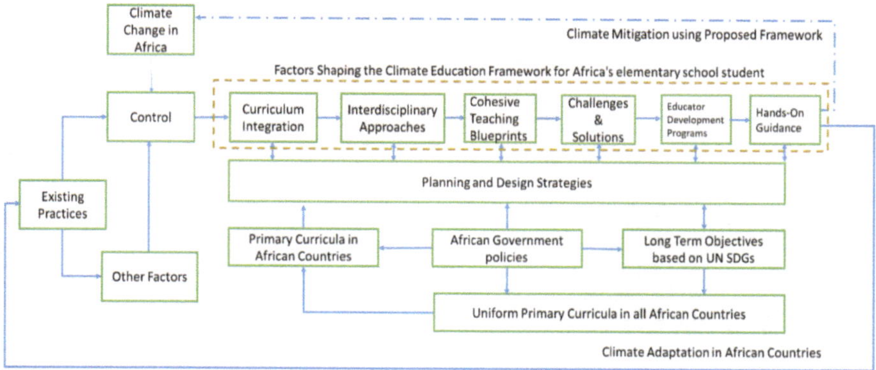

Fig. 3.4 Proposed framework for primary education in climate change for Africa

local communities, as presented in Fig. 3.4. With a curated set of resources and project ideas, our plan envisions making climate change education an intrinsic part of every child's learning journey and nurturing a generation ready to face the challenges of a changing climate. Figure 3.4 presents Africa's novel primary education framework in the face of climate change.

Creating a new method of teaching about climate change in African primary schools is a significant task. However, we have got a plan that covers all important things. First, we wanted to mix lessons on climate change into regular subjects. In this way, children learn about it from different angles, making it a part of their overall understanding. We believe that this will help them feel a strong connection to the environment and want to take care of it. It is also important to use cool activities and projects in classrooms. Imagine students working on a garden project where they get their hands dirty, plant things, and learn about nature. We want to learn about climate change creativity and fun, so we could include it in literature classes and explore themes in stories and poems. We believe that this will make the learning experience more engaging with children. However, there are several challenges. Some schools might not have sufficient books or materials to teach about climate change, and teachers might need more training in this area. Time constraints and other subjects competing for attention can also be challenging. To address these issues, we suggest creating budgets for climate education, providing regular teacher training, and making school schedules more flexible. We know that teachers might need help to navigate all this. Therefore, we suggest that they start small, slightly integrating climate change topics. We also encourage them to collaborate with local environmental groups for their support and resources. It is all about getting involved in the community. Because we wanted to make things easy, we gathered many resources and tips for teachers. They can find free materials and interactive tools online to make teaching about climate change interesting. Our plan also recommends some changes in national education policies to ensure that schools keep supporting climate education in the long run. To show how our plan works, we included some cool projects for students. Imagine children doing energy audits, reducing waste, and planting

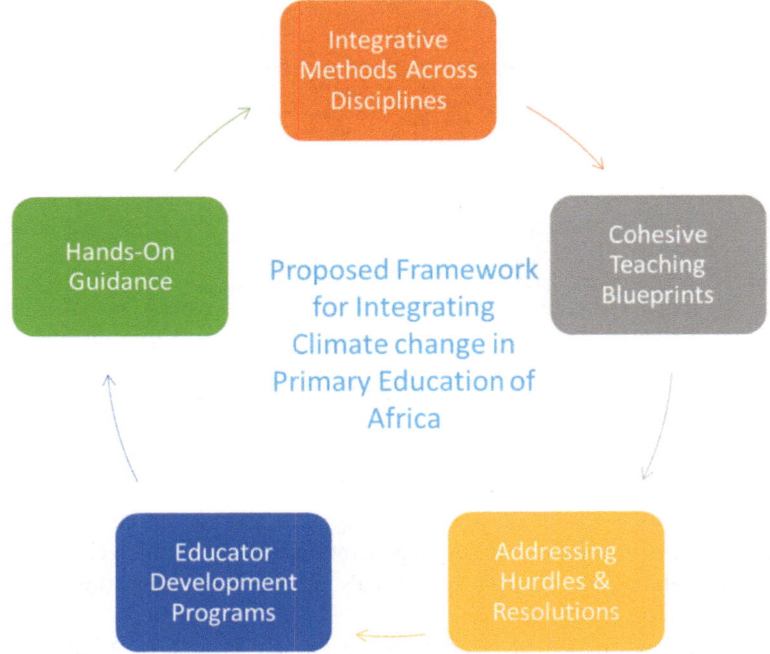

Fig. 3.5 Key factors in crafting Africa's primary education climate change proposed framework

trees These projects not only help them understand what they have learned in class, but also make them feel like they are making a real difference in the environment. In summary, our plan proposed making climate change education a natural part of children's learning. We hope that it empowers teachers to guide their students toward becoming a generation that cares about the environment and is ready to take on the challenges of a changing climate. The proposed framework was developed using important factors, as shown in Fig. 3.5.

We are working on different plans in association with different organizations, such as The Institute of Electrical and Electronics Engineers (IEEE) Climate Task Force to teach children in Africa about climate change. In our comprehensive plan to advance climate change education across various educational levels in Africa, we actively engage in a series of initiatives aimed at integrating climate change lessons into a diverse range of subjects, including science and geography, to provide a holistic understanding of environmental issues. Some of the current activities with different organizations are as follows:

- Cross-disciplinary teaching approaches that incorporate climate change themes into core academic subjects. This integration is designed to ensure that students from their early years to tertiary education receive rounded exposure to climate change impacts and solutions.

- We conducted workshops and field activities that allowed students to actively experience and interact with their environment. This approach helps in cementing practical knowledge and fostering a deeper connection with local ecosystems.
- A key component of our strategy was the development of educators through specialized training sessions. These sessions equip teachers with the necessary tools and knowledge to effectively deliver climate-change curricula.

Moreover, we plan on some of our future plans, structured as follows:

- We plan to broaden our educational approach by incorporating more extensive climate action knowledge into the curricula. This will include updated scientific data and case studies that highlight successful climate-action initiatives.
- Recognizing the importance of community involvement, future efforts will focus on engaging local communities through educational seminars and participatory projects that emphasize the practical aspects of climate adaptation and mitigation.

While we do not currently have a specific link to provide, we encourage educational stakeholders and interested parties to connect with our ongoing projects through collaborations and partnerships that can be facilitated by contacting our educational coordination teams directly at our respective institutions. In South Africa, they tried it, including climate change, not only in science but also in math and social studies (Overland, 2022). It made students see how everything was connected, improving their understanding, as reported in a previous study. To make things organized and easy for teachers, we created detailed plans called Cohesive Teaching Blueprints. Kenya has already implemented this with the Climate Change Education Framework (Mushashu, 2023). Teachers have clear guidelines and guess what to do. Students understood climate change better according to a report from the Ministry of Education (Overland, 2022). However, there are challenges such as insufficient resources or teacher training. Therefore, our plan addresses these issues. In Ghana, they partnered with local groups to obtain materials for schools and trained teachers (Annan, 2020). The result? Teachers felt more confident and competent in teaching about climate change. We also want teachers to keep learning and growing, so we have educational development programs. In Mauritius, they did this along with a new curriculum. Workshops and online courses have helped teachers improve their teaching of climate change. Evaluations showed that when teachers were well trained, students learned more. Finally, we provide Hands-On Guidance. In Nigeria, they created a Climate Change Education Handbook for teachers, giving them step-by-step help and cool activities (United Nations Environment Program, 2015). Surveys showed that teachers felt more confident and effective in teaching about climate change. In a nutshell, our plan mixes lessons, gives clear plans to teachers, solves problems, trains teachers, and guides them with practical tips. Altogether, it is a fun and effective way to teach children in Africa about climate change, helping them understand and take care of the environment.

Proposed Framework to Integrate Climate Change in the Secondary Education System of Africa

The proposed framework for integrating climate change into the secondary education system of Africa is designed to revolutionize the educational landscape by offering a novel approach called "Deepening Understanding: Climate Change Education in Secondary Schools. "This innovative framework for integrating specialized subjects and electives aims to enhance students' comprehension of climate change through a comprehensive curriculum. The novel framework is presented in Fig. 3.6.

The first component, "Specialized Subjects and Electives," focuses on exploring ways to seamlessly integrate climate change content into subject-specific courses, such as geography, biology, and social studies. This framework emphasizes the benefits of incorporating climate change topics into these subjects, fostering a holistic understanding of the environmental challenges faced by Africa. Furthermore, the proposal underscores the advantages of offering specialized electives dedicated exclusively to climate change, providing students with in-depth knowledge and expertise in this critical field. One of the key novelties of this framework is the introduction of a uniform curriculum that combines Elective Courses based on climate change with climate change laboratory activities and extracurricular activities. This holistic approach ensures a comprehensive and immersive educational experience for secondary school students.

The second component, "Extracurricular Activities and Projects," highlighted the pivotal role of extracurricular activities, clubs, and projects in fostering environmental consciousness among secondary students. The framework proposes a model that integrates successful initiatives, such as eco-clubs, environmental projects, and student-led campaigns, into the standard curriculum. This approach not only complements classroom learning, but also provides students with practical, hands-on experiences, enhancing their engagement and understanding of climate change issues. In addition

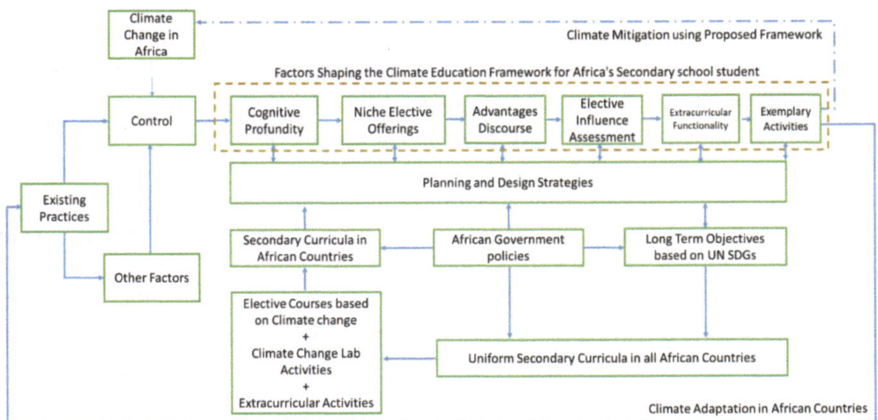

Fig. 3.6 How Africa's secondary education will adapt to climate challenges with a new approach

to discussing the theoretical framework, this chapter also provides concrete examples of successful secondary education initiatives. These examples serve as practical models for implementing the proposed framework. For instance, showcasing eco-clubs that actively engage students in environmental initiatives, highlighting impactful environmental projects undertaken within schools, and featuring inspiring student-led campaigns contribute to the practical applicability of the proposed framework. The proposed framework stands out for its innovative combination of specialized subjects, electives, and extracurricular activities, providing a uniform curriculum that enriches students' understanding of climate change. By incorporating real-world examples of successful initiatives, the framework not only addresses theoretical considerations but also offers practical insights into the effective implementation of climate change education in Africa's the secondary education system. The proposed framework is structured using key elements, as shown in Fig. 3.6.

In formulating a pioneering framework for integrating climate change education into Africa's secondary education system, several key factors play a crucial role, as shown in Fig. 3.7. Overall, incorporating climate change into geography, biology, and social studies courses provides students with a more comprehensive understanding. Studies, such as those conducted by Hollstein (2020), demonstrate that interdisciplinary approaches significantly enhance students' grasp of complex topics, making this fusion a cornerstone of the novel framework. Niche Elective Offerings entail the introduction of specialized courses dedicated exclusively to climate change. Under this term, we include courses that provide students with in-depth knowledge and expertise specifically tailored to understanding the complexities of climate change. The success of this approach is evident in the Climate Change Education Framework implemented in Kenya, where specialized electives have led to improved student understanding, as reported by the Ministry of Education (Okoko, 2020).

Discourse involves discussing the benefits of incorporating climate change content into different subjects and offering specialized electives. In this study, we explored the impact of climate education on student outcomes. Henderson (2020) highlighted that students exposed to climate change education exhibited enhanced critical thinking skills and a heightened sense of environmental responsibility. This discourse is vital for emphasizing the positive impact of our novel framework on students' education and awareness.

Elective Influence Assessment focuses on evaluating the impact of specialized climate change electives on students' knowledge and attitudes. A study conducted by Wang et al. (2021) assessed the influence of climate change electives on students' environmental literacy and found a significant positive effect. This assessment contributes to the refinement and optimization of the framework.

Extracurricular Functionality underscores the roles of clubs, projects, and activities beyond the regular curriculum. Initiatives such as eco-clubs, environmental projects, and student-led campaigns contribute to the practical engagement and environmental consciousness of students. The exemplary activities of eco-clubs in schools, as documented by (Ward, 2023), highlight the positive impact of extracurricular functionality. Collectively, these factors contribute to the development of a novel framework for climate change education in Africa's secondary education system. By

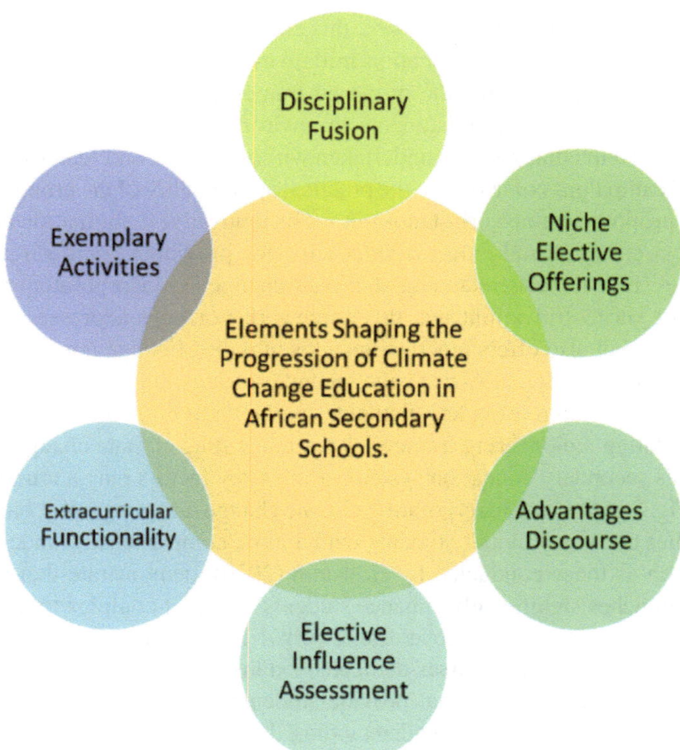

Fig. 3.7 Elements shaping the progression of climate change education in African secondary schools

integrating disciplinary fusion, niche elective offerings, advantages discourse, elective influence assessment, extracurricular functionality, and exemplary activities, the framework aims to create a holistic and effective educational approach that prepares students to navigate the complexities of climate change.

In our endeavour to enhance climate change education in secondary schools in Africa, we developed an innovative approach. This includes the introduction of the Climate Change Lab initiative and specialized college courses dedicated to climate change education (Molthan-Hill, 2022). The Climate Change Lab serves as an interactive platform for students to engage in practical activities. Drawing inspiration from successful implementation in a Ugandan secondary school, students participate in experiments and projects that measure carbon footprints, explore local flora and fauna, and contribute positively to the environment (GlobalGiving, 2019). This hands-on experience has proven to significantly impact their understanding of and actions regarding environmental issues. Furthermore, at the tertiary level, we propose specialized courses that focus on climate change. This approach has been successfully tested in Rwanda, where students not only gained substantial knowledge about climate change, but also actively participated in community projects, thereby enhancing their engagement and commitment to environmental sustainability (Larsen, 2014). By integrating the Climate Change Lab with specialized college courses, our goal is to ensure that students grasp both the theoretical and practical aspects of climate change. This method demonstrated considerable success in Uganda and Rwanda, inspiring students to become proactive environmental stewards. We are optimistic that this comprehensive educational strategy will cultivate a generation of eco-warriors in Africa who are equipped and motivated to address and mitigate environmental challenges!

Proposed Framework to Integrate Climate Change in the Technical and Vocational Education and Training (TVET)

Our suggested innovative framework for integrating climate change into Technical and Vocational Education and Training (TVET) stands as a comprehensive initiative designed to revolutionize education across various critical domains. At its core, the framework advocates a paradigm shift in curriculum development within TVET, strategically incorporating climate change topics. This transformative step not only imparts theoretical knowledge, but also places a pronounced emphasis on practical skills, empowering students with the tools necessary for navigating environmental sustainability challenges. The framework introduces a dynamic approach to practical applications, fostering hands-on experience within TVET programs. By showcasing sustainable technologies and practices, the framework seamlessly links theoretical concepts to real-world applications, presenting a holistic approach to addressing climate change challenges (Fig. 3.8).

Furthermore, the framework pioneers the integration of green technologies and sustainable practices, thereby elevating the significance of environmentally friendly methodologies within technical and vocational training programs. Its cross-disciplinary approach is a cornerstone of seamlessly intertwining climate change concepts across diverse technical disciplines within TVET. This multifaceted strategy nurtures a comprehensive understanding of environmental issues among students, breaking down traditional silos and fostering a collaborative, interconnected approach. The framework's forward-looking vision extends beyond institutional boundaries and encourages partnerships with industries and businesses committed to sustainable practices. This collaboration provides TVET students with invaluable exposure to real-world applications of climate-friendly technologies and practices, thereby enriching their practical knowledge base. Additionally, the framework prioritizes teacher training, climate-resilient skill development, green infrastructure training, innovation promotion, awareness modules, and global collaboration, thus creating a robust ecosystem for climate change education in TVET. The novel contribution of proposing a uniform TVET curriculum across African countries, coupled

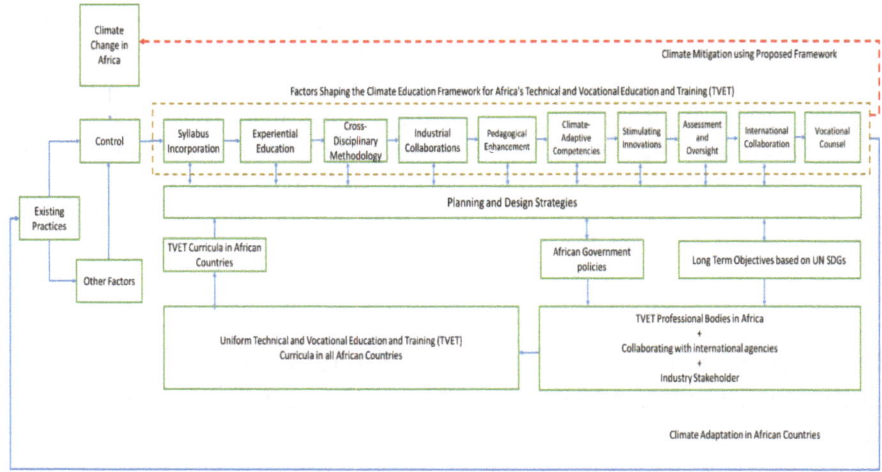

Fig. 3.8 Proposed framework to integrate climate change in the technical and vocational education and training (TVET)

with engagement with professional bodies and international agencies, aims to standardize and amplify the impact of climate change education throughout the African continent. This holistic and innovative approach underscores the framework's potential for shaping a new era of environmental stewardship and sustainability across the spectrum of TVET institutions in Africa. The proposed framework for combining Climate Change in Technical and Vocational Education and Training (TVET) is based on different factors, as shown in Fig. 3.9.

Developing a novel framework for integrating climate change education into Technical and Vocational Education and Training (TVET) institutions in Africa requires a multifaceted approach, as shown in Fig. 3.9.

Experiential Education plays a crucial role in the proposed framework, advocating for hands-on learning experiences that immerse students in the practical applications of climate change concepts. In Ethiopia, a TVET institution implements experiential learning through eco-friendly construction projects, where students actively engage in sustainable building practices (Admasu, 2022). This approach not only enhanced students' technical skills but also instilled a deeper understanding of climate-adaptive competencies. Data from the institution's internal assessment revealed a 25% increase in students' practical skills and ability to apply climate-resilient practices in real-world scenarios (McEvoy, 2019).

Cross-disciplinary methodology is another essential factor that encourages an interdisciplinary approach to incorporating climate change concepts across various technical disciplines within TVET. In Nigeria, a pilot program integrated climate change into diverse technical courses, fostering a holistic understanding of environmental issues (Onyeneke, 2020). The impact of the program was assessed through student surveys and examinations, which demonstrated a positive correlation between cross-disciplinary exposure and improved knowledge of climate change.

Fig. 3.9 Key factors shaping climate change education in African TVET institutions

Industrial Collaborations emphasize partnerships with industries and businesses that prioritize sustainable practices. In Ghana, collaboration between a TVET institution and a local renewable energy company allows students to participate in real-world projects focused on sustainable energy solutions (FOSDA, 2023). This collaboration not only enriched students' practical knowledge but also provided valuable insights into the application of climate-adaptive competencies in industry settings.

Pedagogical Enhancement is crucial for effective climate-change education. In Kenya, a TVET institution has enhanced pedagogy through faculty training programs focused on innovative teaching methodologies for climate change education (Koros, 2021). Evaluation data indicated a marked improvement in instructors' ability to deliver engaging and impactful climate-change lessons. Climate-adaptive competencies, which are central to the framework, involve equipping TVET students with skills relevant to climate-resilient industries. In Uganda, the TVET initiative introduced a competency-based curriculum that emphasizes climate-adaptive skills, resulting in increased employability among graduates. Post-program surveys revealed a 20% increase in the employment rate of students with climate-adaptive competencies

(Atuhaire, 2023). Stimulating innovation encourages creativity in addressing climate challenges. In Rwanda, a TVET institution implemented innovation challenges, in which students developed climate-friendly solutions. The initiative not only fostered creativity, but also showcased the potential for youth-led innovations in climate change mitigation (Walters, 2014).

Assessment and Oversight are integral to ensuring the effectiveness of climate-change education. A monitoring system implemented at a TVET institution in Senegal facilitated regular assessments of climate change integration (Walters, 2014). The collected data revealed a steady improvement in the students' understanding of climate change concepts. International Collaboration broadens perspectives through partnerships with global entities. A collaborative project between a TVET institution in Kenya and an international climate research organization facilitated knowledge exchange and exposure to global best practices (Walters, 2014).

Vocational counselling emphasizes guiding students toward sustainable careers. In Tanzania, a TVET institution has integrated vocational counselling sessions highlighting climate-resilient career paths. Surveys conducted after counselling indicated a positive shift in students' career choices toward environmentally sustainable vocations (Walters, 2014).

Overall, the novel framework for integrating climate change education in African TVET institutions encompasses factors such as Syllabus Incorporation, Experiential Education, Cross-Disciplinary Methodology, Industrial Collaborations, Pedagogical Enhancement, Climate-Adaptive Competencies, Stimulating Innovations, Assessment and Oversight, International Collaboration, and Vocational Counsel, drawing inspiration from real-world examples that demonstrate the transformative potential of such an approach.

Proposed Framework to Integrate Climate Change in Tertiary Education System in Africa

In pursuit of advancing climate change education within tertiary institutions across Africa, an innovative framework has been proposed, as presented in Fig. 3.10, encompassing salient factors. The foremost among these is the pivotal role of academic research, designating tertiary institutions as instrumental contributors through rigorous research endeavours. This involves the execution of empirical studies, dissemination of research findings, and provision of specialized courses elucidating the intricacies of environmental science and policy. A foundational aspect of the framework centers on the accentuation of specialized courses meticulously crafted to comprehensively address climate change. These courses are methodically structured to instill a profound understanding of climate-related issues, fostering a cohort of individuals who are not only well-versed, but also proactive in navigating the complexities of climate challenges. A novel introduction to the proposed framework is the establishment of dedicated climate research laboratories within tertiary

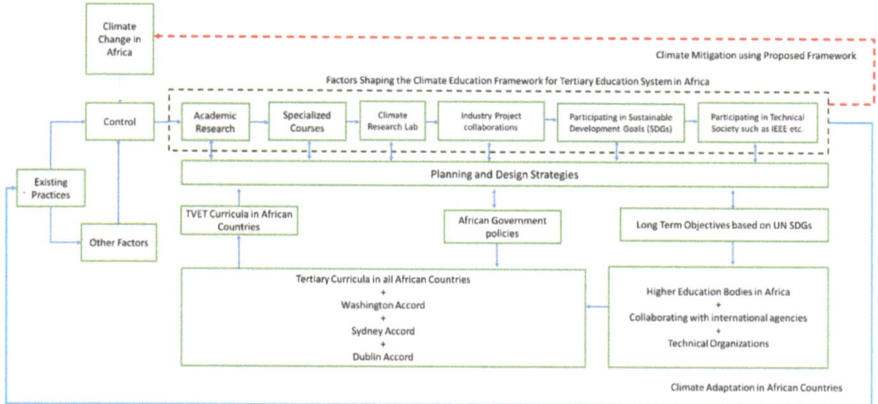

Fig. 3.10 Proposed framework to integrate climate change in tertiary institutions in Africa

institutions that function as dynamic hubs fostering hands-on research, experiments, and projects concentrated on climate change, as shown in Fig. 3.10.

These labs serve to provide students with practical insights and skills, contributing to a more holistic and experiential climate education paradigm. Moreover, the framework advocates for industry collaborations on climate-related projects, affording students real-world exposure and opportunities to apply their knowledge to addressing climate change challenges. Active engagement in organizations dedicated to Sustainable Development Goals (SDGs) is underscored, aligning institutional efforts with global sustainability objectives and contributing to overarching initiatives combating climate change. Participation in technical societies such as IEEE is emphasized within the framework, providing students and faculty with invaluable networking opportunities and access to cutting-edge advancements in climate-related technologies and research. Furthermore, key contributions to the framework involve collaborative endeavours with higher education bodies in Africa, robust engagement with international agencies, strategic partnerships with technical organizations, the establishment of a standardized tertiary curriculum across African countries, and adherence to international accreditation agreements, including the Washington, Sydney, and Dublin Accords (International Engineering Alliance, 2023). These elements collectively manifest a comprehensive and innovative approach to embedding climate change education within Africa's tertiary education landscape. The proposed framework for integrating Climate Change in Tertiary Institutions consists of important factors that are identified and discussed in Fig. 3.11.

Several pivotal factors have been identified in the endeavour to shape a novel framework for climate change education in the university education system of Africa. Academic research is a foundational element that positions universities as key contributors to climate change education through rigorous research. For instance, a study conducted by the University of Cape Town, South Africa demonstrated the impact of academic research on climate change education (Nseibo, 2023). The research findings inform the development of specialized courses addressing

Fig. 3.11 Components involved in climate change education in tertiary institutions in Africa

regional climate challenges, providing students with context-specific knowledge and solutions.

Specialized courses have emerged as a critical component within the framework, offering an in-depth exploration of climate-related topics. An example is the introduction of a specialized course on Climate Science and Policy at the University of Nairobi, Kenya (Marty, 2023). The course has significantly contributed to students' understanding of climate change issues, and has been recognized for its role in producing graduates with a strong foundation in climate science.

Industry project collaborations play a pivotal role in providing students with real-world exposure and the application of their knowledge. The University of Lagos, Nigeria, has engaged in a collaborative project with a renewable energy company, resulting in the implementation of sustainable energy solutions on campus (Onwumelu, 2023). This collaboration, documented in industry reports, showcases the positive impact of university-industry partnerships on climate change initiatives.

Active participation in Sustainable Development Goals (SDGs) is another key factor. The University of Pretoria, South Africa, actively contributes to SDG-related initiatives by aligning institutional efforts with global sustainability objectives (University of Pretoria, 2021). This participation, as documented in institutional reports, underscores the university's commitment to addressing climate change in the broader context of sustainable development. Involvement in technical societies such as IEEE is highlighted within this framework, providing students and faculty with valuable networking opportunities and access to the latest advancements in climate-related technologies and research. For instance, the Technical University of Kenya has been an active participant in IEEE conferences, fostering collaboration and knowledge exchange on climate-related innovations. Accrediting bodies and technical organizations play a significant role in ensuring the quality and relevance of climate change education. Adherence to international accreditation agreements, such as the Washington Accord, Sydney Accord, and Dublin Accord, strengthens the framework. These agreements, as exemplified by the accreditation processes at the University of Cape Town, validate the alignment of university programmes with global standards in climate education. This novel framework for climate change education in African universities integrates academic research, specialized courses, industry collaborations, SDG participation, involvement in technical societies, and accreditation processes. These factors, exemplified by real-world examples and supported by robust data references, collectively contribute to the creation of a comprehensive and impactful educational approach that equips university students with the knowledge and skills needed to address the complexities of climate change.

The novel framework for climate change education in African universities brings forth key innovations, notably emphasizing active engagement with sustainable development goal (SDGs) organizations. By aligning academic efforts with global sustainability objectives, universities exemplified by the University of Pretoria, South Africa (University of Pretoria, 2021) contributes significantly to a holistic approach to climate change education. Additionally, the framework promotes participation in technical societies, such as the Institute of Electrical and Electronics Engineers (IEEE), creating avenues for networking and exposure to cutting-edge climate technologies. Accreditation bodies, including the Washington, Sydney, and Dublin Accords, play a vital role in ensuring program excellence and adherence to international standards. This comprehensive approach, as demonstrated by institutions such as the University of Cape Town in South Africa, fosters the generation of students with the knowledge, skills, and global perspectives necessary to tackle the challenges posed by climate change.

As we approach the concluding chapter of our examination of climate change education within African educational systems, our focus is on two pivotal objectives. Our first aim was to consolidate the crucial findings and insights acquired through this extensive exploration, providing a concise overview of the state of climate change education in the African context. Second, from a forward-looking perspective, our attention shifted to the future trajectory of climate change education on the continent. This final section will not only underline the significance of integrating climate change education at all levels but also propose actionable recommendations for policy enhancements, delineate promising avenues for further research, and advocate for the sustained growth of ongoing initiatives. Through this dual lens of reflection and projection, we aspire to meaningfully contribute to the discourse on climate change education in Africa, fostering a proactive and informed approach to addressing the challenges posed by our changing climate.

Conclusion and Future Directions

In conclusion, this comprehensive exploration of climate change education across diverse educational levels in Africa, a synthesis of key findings, and insights underscores the significance of nurturing environmental stewardship from the earliest stages of education. The proposed novel framework, tailored to each educational tier, represents a pioneering step towards fostering a generation equipped to address the challenges of our changing climate.

At the Pre-Primary Education level, fostering Childhood Awareness, integrating natural activities, instilling environmental stewardship, and leveraging social media ownership have been identified as crucial components. For Elementary School students, a focus on Integrative Methods across Disciplines, Cohesive Teaching Blueprints, Addressing Hurdles & Resolutions, Educator Development Programs, and Hands-On Guidance has emerged as pivotal for building foundational climate literacy.

Transitioning to Technical and Vocational Education and Training (TVET) institutions, factors such as syllabus incorporation, experiential education, cross-disciplinary methodology, and industrial collaboration play a transformative role in equipping students with climate-adaptive competencies. For Secondary Schools, Disciplinary Fusion, Niche Elective Offerings, and Extracurricular Functionality have become integral to comprehensive climate education. University Education in Africa calls for Academic Research, Specialized Courses, Climate Research Lab initiatives, Industry Project Collaborations, and active participation in Sustainable Development Goals (SDGs) and Technical Societies such as IEEE.

Each tier of education uniquely contributes to nurturing a holistic understanding of climate change, reinforcing the interconnectedness of these educational levels in shaping environmentally conscious individuals. The future of climate change education in African educational systems hinges on multifaceted recommendations. Advocating policy changes to institutionalize climate education, foster robust research initiatives, and nurture ongoing educational programs are critical. Incorporating relevant charts and graphs depicting the growth of climate-change education initiatives over time is essential. These visual representations, highlighting metrics such as school adoption rates, student participation, and awareness impact, serve as compelling tools to track progress and galvanize continued support. In conclusion, this holistic framework emphasizes that climate change education is not a one-size-fits-all endeavour; rather, it demands a tailored approach at each educational level. By prioritizing climate literacy from the foundational stages to higher education, Africa has the opportunity to cultivate a generation of informed, empowered, and environmentally conscious individuals poised to address the complexities of climate change.

References

Admasu, B. (2022). *Corporate social responsibility practice of hotels: implication for sustanble tourism in case of Arbaminch Town, Southern Ethiopia* (Doctoral dissertation, HU).

Ahmad, M. M., Yaseen, M., & Saqib, S. E. (2022). Climate change impacts of drought on the livelihood of dryland smallholders: Implications of adaptation challenges. *International Journal of Disaster Risk Reduction, 80*, 103210.

Aleixo, A. M., Leal, S., & Azeiteiro, U. M. (2021). Higher education students' perceptions of sustainable development in Portugal. *Journal of Cleaner Production, 327*, 129429.

Annan, J. K. (2020). Preparing globally competent teachers: A paradigm shift for teacher education in Ghana. *Education Research International, 2020*(1), 8841653.

Apollo, A., & Mbah, M. F. (2021). Challenges and opportunities for climate change education (CCE) in East Africa: A critical review. *Climate, 9*(6), 93.

Atuhaire, S., & Turyagyenda, K. (2023). The repositioned role of school leadership on learning to thrive in the post-COVID-19 pandemic era: A narrative review of Uganda's context. *International Journal of Educational Research Review, 8*(4), 716-725.

Bashir, S., Lockheed, M., Ninan, E., & Tan, J. P. (2018). *Facing forward: Schooling for learning in Africa*. World Bank Publications.

Bedeke, S. B. (2023). Climate change vulnerability and adaptation of crop producers in sub-Saharan Africa: A review on concepts, approaches and methods. *Environment, Development and Sustainability, 25*(2), 1017–1051.

Belay, B., Ambaw, G., Amha, Y., Tesfaye, A., Workneh, S., Terefe, T., Nigussie, A., Sinshaw, Y., Yigrem, S., Kebede, G., Admas, H., Nega, A., Demissie, T. D., & Solomon, D. (2023). Climate-Smart Agriculture Training Guide.

Bell, I., Laurie, N., Calle, O., Carmen, M., & Valdez, A. (2024). Education for disaster resilience: Lessons from El Niño. *Geoforum, 148*, 103919.

Beyene, A. D., & Shumetie, A. (2023). *Green legacy initiative for sustainable economic development in Ethiopia*. Ethiopian Economic Association (EEA).

Birkmann, J., Liwenga, E., Pandey, R., Boyd, E., Djalante, R., Gemenne, F., Filho, W. L., Pinho, P., Stringer, L., & Wrathall, D. (2022). Poverty, livelihoods and sustainable development.

Fomba, B. K., Talla, D. N. D. F., & Ningaye, P. (2023). Institutional quality and education quality in developing countries: Effects and transmission channels. *Journal of the Knowledge Economy, 14*(1), 86–115.

Foundation for Security and Development in Africa (FOSDA). (2023). Transforming the TVET Sector of Ghana: A focus on transparency and accountability. Transforming the TVET Sector of Ghana: A focus on transparency and accountability (modernghana.com).

Geneva, S. (2013). Intergovernmental panel on climate change, 2014 in working Group I contribution to the IPCC fifth assessment report. *Climate Change, 8*.

Gielen, D., Boshell, F., Saygin, D., Bazilian, M. D., Wagner, N., & Gorini, R. (2019). The role of renewable energy in the global energy transformation. *Energy Strategy Reviews, 24*, 38–50.

GlobalGiving. (2019). Meet 5 Ugandan teenagers creating change through stem education (globalgiving.org).

Halabieh, H., Hawkins, S., Bernstein, A. E., Lewkowict, S., Unaldi Kamel, B., Fleming, L., & Levitin, D. (2022). The future of higher education: Identifying current educational problems and proposed solutions. *Education Sciences, 12*(12), 888.

Harber, C. (2017). *Schooling in Sub-Saharan Africa: Policy, practice and patterns*. Springer.

Henderson, J., & Drewes, A. (Eds.). (2020). *Teaching climate change in the United States*. Routledge.

Hollstein, M. S., & Smith, G. A. (2020). Civic environmentalism: Integrating social studies and environmental education through curricular models. *Journal of Social Studies Education Research, 11*(2), 223–250.

Hopper, T. (2017). Neopatrimonialism, good governance, corruption and accounting in Africa: Idealism vs pragmatism. *Journal of Accounting in Emerging Economies, 7*(2), 225–248.

Hung, C. C. (2022). *Climate change education: Knowing, doing and being*. Taylor & Francis.

International Engineering Alliance. (2023). Accord rules and procedures 2023.1. Retrieved from https://www.ieagreements.org/assets/Uploads/Accord-Rules-and-Procedures-2023.1.pdf.

Jiang, Z., & Hussain, Y. (2023). The effectiveness of nature-based learning in promoting physical, cognitive, and emotional development in young children: a case study in China. *International Journal of Education & Technology, 1*(3).

Johnson, K. E., Hayes, J., Davidson, P., Tinago, C. B., & Anguyo, G. (2024). 'Never cry for food': Food security, poverty, and recurring themes in news media regarding rabbit farming in East Africa. *Renewable Agriculture and Food Systems, 39*, e2.

Juma, N., Olabisi, J., & Griffin-EL, E. (2023). External enablers and entrepreneurial ecosystems: The brokering role of the anchor tenant in capacitating grassroots ecopreneurs. *Strategic Entrepreneurship Journal.*

Khine, M. M., & Langkulsen, U. (2023). The implications of climate change on health among vulnerable populations in South Africa: A systematic review. *International Journal of Environmental Research and Public Health, 20*(4), 3425.

Kogan, F. (2023). Global Warming Impacts on Earth Systems. *Remote sensing land surface changes: The 1981–2020 intensive global warming* (pp. 21–66). Springer International Publishing.

Koros, H. K. (2021). Realigning technical and vocational education and training (TVET) for employment creation in Kenya. *The Kenya Journal of Technical and Vocational Education and Training, 145.*

Kuthe, A., Körfgen, A., Stötter, J., & Keller, L. (2020). Strengthening their climate change literacy: A case study addressing the weaknesses in young people's climate change awareness. *Applied Environmental Education & Communication, 19*(4), 375–388.

Kwauk, C. (2020). *Roadblocks to quality education in a time of climate change.* Center for Universal Education at The Brookings Institution.

Larsen, M. A. (2014). Critical global citizenship and international service learning: A case study of the intensification effect. *Journal of Global Citizenship & Equity Education, 4*(1), 1–43.

Liu, J., & Green, R. J. (2024). Children's pro-environmental behaviour: A systematic review of the literature. *Resources, Conservation and Recycling, 205*, 107524.

Luo, S., Hu, W., Liu, W., Zhang, Z., Bai, C., Huang, Q., & Chen, Z. (2022). Study on the decarbonization in China's power sector under the background of carbon neutrality by 2060. *Renewable and Sustainable Energy Reviews, 166*, 112618.

Marty, E., Bullock, R., Cashmore, M., Crane, T., & Eriksen, S. (2023). Adapting to climate change among transitioning Maasai pastoralists in southern Kenya: An intersectional analysis of differentiated abilities to benefit from diversification processes. *The Journal of Peasant Studies, 50*(1), 136–161.

Maswabi, N. L. (2022). *Factors affecting the implementation of the 2013 revised curriculum in selected primary schools in Mumbwa district of Zambia: an educational management perspective* (Doctoral dissertation, The University of Zambia).

McEvoy, D., Iyer-Raniga, U., Ho, S., Mitchell, D., Jegatheesan, V., & Brown, N. (2019). Integrating teaching and learning with inter-disciplinary action research in support of climate resilient urban development. *Sustainability, 11*(23), 6701.

McGrath, S., & Russon, J. A. (2023). TVET SI: Towards sustainable vocational education and training: Thinking beyond the formal. *Southern African Journal of Environmental Education, 39.*

Molthan-Hill, P., Blaj-Ward, L., Mbah, M. F., & Ledley, T. S. (2022). Climate change education at universities: Relevance and strategies for every discipline. *Handbook of climate change mitigation and adaptation* (pp. 3395–3457). Springer International Publishing.

Mukwawaya, O. Z., Proches, C. G., & Green, P. (2022). Perceived challenges of implementing an integrated talent management strategy at a tertiary institution in South Africa. *International Journal of Higher Education, 11*(1), 100–107.

Nand, K., Baghaei, N., Casey, J., Barmada, B., Mehdipour, F., & Liang, H. N. (2019). Engaging children with educational content via Gamification. *Smart Learning Environments, 6*, 1–15.

Nepraš, K., Strejčková, T., & Kroufek, R. (2022). Climate change education in primary and lower secondary education: Systematic review results. *Sustainability, 14*(22), 14913.

Ngcamu, B. S. (2023). Climate change effects on vulnerable populations in the global South: A systematic review. *Natural Hazards, 118*(2), 977–991.

Nseibo, K., Samuels, C., McKenzie, J., Small, J., Karisa, A., Butler, L., & van Tonder, K. (2023). Redesigning blended courses using the universal design for learning framework: A case of disability studies in an education short course at the University of Cape Town. In *Developing Inclusive Environments in Education: Global Practices and Curricula* (pp. 34–52). IGI Global.

Oketch, M. (2023). What is the appropriate higher education finance model for Africa? Some reflections. *South African Journal of Higher Education, 37*(6), 131–152.

Okoko, J. M. (2020). Framing school leadership preparation and development for Kenya: Context matters. *Educational Management Administration & Leadership, 48*(2), 396–413.

Oluwatimilehin, I. A., & Ayanlade, A. (2023). Climate change impacts on staple crops: Assessment of smallholder farmers' adaptation methods and barriers. *Climate Risk Management, 41*, 100542.

Onyeneke, R. U., Nwajiuba, C. U., Tegler, B., & Nwajiuba, C. A. (2020). Evidence-based policy development: National adaptation strategy and plan of action on climate change for Nigeria (NASPA-CCN). In *African handbook of climate change adaptation* (pp. 1–18).

Onwumelu, D. C. (2023). Biomass-to-power: Opportunities and challenges for Nigeria. *World Journal of Advanced Research and Reviews, 20*(2), 001–023.

Overland, I., Fossum Sagbakken, H., Isataeva, A., Kolodzinskaia, G., Simpson, N. P., Trisos, C., & Vakulchuk, R. (2022). Funding flows for climate change research on Africa: Where do they come from and where do they go? *Climate and Development, 14*(8), 705–724.

Rooney, T., & Blaise, M. (2022). *Rethinking environmental education in a climate change era: Weather learning in early childhood*. Taylor & Francis.

Rosenstock, T. S., Dawson, I. K., Aynekulu, E., Chomba, S., Degrande, A., Fornace, K., Jamnadass, R., Kimaro, A. A., Kindt, R., Lamanna, C., Malesu, M. M., Mausch, K., McMullin, S., Murage, P., Namoi, N., Njenga, M., Nyoka, B. I., Valencia, A. M. P., Sola, P., Shepherd, K. D. & Steward, P. (2019). A planetary health perspective on agroforestry in Sub-Saharan Africa. *One Earth, 1*(3), 330–344.

Seddon, N., Smith, A., Smith, P., Key, I., Chausson, A., Girardin, C., House, J. I., Srivastava, S., & Turner, B. (2021). Getting the message right on nature-based solutions to climate change. *Global change biology, 27*(8), 1518–1546.

Shanahan, K. (2022). *Putting children first: a case study exploring the perspectives toward child protection and safeguarding in thirteen North Dublin primary schools* (Doctoral dissertation, Dublin City University).

Singh, R. L., & Singh, P. K. (2017). Global environmental problems. In *Principles and applications of environmental biotechnology for a sustainable future* (pp. 13–41).

Torto, M. S., Smith, D. T., McKnight, L. W., & Ghosh, P. K. (2022). The internet backpack: Transforming STEM education, agriculture and economic development in Liberia, West Africa. In *2022 IEEE International Symposium on Technology and Society (ISTAS)* (Vol. 1, pp. 1–5). IEEE.

Trisos, C. H., Merow, C., & Pigot, A. L. (2020). The projected timing of abrupt ecological disruption from climate change. *Nature, 580*(7804), 496–5010.

Union, A. (2015). Continental education strategy for Africa. In *Addis Ababa: African Union.*

University of Pretoria. (2021). UP Institutional Advancement SDG Report. Retrieved from https://issuu.com/universityofpretoria/docs/13170_up_institutional_advancement_8cde5f05081767.

United Nations Environment Programme. (2015). Climate change toolkit for teachers. Retrieved [23 June 2024], from https://www.unep.org/resources/toolkits-manuals-and-guides/climate-change-toolkit-teachers.

Walters, S., Yang, J., & Roslander, P. (2014). *Key Issues and Policy Considerations in Promoting Lifelong Learning in Selected African Countries: Ethiopia, Kenya, Namibia, Rwanda and Tanzania. UIL Publication Series on Lifelong Learning Policies and Strategies. No. 1.* UNESCO Institute for Lifelong Learning. Feldbrunnenstrasse 58, 20148 Hamburg, Germany.

Wang, R., Jia, T., Qi, R., Cheng, J., Zhang, K., Wang, E., & Wang, X. (2021). Differentiated impact of politics-and science-oriented education on pro-environmental behavior: A case study of Chinese university students. *Sustainability, 13*(2), 616.

Ward, K., Birch, R., MacDonald, T., Beresford-Dey, M., Lakin, L., Purcell, M., & Searle, B. (2023). Learning for sustainability: Young people and practitioner perspectives.

Yiga, S. (2022). Assessment methodologies and determinants of employability and skills level among Technical and vocational education training (TVET) graduates in Central Uganda. *International Journal of Vocational and Technical Education, 14*(2), 40–47.

You, D., Hug, L., & Anthony, D. (2014). *Generation 2030/Africa.* UNICEF. 3 United Nations Plaza, New York, NY 10017.

Zbieć, M., Franc-Dąbrowska, J., & Drejerska, N. (2022). Wood waste management in Europe through the lens of the circular bioeconomy. *Energies, 15*(12), 4352.

Chapter 4
Conceptual Premises for Climate Change Adaptation Education in African Universities

Marcellus Forh Mbah, Brian Harman, and Petra Molthan-Hill

Abstract Africa is especially vulnerable to the deleterious effects of climate change. Unless there is a significant shift in current trends, many African countries are likely to continue facing extreme weather events that will threaten their food security, water resources, human health, and biodiversity. Consequently, there is a pressing need to ensure that climate change mitigation strategies (strategies to reduce carbon emissions) and climate change adaptation strategies (strategies to circumvent the deleterious effects of climate change) are rolled out across all courses at African universities. Universities serve as both knowledge hubs and vehicles for societal change. However, African universities have traditionally adopted a Eurocentric approach to education that delegitimises Indigenous knowledge and reinforces colonial narratives. To overcome these historical shortcomings, African universities must engage with local populations and leverage Indigenous knowledge systems to co-create place-based climate solutions that provide transformative change for all. In this chapter, we call for African universities to reposition their orientation by reconsidering their conceptualisation of climate change education.

Keywords Climate change education · African universities · Epistemological plurality · Polycentrism · Knowledge co-creation

M. F. Mbah (✉)
School of Environment, Education and Development, University of Manchester, Manchester, England
e-mail: marcellus.mbah@manchester.ac.uk

B. Harman · P. Molthan-Hill
Nottingham Business School, Nottingham Trent University, Nottingham, England
e-mail: brian.harman@ntu.ac.uk

P. Molthan-Hill
e-mail: petra.molthan-hill@ntu.ac.uk

M. F. Mbah et al. (eds.), *Practices, Perceptions and Prospects for Climate Change Education in Africa*, https://doi.org/10.1007/978-3-031-84081-4_4

Introduction

The all-encompassing magnitude of the existential threat posed by climate change constitutes a "wicked," problem (Cross & Congreve, 2021). Indeed, predictions of the dire downstream consequences associated with average temperature rises above 1.5 degrees °C are already well documented (IPCC, 2018). Climate change education (CCE) is considered an essential tool in countering environmental degradation and redressing societal imbalances (UNESCO, 2020). Mainstreaming CCE within all levels of education is critical to achieving a sustainable future (Molthan-Hill et al., 2022). However, CCE must adopt multidisciplinary, interdisciplinary, and transdisciplinary approaches to help shift societal norms and individual mindsets towards sustainable practice (Dupigny-Giroux, 2010).

Traditionally, CCE within universities has focused on teaching climate change science education. This reliance on science education has come at the expense of the other two aspects of CCE, namely climate change mitigation education (CCME) and climate change adaptation education (CCAE). There is now a pressing need to ensure that climate change mitigation strategies (strategies to reduce carbon emissions) and climate change adaptation strategies (strategies to circumvent the deleterious effects of climate change) are rolled out across all university courses (Molthan-Hill et al., 2022). CCE promises the potential of a "multiplier effect" through the broad diffusion of knowledge across society and between disparate communities (Mochizuki & Bryan, 2015). However, the lack of training relating to climate change mitigation and climate change adaptation currently limits the potential for widespread societal change.

Within the Global South, Climate Change Adaptation (CCA) strategies are especially important because countries in this region are often forced to confront extreme climate change. CCA strategies are therefore vital to help secure the lives and livelihoods of those living in the Global South. Importantly, research suggests that CCA strategies are scalable and transferable across different contexts (Paytan et al., 2017). Nevertheless, for CCA strategies to be effective, they must align with the culture of the resident community (Johnson et al., 2022). However, to date, the heavy reliance on climate change science education within CCE has limited the scope of societal response. More broadly, CCE has often undermined the agency of communities within the Global South due to the expansionist, Eurocentric orientation that unpins its reductive, science-based stance. Indeed, "colonially induced environmental changes (have) altered the ecological conditions that supported Indigenous peoples' cultures, health, economies, and political self-determination" (Whyte, 2017 p. 154). Recent international resolutions at COP meetings (COP 26, 27, 28) has renewed a sense of determination to democratise CCE for a global community. However, in Africa, the roll out of CCA and CCM strategies remains painfully slow. This brings us to an important point.

In common discourse, climate change mitigation and climate change adaptation are often misunderstood as being mutually exclusive routes to climate solutions. This misconception is problematic since it presupposes a false dichotomy. It is important to understand that effective climate solutions can serve both mitigation and adaptation goals simultaneously. So, while CCA strategies in Africa may aim to help indigenes adapt to climate change, they may also act as climate mitigation strategies that help to reduce carbon emissions. For example, using solar panels as a power source to refrigerate food satisfies both climate adaptation and climate mitigation goals. Refrigeration allows people to store food thereby making them more adaptable and resilient to climate change. Crucially though, this particular CCA strategy supports climate change mitigation goals by leveraging clean, renewable solar power. In doing so, this CCA strategy cuts carbon emissions while also reducing the potential for food waste (another major contributor of greenhouse gases). This example demonstrates the interconnected nature of climate change solutions and the importance of adopting a holistic approach to problem solving. When evaluating competing CCA strategies, it is therefore important to consider the complimentary climate mitigation potential of these interventions. For example, planting trees can represent an adaptive response to extreme heat (i.e. provides shade and cooling effects) while also supporting climate change mitigation (by converting carbon dioxide into oxygen). Alternative CCA strategies to deal with extreme heat (e.g. the construction of concrete shelters) are suboptimal since the embedded carbon in the construction of the shelters undermines, rather than supports climate mitigation goals. Where possible, CCA strategies should be evaluated holistically to assess their overall impact as a climate solution. In this light, CCE can play an important role in helping communities design climate solutions that best address their needs. In the ensuing sections, we consider the climate change context in Africa and the role of the university in reshaping the status-quo. Central to the role of any progressive university is the widespread operationalisation of CCE. However, for African universities in particular, the onerous task of achieving widespread CCE is complicated further by historical and societal factors. In this chapter, we argue for the decolonisation of CCE and for a recognition of Indigenous knowledge systems (IKS) as a vehicle for promoting knowledge co-creation through transformative social learning. We do so in the pursuit of epistemological plurality and polycentrism for climate solutions.

Climate Change in Africa and the Role of the University

Climate change results in extreme weather events that expose and exacerbate Africa's longstanding vulnerabilities in key areas relating to food security, water resources, human health and biodiversity (Apollo & Mbah, 2021). Indeed, Africa is at the forefront of the climate emergency and is predicted to suffer some of the worst effects of climate change (IPCC, 2018). Despite contributing less than 10% of global greenhouse gases, Africa is especially vulnerable to the effects of climate change. In 2022, climate change cost the continent over US$ 8.5 billion in economic damages and directly affected the lives of over 110 million Africans (Reliefweb, 2023).

It seems reasonable to assume that the deleterious effect of climate change in Africa can be reduced if climate literacy is improved. Climate literacy can be defined as "an awareness of climate change and its anthropogenic causes" (Simpson et al., 2021 p. 937). Note, awareness of climate change alone (perhaps by observing changing weather patterns) may simply engender passivity or maladaptive responses (Eriksen et al., 2012). It is only when a person understands the anthropogenic causes of climate change that remedial and combative action can take place. Simply put, climate literacy "underpins informed mitigation and adaptation responses" that promote climate action (Simpson et al., 2021 p. 937). Sadly, climate literacy rates in Africa vary widely among different states and between different regions. Simpson et al. (2021) found that climate literacy rates ranged from 23–66% across the 33 African countries surveyed. What's more, even larger variances in climate literacy rates were witnessed in states within these African countries (e.g. 5–71% in Nigeria).

Research suggests that African communities are often misinformed about the causes of climate change (Silvestri et al., 2012). Thus, there is a compelling case to be made for striving towards greater climate literacy in Africa since a basic understanding of climate change is required before climate change adaptation and climate change mitigation strategies can be pursued. In addition to promoting adaptive climate change solutions, CCE can also be used to highlight the dangers of maladaptive, carbon producing activities (e.g. mining) that accelerate climate change. In the coming years, fossil fuel extraction in Africa is set to quadruple (Earth Insight, 2022). This is especially worrisome when we consider that 90% of the land earmarked for mining, oil, and gas reserves lies within tropical rainforests (Earth Insight, 2022). Future fossil fuel extraction in Africa is therefore doubly destructive since it fuels further carbon emissions while irreparably damaging the much-needed carbon sinks (forests) that absorb carbon emissions. It is imperative that local communities understand not only the benefits of climate adaptation but also the costs associated with environmental degradation. It is to this end that universities must apply themselves.

Universities serve as "knowledge hubs" and "training centres" (Ssekamatte, 2022) that promote sustainable development within local communities (Blum et al., 2013). In addition to striving towards carbon neutrality themselves (see Udas et al., 2018), many universities are extending this ethos outwards to promote climate adaptation and mitigation measures within their broader communities (Filho et al. 2021). In doing so they are expanding the "societal carbon brainprint by teaching knowledge and skills in the area of carbon neutral practices" (Filho et al. 2021 p. 2). This outward looking, forward-facing perspective chimes with past calls for universities to take up leadership roles within society. Universities should actively shape society by being "proactive leaders in promoting societal change" rather than simply be "indicators of (societal) change" that reflect the zeitgeist (Virtenen, 2010 p. 232). To proactively lead climate action, universities must adopt different climate related roles and goals within society. Specifically, universities must be instrumental in "generating scientific knowledge through scientific research; providing training and capacity building; carrying out sensitizations and providing guidance to communities and policy makers" (Ssekamatte, 2022 p. 12). Their role must also extend to engaging with local communities to create place-based climate change mitigation and climate change adaptation solutions (Ssekamatte, 2022 p. 12). Indeed, recent research has outlined how CCE can be integrated into all university courses (Molthan-Hill et al.,

2022). This mainstreaming and integration of CCE across all university courses is an important step in addressing climate change (see Boateng & Boateng, 2015; Buckland et al., 2018; Reza, 2016).

Climate Change Education in Africa

CCE provides the means to address the current knowledge deficits that exist within rural African communities. As thought leaders within their communities, African universities are uniquely placed to deliver the CCAE and climate change mitigation education (CCME) that Africa so desperately needs. Since climate change will influence all aspects of daily life, it stands to reason that CCE should be a cornerstone of a university student's education. Indeed, since universities are beneficiaries of taxpayer money, it seems wholly appropriate that they should serve the communities in which they reside. Indeed, some would argue that universities are morally obliged to cascade down important information that is relevant to the future lives of their students (Nussbaum et al. 2015).

This renewed focus on CCE has prompted some to assert that "the way forward for universities is to dynamically reposition" (Filho et al., 2021 p. 2). CCE can be incorporated into a wide variety of (in)formal learning scenarios and contexts if operational and regulatory inertia can be overcome within the university sector (Molthan-Hill, 2019). Indeed, CCE provides new opportunities for universities to transcend traditional barriers and mobilise transformative change within society (Apollo & Mbah, 2021; Ssekamatte, 2020). However, research suggests that a lack of commitment among university officials, shortsighted priorities, knowledge silos and a general ignorance of sustainability agendas all act as barriers to CCE within universities (see Larrán et al., 2016; Tilbury, 2011; Lotz-Sisitka, 2011). It must also be acknowledged that the transdisciplinary nature of CCE poses challenges for educators who must grapple with their own knowledge deficits when navigating the complex, interdisciplinary climate change literature (Berger et al., 2015; Pruneau et al., 2010). However, research suggests that educators are keen to embrace this challenge if given training opportunities to address their knowledge deficits (Apollo & Mbah, 2021). While CCE is taking root within at least some Africa universities (see Apollo & Mbah, 2021; Mbah & Ezegwu, 2024) serious consideration still needs to be given to the factors that help and hinder its further expansion. The green shoots of this greener education system will only grow if CCAE strategies are supported by decolonisation efforts (Mbah & Ezegwu, 2024) that undermine the prevailing Eurocentric perspective on climate change.

Decolonising Climate Change Education

"The colonial invasion that began centuries ago caused anthropogenic environmental changes that rapidly disrupted many Indigenous peoples," (Whyte, 2017 p. 155). The aftermath of this invasion has been pollution, deforestation and soil degradation. Today, the exportation of CCE from first world nations to the global south represents "climate colonialism" (Sultana, 2022). Eurocentric education systems have sought to inculcate colonised people with value systems and perspectives that are not in alignment with their own traditional practices (Ajaps & Mbah, 2022). By delegitimizing the value of Indigenous knowledge, exported education systems have marginalised communities by robbing them of important cultural components (e.g. language, religion, myth, traditions, rituals, songs etc.). The marginalising effects of occupation and subsequent indoctrination undermine community efforts to seek local solutions to local problems (Asante, 2008). Indeed, the "epistemic violence" perpetrated by colonial forces has incarcerated the minds of indigenes within "cognitive prisons" (Cajete, 2005). The devaluation of Indigenous Knowledge Systems (IKS) reinforces the prison bars of these cognitive prisons through a process of marginalisation. To stem this tide of marginalisation, curricular reform is required. According to Adebisi (2016), the decolonisation of curriculums can be viewed as an evolving process that shrugs off past colonial influence through the restoration of IKS. "The foundational intent of decolonisation is to equip students with "diverse academic learning environments, curricula and approaches to research within which Indigenous cultures, histories, and knowledge are embedded" (Waghid & Hibbert, 2018 as cited by Lumadi, 2021, p. 2). Decolonising education is an issue that has risen to prominence in recent years (Zembylas, 2018). Decolonising CCE calls for the widespread adoption of transdisciplinary teaching approaches. "Transdisciplinary education goes beyond interdisciplinary content" (Newberry & Trujillo, 2018 p. 205) and fosters knowledge exchange, and problem solving between academics and practitioners (Williams et al., 2016). The expansionist perspective afforded by transdisciplinary education permits input from those within and those outside the different scientific communities. While structural fragmentation and ever-increasing specialisation characterise the traditional sciences, transdisciplinary education seeks to withdraw the lines of demarcation between disciplines and domains. In doing so, it provides a remedy to the shortcomings of western science (Aldunce et al, 2016) that can be exclusive rather than inclusive in its outlook. Accordingly, Chao and Enari (2021 p. 32) call for transdisciplinary, experimental and decolonial imaginations" grounded in an ethos of inclusivity, participation and humility" to "destabilise the prevailing hegemony of secular science". In doing so, they invoke different types of imagination to combat climate change; relational imagination, storied imagination, beyond-human imagination, multi-sensory imagination, reflective imagination, emplaced imagination and transdisciplinary imagination. While climate change has been conceptualised as "an incredible failure of imagination" (Wallace-Wells, 2019), the authors assert that utilising these complementary imaginations will lead to a "decolonised imagination". The reductionist perspective of westernised climate science fails to capture the web of social complexity that underpins the climate emergency. African climate change adaptation education must therefore be customised to conform to the lived experiences of those within the community (Shava & Nkopodi, 2020). To achieve this goal, CCAE strategies must leverage the Indigenous knowledge (IK) of the local community. This view appears to be shared by university students. Mampane, Omidir and Aluko (2018) found that postgraduate students perceived glocal initiatives to be an essential component of decolonisation efforts. These students also maintained that technology did not have to be relinquished to achieve decolonisation. However, the students did believe that a decolonised education system should be "foregrounded in Indigenous knowledge" (Mampanne et al. 2018 p. 1).

The Importance of Indigenous Knowledge Systems (IKS)

Indigenous Knowledge Systems (IKS) encapsulate the local skills, knowledge, cultural components and inter-generational traditions that allow indigenes to work in concert with nature (Ubisi, Kolanisi, & Jiri, 2019). IKS's are especially relevant to communities in the Global South where the intergenerational transmission of knowledge is commonplace (Greenwood, and Lindsay, 2019). IKS provide "ground truth" (Praskievicz, 2022) and may be considered integral to securing sustainable development in rural communities (Thaman, 2002). IK is "holistic, synthetic and multi-contextual" (Newberry & Trujillo, 2018 p. 204). These properties of IK make it particularly responsive to finding consensus among the republic of stakeholders that are invested in finding local climate solutions. Thus, by fortifying CCAE strategies with IK, the needs for communal inclusivity and individual self-determination are addressed. In contrast, misguided climate policies based solely on reductionist climate change science are unlikely to gain traction within rural communities. CCAE strategies must therefore embrace resident cultures and speak to the lived experiences of those within the community. To achieve their goals, CCAE strategies will require the decolonisation and indigenisation of university curriculums. Research suggests that the adoption of IKS's within CCE is linked to successful CCE programmes within the Global South (Johnson et al., 2022). However, little research has addressed the need to integrate IKS within university based CCE interventions (Ulmer & Wydra, 2020). Indeed, "western conceptions of the sustainability discourse alienate and remove the socio-cultural specificities in sustainability" (Kumalo, 2017, p.19). Past research has demonstrated that Indigenous students are more likely to suffer from alienation and a loss of identity if their curriculum undermines the Indigenous knowledge that has been passed down to them. As Newberry and Trujillo (2018) note, higher education is largely designed for 'low' context learners (also see Ibarra, 2001). 'Low' context learners are comfortable with procedural learning and can easily compartmentalise and assimilate information devoid of a social context. In contrast, 'high' context learners require information to be socially constructed and situated within their lived experience. Here, "demonstration, application, and experience" are required for learners to fully assimilate the information being prescribed. To service both sets of learners Cajete (2005) suggests that learners should be exposed to "engaged civic learning" that involves authentic problem-solving scenarios. University CCE programmes should therefore be relatable and relevant to indigenes. Accordingly, CCE programmes should be characterised by localised curriculums and tailored outreach programmes that are both malleable and sensitive to local concerns (Mignolo, 2011). This simple logic underpins the arguments for the customisation, decolonisation and indigenisation of higher education within the Global South. However, achieving these aims requires active and continual collaboration between the university and the local population (Mbah, 2019).

Indigenous knowledge has stood the test of time and increases a community's potential for resiliency. Its longevity is testament to the depth of knowledge that can exist between people and place. Furthermore, a recent systematic review finds that IKS can shape influential CCAE strategies that are transferable across regions (Schlingmann et al., 2021). However, scientists need to exercise restraint and ensure these IKS are not exploited, monetised, or misused (Latulippe & Klenk, 2020). The exploitation of IKS by non-Indigenous scientists simply reinstates a new brand of colonialism (Chavez & Gavin, 2018).

Research suggests that embedding IKS within CCAE strategies presents exciting opportunities for impactful knowledge co-creation (Mbah, 2019). What's more, universities have an important and expanding role in progressing sustainability goals through co-creation activities (Stein, 2023). By opening the channels of communication among stakeholders this participative approach allows for the co-creation of shared solutions through the distillation of a negotiated wisdom. However, this is contingent upon affording an equal weighting to formal (scientific) knowledge and informal (Indigenous) knowledge systems. Indeed, community based, knowledge creation within universities appears to be the exception rather than the rule. Ulmer and Wydra (2020) argue that the longstanding absence of co-creation activities between universities and local communities may be due to language barriers and cultural differences. To surmount these barriers, universities need to embrace the principles of transformative social learning if co-created, glocal solutions are to be realised (Mampanne et al. 2018).

Co-Creation of Climate Change Adaptation Education (CCAE) via Transformative Social Learning

Ensor and Harvey (2015) define social learning as a process "emerging through practices that facilitate knowledge sharing, joint learning, and co-creation of experiences between stakeholders around a shared purpose in ways that: 1) take learning and change beyond the individual to communities, networks, or systems; and 2) enable new shared ways of knowing to emerge that lead to changes in practice" (p. 510). Indeed, CCAE has been reformulated as a form of social learning that has transformative properties (Collins & Ison, 2009). Transformation learning is characterised by an "emancipatory, participatory, value laden, transgressive co-engagement with complex matters of concern" (Macintyre et al., 2018 p.85). As such, it allows different actors to "co-define" the "matters of concern". This flexible approach facilitates cooperation and inclusivity among a broad "republic of stakeholders" by abandoning the hegemony of a purely science approach. As such, transformative social learning leaves space to develop emerging solutions through reflexive means and dialogical processes. Arguably, this approach lends itself well to addressing complex problems such as climate change. Interestingly, transformative learning has also been used in tandem with decolonizing pedagogy to progress climate change solutions (Mackinlay & Barney, 2014; Zembylas, 2018). Zembylas (2018) suggests that applying a humanist lens to "decolonial thinking" (Mignolo, 2011) provides solutions to the questions surrounding the future format of CCE. These complementary approaches lead to greater cooperation and knowledge sharing among stakeholders. Indeed, research suggests that cooperation and knowledge sharing between different cohorts permits the "co-creation" of knowledge that forms the bedrock of effective CCA strategies (Utter et al., 2021).

The co-creation of knowledge is "a collaborative process involving two or more actors, who are intentionally integrating their knowledge and learning, resulting in the development of insights and solutions that would not otherwise be reached independently" (Utter et al., 2021 p. 1). Co-creation within agroecology is characterised by bottom-up, participatory action research that focuses on the creation of novel knowledge. Interestingly, Utter et al. (2021) suggests that this novel knowledge is often "appropriated and co-opted by academics and relabelled as "new knowledge". Herein lies a problem. Undertaking research on IK can become an extractive process whereby academics assimilate and classify this wisdom through a scientific lens (Latulippe & Klenk, 2020). In short, researchers must appreciate the "epistemology of the south". Renouncing the binary classifications of Indigenous knowledge and scientific knowledge allows scholars to discern the similarities and shared ground between both these knowledge systems (Agrawal, 1995). Going forward, academics must be willing to embrace a "pluralistic" approach to CCE and climb down from their westernised ivory towers. They must move beyond the cosseted environs of the university and seek knowledge within their local communities. By relocating to the broad church of pluralism, universities and their staff will be able to fulfil their new mandate of co-creating a relatable, CCAE knowledgebase with Indigenous knowledge holders.

Reorienting Knowledge Creation to Capture Epistemological Plurality

The unwanted legacy of a colonial past is today being dismantled through indigenisation and decolonisation initiatives within higher education (Adebisi, 2016). To this end, Ajaps and Mbah (2022) advocate applying "epistemological plurality" to CCE. The plurality of knowledge finds itself occupying the middle ground between competing schools of thought, such as Afrocentrism and cosmopolitanism. Advocates of Afrocentrism hold that education within Africa must be uniquely tailored to the continental context (Royster, 2020). Furthermore, Afrocentrism asserts that education should use African concerns as the focal point for education. As the name suggests, this perspective endorses an "Africa" centric perspective whereby all ancillary concerns are pushed to the periphery. Such a perspective assumes that Africa can be both self-sufficient and self-sustaining on its own terms. This insular perspective shuns the notion of global input and seeks to ignore the unyielding power of globalism. While some might view Afrocentrism as a path to reinvigorating Indigenous knowledge systems it may also be viewed as a path to isolationism. The alienation and marginalisation suffered by Indigenous populations of the past may well be perpetuated again if they are cut off from the outside world. Afrocentrism also assumes a concept of Africa as a unity of measure which seems highly problematic given the scale and complexity that characterise the continent. Such generalisations are unhelpful and likely to be regressive.

In contrast, cosmopolitanism asserts that Africa is part of a global community and should therefore tailor its education system to embrace pluralistic and globalist ideals. Cosmopolitism holds that Africa is part of a global community and, as such, must remain fully open to outside influences. Arguably, influences from the west and more recently, influences from the east, allow Africa to find its place in the world. This perspective seeks to centralise a global perspective and pushes this agenda to the fore at the expense of local and regional concerns.

Negotiating a more conciliatory and centrist position, pluralism seeks to reconcile these binary propositions by endorsing the use of both local and glocal responses to CCE. By weighting all knowledge systems equally, pluralism advocates utilising both IKS and western sciences to arrive at equitable, localised, co-created, climate solutions. Importantly, the inclusion of localised knowledge and IKS serves to undermine the Eurocentric hegemony that currently characterises African education. Pluralism therefore champions both place-based pedagogy (Sobel, 2005) and critical pedagogy (De Sousa Santos, 2014) in its aim to renounce colonial narratives and progress social justice. Critical pedagogy renounces imported doctrine and embraces the social situations of the local people. This teaching approach champions a social justice agenda which may be viewed as a recurring theme within pedological research. Indeed, this theme has been repackaged under various guises and could hardly be considered contentious. For example, while Banks and Banks (1995) speak of multicultural education and Kumashiro (2002) of anti-oppressive education, the roots of this school of thought can be traced back to transformative learning (Mezirow, 1978) and the democratic pedagogy espoused by Freire (1996). By restoring the inextricable link between people and place, pluralism ensures that the full "ecology of knowledges" are represented in problem solving and co-creation activities (Ajaps & Mbah, 2022). The aim of pluralism is therefore to achieve parity between different knowledge systems in an effort to deconstruct the prevailing Eurocentric hegemony that is both persistent and frequently pernicious. As depicted in Fig. 4.1, the ecology of knowledges should be evident in different areas of a university's mission, namely teaching, research and outreach activities.

In real terms, the ecology of knowledges involves an emancipatory engagement with all stakeholders. Solutions are therefore derived through a process of thoughtful consultation and co-creation between all vested interests. Learners within the education system can therefore be viewed as active contributors. As such, these learners should have the power to bend a malleable curriculum to their will in order to address and redress the historical repressions of thought and action.

Hegger et al (2012) suggests that seven conditions need to be met to facilitate co-creation. Co-creation activities require a "broad church" where all actors find a voice in the problem definition and goal setting stages. The differing perspectives and competencies of stakeholders must be acknowledged, and the resources

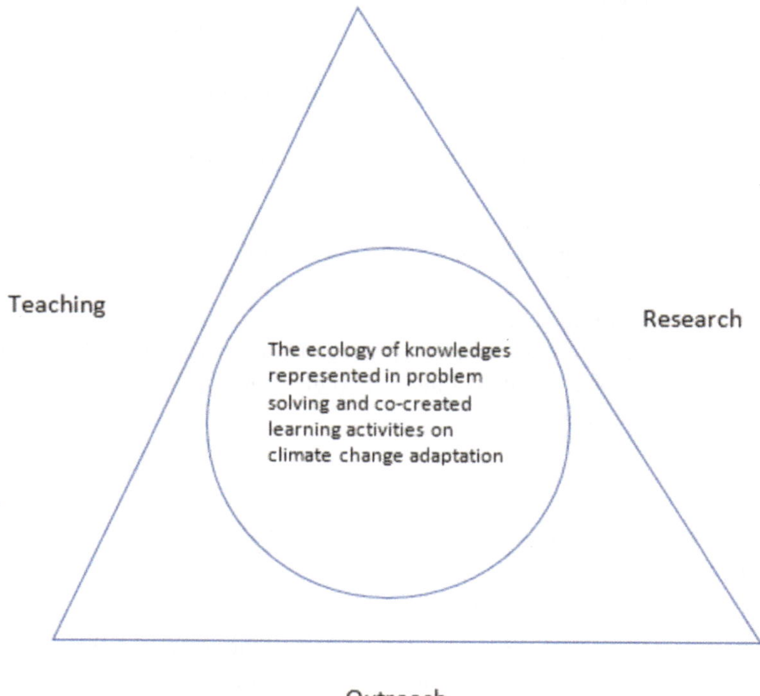

Fig. 4.1 The relationship between the ecology of knowledges and different missions of the university

and facilities required to support co-creation activities, carefully considered. Meaningful incentives and rewards need to be put in place to overcome inertia and help promote innovation. Sadly, a recent systematic review of co-creation activities within agroecology finds that only 6% (of the 69 cases reviewed) resulted in positive outcomes (Cartagena, 2019). The review concluded that co-creation activities that yielded successful impacts were those that solicited a sense of attachment to the project and those that were personally relevant to the farmers. External factors such as resource limitations, power asymmetries and obstructive regulation were found to hinder co-creation efforts. Past research on agroecology co-production also finds that shifting power dynamics between stakeholders greatly affects engagement and interaction levels (Carolan, 2006). This is an important consideration for future university outreach programmes. Outreach programmes that aim to co-create knowledge between academics and farmers must foster a mutual respect for both people and place (Scannell & Gifford, 2010). Relationship building is therefore fundamental to the success of outreach programmes (White & Utter, 2021). Humility, active listening, and empathy are required to gain the trust of farmers who may feel wary of imported doctrines that are incompatible with their local culture (De Sousa Santos, 2014). Differences in perspective, interests and goals can also drive a wedge between stakeholders unless they are properly managed (Ayala-Orozco, et al., 2018). Indeed, the importing of solutions from the Global North represents a form of techno-solutionism that arises from climate colonialism (Stein, 2023). Nevertheless, transdisciplinary outreach activities focused on co-creation have the potential to support transformational social learning within communities (Macintyre et al., 2018).

The inclusivity of transdisciplinary teaching approaches holds the promise of meaningful dialogue between local communities and scientists. This participative approach allows for the co-creation of optimised and localised climate solutions. Couched within the resident culture and nested within nuance, transdisciplinary approaches have yet to be fully realised. Parity between science and non-science inputs appears difficult to achieve. However, scientists need to realise that effective climate solutions all inherently require a polycentric approach and understanding of the behavioural complexities that contribute to the problem. In this regard, western science alone comes up short.

Pursuing Polycentricity within the University to Address Climate Solution

Changes in perceptual and behavioural patterns are required across different facets of a university's mission to facilitate climate solutions. Solutions that fail to leverage the generational wisdom embedded with communities are likely to fail. While western science has served society well, it must adopt a more conciliatory stance with regard to Indigenous knowledge. Scientists within the hard sciences must listen to their colleagues within the softer social sciences who are more receptive to the wisdoms contained within culture. The epistemological implications may be uncomfortable for some who have spent their lives asserting the supremacy of western sciences. Since the age of the enlightenment (Pinker, 2018), reason and logic have become the mainstays of this domain. The unbounded success of the scientific method as a means of progressing development has led to an epistemology that reinforces the narrative that science in isolation can overcome all challenges. However, now science faces its most pressing challenge, one that cannot arguably be overcome without transdisciplinary input. Science must extend an olive branch to those who seek to contribute to climate solutions from outside academia and outside the traditional sciences (Newberry & Trujillo, 2018). It is worth considering that academics have been indoctrinated with the scientific method. Thus, training and resources may be needed to help facilitate their transition to a more inclusive modus operandi that considers the ecology of knowledges. The vaulted position that academics occupy within society furnishes them with the power to influence vast swathes of people. However, perhaps the issue is that the self-sustaining culture within academia is at loggerheads with the culture that exists outside of the university gates. Typically, it is environmental evangelists and politicians who extoll the virtues of a data driven approach to climate change solutions. These prominent thought leaders often set the agenda and frame the messaging on climate change solutions. In stark contrast, the voices of those who are most affected by climate change are often muted or unheard. Going forward, universities have an important role to play in giving voice to these underrepresented communities. Since universities should serve their communities, they must consider how the Indigenous knowledge of the region can be incorporated into their curricula to comunially transform local and national responses to climate change (Mbah et al., 2021).

We assert that the African university of tomorrow should embrace a placed-based focus when delivering CCE. Such a university would aim to capture the full ecology of knowledges through its teaching, research, and outreach programmes (see Fig. 4.1). In short, African universities should adopt a polycentric approach to CCE (Ajaps & Mbah, 2022). This approach should be characterised by a tripartite mission focus that embeds IK within all teaching, research and outreach activities within the university. The role of academics within this approach is to actively raise the profile of IK among students, the private sector and the government. Table 4.1 summarises how research, teaching, and outreach/fieldwork can be used to pursue polycentricity among these different cohorts. Below, we provide an example of how a polycentric approach might be used as a framework for operationalising a climate solution (clean cooking).

Rigorous research by an international team of scientists has identified 93 of the most impactful technologies and practices that can be employed to reduce concentrations of greenhouse gases (Project Drawdown, 2022a). Highly ranked in this list of climate solutions is the practice of "clean cooking". Indeed, "clean cooking is one of the most potent solutions to reducing global greenhouse gases" (Project Drawdown, 2022a). Unlike traditional means of cooking, clean cooking does not involve burning materials that significantly contribute to greenhouse gases. Instead, clean cooking uses renewable fuel (e.g. solar power) or low impact fuels (e.g. liquid petroleum

Table 4.1 A tripartite approach to pursuing polycentricity within the university

Research	As intermediaries between government and local communities, universities should engage in research that seeks to build bridges between rural communities and national government. Transdisciplinary and interdisciplinary research should be employed to promote curriculum decolonisation (Newberry & Trujillo, 2018). The findings of the research should inform future government policy on climate action
	More funding should be provided to support local outreach programmes aimed at achieving glocal solutions (Mampane et al., 2018). Government policies should seek to nurture farmers' links with the land (Ajaps & Mbah, 2022) by encouraging cross pollination between academia and agrarian practice
Teaching	Universities should employ place-based pedagogy and outreach work to prioritise and normalise transformative social learning. These activities should foster continuous engagement with the local community to create stronger links between the university and the community (Mbah, 2019)
	A flexible, interactive and innovative approach to curriculum design should be adopted. Students should be actively involved in problematising and problem solving local issues through the lens of a polycentric approach (Ajaps & Mbah, 2022; Mignolo, 2011). Mainstreaming CCE within all levels of higher education is critical to achieving a sustainable future. However, future CCE efforts need to focus on teaching tailored climate solutions that simultaneously serve both climate mitigation and climate adaptation goals (Molthan-Hill et al., 2022). The route to achieving this goal is through co-created, glocal solutions that combine the full ecology of knowledges
Outreach and fieldwork	Best practices need to be adopted when undertaking fieldwork to avoid conflict (Chavez & Gavin, 2018) and to ensure that the research being undertaken is directly applicable to the indigenes (Olesen & Nordentoft, 2018). Students and academics who undertake outreach work should take a collaborative approach to fieldwork. In the spirit of restorative justice, researchers should adopt a humble, open-minded outlook that seeks to redress traditional power asymmetries
	Relinquishing incompatible goals and the allure of techno-solutionism (Stein, 2023) will ameliorate relations with the indigenes. Lecturers and other stakeholders who enjoy privileged positions of power need to remain open-minded and magnanimous during co-creation activities. Knowledge systems should be afforded equal weight if the stereotypes of ill-fated past encounters are to be avoided (Briley et al., 2015)

gas) that create much less greenhouse gases than traditional solid fuel stoves. Clean cooking may also be viewed as a CCM strategy but also as a CCA strategy since it potentially allows people to cook food in the absence of dwindling resources such as firewood. The potential impact of educating people on the benefits of clean cooking are huge when we consider that "worldwide, billions of people mainly cook with polluting fuels and technologies" (Project Drawdown, 2022b).

."As of 2020, an estimated 43% of families in low and middle-income countries were mainly using cookstoves fuelled by traditional wood or coal stoves for cooking" (Project Drawdown, 2022b). What's more, research suggests that 31% of the global population will still be using solid fuel stoves by 2030 (Stoner et al., 2021). While clean cooking is an important climate solution it nevertheless represents an externally sourced, technocratic intervention to reduce carbon emissions. If widespread behaviour change is to be adopted by Indigenous communities then this departure from traditional cooking methods must be sympathetic to the existing Indigenous knowledge base.

Similarly, universities can adopt a polycentric approach to CCE as itemised in Table 4.1, touching on their missions: (1) Research, (2) Teaching, and (3) Outreach and fieldwork.

Specifically, for clean cooking to be widely adopted, significant research is first required to ascertain the type of stove that would work best in each region. Weather conditions, infrastructure and the availability of renewable materials are key determinants. University researchers could assess the viability and suitability of the different options available in the area. Importantly, researchers would also need to work with local communities to understand the barriers and constraints that might hinder the adoption of this new practice. This information could then be fed back to government bodies to inform future public policy. In doing so, researchers could act as intermediaries between local communities and legislative bodies to help create the social conditions conducive to the adoption of clean cooking practices.

After the research has established the best stove, it could then fall to educators to disseminate this knowledge widely within their communities. Here, the role of the educator is to cascade the information down to the grassroot level within communities. As respected thought leaders within their communities, academics also have the power to sway industry leaders and local influencers within the community.

Finally, outreach and field research could be used to demonstrate the utility of the clean cooking stoves in real world conditions. This type of outreach work would extend the university's conservation ethos outwards. It would bring the message to hard-to-reach communities that traditionally lay beyond the reach of the university. By adopting a polycentric approach to CCE, researchers and educators can begin to redress the knowledge gap between the "town and the gown". In doing so, they can develop ever more impactful climate change adaptation and climate change mitigation strategies within their resident communities.

Conclusion

IK has an important role to play in adaptation practices in Africa. However, greater recognition of IKS within higher education and governmental spheres is required to co-create CCA solutions that are relevant and relatable to rural farmers. The lack of recognition of IK within governmental quarters seems to undermine the utility of IK at university level, thereby delegitimising this knowledge base. It appears that colonial narratives still shape education within Africa. We propose that future IK research should challenge scientists to find the humility to relinquish the power advantage bestowed unto them by their datasets. Similarly, we propose that outreach programmes should promote genuine co-creation opportunities rather than simply be used as fortuitous encounters to harvest Indigenous wisdom or propagate the scientific faith. The goal of the epistemologically plural or polycentric university is not to integrate knowledge but rather to co-create it. This constructionist perspective challenges the positivist orthodoxy embedded within academia and governments. Government policy and university curriculums must strive to accommodate the place-based attachments that bind people to places. It is clear that IK must form an integral part of climate mitigation and climate adaptation strategies in Africa.

The conceptual premises advanced in this chapter are intended to support the transdisciplinary research and co-creation activities needed to deliver the climate solutions of tomorrow. Alas, if the promise of an epistemological plural or polycentric university is to be realised then the cultural barriers of climate colonisation and techno-solutionism must first be overcome. A departure from a neo-liberal, market-driven education system is required to secure transformative education (Odell et al., 2019). Furthermore, transformative education that seeks to address all the Sustainable Development Goals must "restructure power and the embedded values within society" (Odell et al., (2019 p. 3). To promote the required "third order changes" to learning, Sterling (2011) asserts that we must engage in "seeing things differently". An epistemological plural or polycentric approach to CCE in Africa will help to achieve these transformative changes by promoting core sustainability competencies that empower learners to integrate sustainability into their everyday lives (see Wiek et al., 2011). Empowering learners to tackle climate change will require them to engage in different modes of thinking; (1) Systems thinking competency, (2) anticipatory or future thinking competency, (3) normative or value thinking competency, (4) strategic thinking or action-orientated competency and (5) interpersonal or collaborative competency. Tailored CCE can induce these modes of thinking that are needed to foster meaningful and impactful changes to thinking and behaviour. As such, CCE has the potential to empower a new generation of Africans to take ownership of their own futures by facing down the threat of climate change on their own terms.

References

Adebisi, F. I. (2016). Decolonising education in Africa: Implementing the right to education by re-appropriating culture and indigeneity. *Northern Ireland Legal Quarterly., 67*, 433–451.

Agrawal, A. (1995). Dismantling the divide between indigenous and scientific knowledge. *Development and Change, 26*(3), 413–439.

Ajaps, S., & Mbah, F. M. (2022). Towards a critical pedagogy of place for environmental conservation. *Environmental Education Research, 28*(4), 508–523. https://doi.org/10.1080/13504622.2022.2050889

Aldunce, P., Bórquez, R., Adler, C., Blanco, G., & Garreaud, R. (2016). Unpacking resilience for adaptation: Incorporating practitioners' experiences through a transdisciplinary approach to the case of drought in Chile. *Sustainability, 8*(9), 905.

Apollo, A., & Mbah, M. F. (2021). Challenges and opportunities for climate change education (Cce) in East Africa: A critical review. *Climate, 9*(6), 93.

Asante, M. K. (2008). Afrocentricity: Toward a new understanding of African thought in the world. In M. Kete, Y. M. Asante, J. Yin (Eds.) *The Global Intercultural Communication Reader* (pp. 101–110). New York, USA: Routledge. ISBN 978-0-415-95812-7.

Ayala-Orozco, B., Rosell, J. A., Merçon, J., Bueno, I., Alatorre-Frenk, G., Langle-Flores, A., & Lobato, A. (2018). Challenges and strategies in place-based multi-stakeholder collaboration for sustainability: Learning from experiences in the global South. *Sustainability, 10*(9), 3217.

Banks, C. A. M., & Banks, J.A. (1995). Equity pedagogy: An essential component of multicultural education. *Theory Into Practice*, 152–158.

Berger, P., Gerum, N., & Moon, M. (2015). Roll up your sleeves and get at it! climate change education in teacher education. *Canadian Journal of Environmental Education, 20*, 154-172.

Blum, N., Nazir, J., Breiting, S., Goh, K. C., & Pedretti, E. (2013). Balancing the tensions and meeting the conceptual challenges of education for sustainable development and climate change. *Environmental Education Research, 19*(2), 206-217.

Boateng, C. A., & Boateng, S. D. (2015). Tertiary institutions in Ghana curriculum coverage on climate change; implications for climate change awareness. *Journal of Education and Practice, 6*(12), 99–106.

Briley, L., Brown, D., & Kalafatis, S. E. (2015). Overcoming barriers during the co-production of climate information for decision-making. *Climate Risk Management, 9*, 41–49.

Buckland, P., Goodstein, E., Alexander, R., Muchnick, B., Mallia, M. E., Leary, N., et al. (2018). The challenge of coordinated civic climate change education. *Journal of Environmental Studies and Sciences, 2018*(8), 169–178.

Cajete, G. A. (2005). American Indian epistemologies. *New Directions for Student Services, 2005*(109), 69–78.

Carolan, M. S. (2006). Sustainable agriculture, science and the co-production of 'expert' knowledge: The value of interactional expertise. *Local Environment, 11*(4), 421–431.

Cartagena, B. L. (2019). Bridging the gap between theory and practice in agroecological farming: Analyzing knowledge co-creation among farmers and scientific researchers in Southern Spain [Unpublished master's thesis]. Murcia, Spain: Spain Regeneration.

Chao, S., & Enari, D. (2021). Decolonising climate change: A call for beyond-human imaginaries and knowledge generation. *ETropic, 20*(2), 32–54. https://doi.org/10.25120/etropic.20.2.2021.3796

Chavez, D. M., & Gavin, M. C. (2018). A global assessment of Indigenous community engagement in climate research. *Environmental Research Letters, 13*(12), 123005.

Collins, K., & Ison, R. (2009). Jumping off Arnstein's ladder: Social learning as a new policy paradigm for climate change adaptation. *Environmental Policy and Governance, 19*(6), 358–373.

Cross, I. D., & Congreve, A. (2021). Teaching (super) wicked problems: Authentic learning about climate change. *Journal of Geography in Higher Education, 45*(4), 491–516.

De Sousa Santos, B. (2014). *Epistemologies of the South: Justice against Epistemicide.* Oxon, UK: Routledge.

Dupigny-Giroux, L. A. L. (2010). Exploring the challenges of climate science literacy: Lessons from students, teachers and lifelong learners. *Geography Compass, 4*(9), 1203–1217.

Earth Insight. (2022). Congo in the Crosshairs: New Oil & Gas Expansion Threats to Forests and Communities. Retrieved 08 Jan 24, from https://www.earth-insight.org/wp-content/uploads/2022/11/Congo-in-the-Crosshairs-November-2022-English-Report-small.pdf.

Ensor, J., & Harvey, B. (2015). Social learning and climate change adaptation: Evidence for international development practice. *Wiley Interdisciplinary Reviews: Climate Change, 6*(5), 509–522.

Eriksen, S., Aldunce, P., Bahinipati, C.S., Martins, R.D.A., Molefe, J.I., Nhemachena, C., O' Brien, K., Olorunfemi, F., Park, J., Sygna, L., & Ulsrud, K. (2012). When not every response to climate change is a good one: Identifying principles for sustainable adaptation. In *Sustainable Adaptation to Climate Change* (pp. 7–20). Routledge.

Filho, W. L., Sima, M., Sharifi, A., Luetz, J. M., Salvia, A. L., Mifsud, M., Olooto, F. M., Djekic, I., Anholon, R., Donkor, F. K., Dinis, M. A. P., Klavins, M., Finnveden, G., Chari, M. M., Molthan-Hill, P., Mifsud, A., Sen, S. K., & Lokupitiya, E. (2021). Handling climate change education at universities: An overview. *Environmental Sciences Europe, 33*, 1–19.

Freire, P. (1996). *Pedagogy of the Oppressed (Revised)*. Continuum.

Greenwood, M., & Lindsay, N. M. (2019). A commentary on land, health, and indigenous knowledge (s). *Global Health Promotion, 26*(3_suppl), 82–86.

Hegger, D., Lamers, M., Van Zeijl-Rozema, A., & Dieperink, C. (2012). Conceptualising joint knowledge production in regional climate change adaptation projects: Success conditions and levers for action. *Environmental Science & Policy, 18*, 52–65.

Ibarra, R. (2001). *Beyond affirmative action: Reframing the context of higher education*. The University of Wisconsin Press.

IPCC. (2018). *Global Warming of 1.5°C, an IPCC special report on the impacts of global warming of 1.5°C above pre-industrial levels and related global greenhouse gas emission pathways, in the context of strengthening the global response to the threat of climate change, sustainable development, and efforts to eradicate poverty*. Retrieved Feburary 14, 2023, from http://www.ipcc.ch/report/sr15/.

Johnson, D. E., Parsons, M., & Fisher, K. (2022). Indigenous climate change adaptation: New directions for emerging scholarship. *Environment and Planning E: Nature and Space, 5*(3), 1541–1578.

Kumalo, S. H. (2017). Problematising development in sustainability: Epistemic justice through an African ethic. *Southern African Journal of Environmental Education, 33*, 14–24.

Kumashiro, K. K. (2002). Troubling education: Queer activism and anti-oppressive pedagogy. *Psychology press*.

Larrán Jorge, M., Herrera Madueño, J., Calzado, Y., & Andrades, J. (2016). A proposal for measuring sustainability in universities: A case study of Spain. *International Journal of Sustainability in Higher Education, 17*(5), 671–697.

Latulippe, N., & Klenk, N. (2020). Making room and moving over: Knowledge co-production, Indigenous knowledge sovereignty and the politics of global environmental change decision-making. *Current Opinion in Environmental Sustainability, 42*, 7–14.

Lotz-Sisitka, H. (2011). The "event" of modern sustainable development and universities in Africa. *Sustainable Development, 2*(1), 41–57.

Lumadi, M. W. (2021). The pursuit of decolonising and transforming curriculum in higher education. *South African Journal of Higher Education, 35*(1), 1–3.

Macintyre, T., Lotz-Sisitka, H., Wals, A., Vogel, C., & Tassone, V. (2018). Towards transformative social learning on the path to 1.5 degrees. *Current Opinion in Environmental Sustainability, 31*, 80–87.

Mackinlay, E., & Barney, K. (2014). Unknown and unknowing possibilities: Transformative learning, social justice, and decolonising pedagogy in indigenous Australian studies. *Journal of Transformative Education, 12*(1), 54–73

Mampane, R. M., Omidire, M. F., & Aluko, F. R. (2018). Decolonising higher education in Africa: Arriving at a glocal solution. *South African Journal of Education, 38*(4). https://doi.org/10.15700/saje.v38n4a1636

Mbah, M. (2019). Can local knowledge make the difference? Rethinking universities' community engagement and prospect for sustainable community development. *The Journal of Environmental Education, 50*(1), 11–22.

Mbah, M., Ajaps, S., & Molthan-Hill, P. (2021). A systematic review of the deployment of indigenous knowledge systems towards climate change adaptation in developing world contexts: Implications for climate change education. *Sustainability, 13*(9), 4811.

Mbah, M. F., & Ezegwu, C. (2024). The Decolonisation of climate change and environmental education in Africa. *Sustainability, 16*(9), 3744.

Mignolo, W. (2011). *The darker side of western modernity: Global futures, decolonial options*. Duke University Press.

Mochizuki, Y., & Bryan, A. (2015). Climate change education in the context of education for sustainable development: Rationale and principles. *Journal of Education for Sustainable Development, 9*(1), 4–26.

Molthan-Hill, P., Blaj-Ward, L., Mbah, M. F., & Ledley, T. S. (2022). Climate change education at universities: Relevance and strategies for every discipline. *Handbook of Climate Change Mitigation and Adaptation* (pp. 3395–3457). Springer International Publishing.

Molthan-Hill, P., Worsfold, N., Nagy, G. J., Filho, W. L., & Mifsud, M. (2019). Climate change education for universities: A conceptual framework from an inter-national study. *Journal of Cleaner Production., 226*, 1092–1101. https://doi.org/10.1016/j.jclepro.2019.04.053

Mezirow, J. (1978). Perspective transformation. *Adult Education, 28*(2), 100–110.

Newberry, T., & Trujillo, O. V. (2018). Decolonizing education through transdisciplinary approaches to climate change education. In *Indigenous and decolonizing studies in education* (pp. 204–214). Routledge.

Nussbaum, E. M., Owens, M. C., Sinatra, G. M., Rehmat, A. P., Cordova, J. R., Ahmad, S., et al. (2015). Losing the Lake: Simulations to promote gains in student knowledge and interest about climate change. *International Journal of Environment & Science Education, 10*(6), 789–811.

Odell, V., Molthan-Hill, P., Martin, S., and Sterling, S. (2019). Transformative Education to Address All Sustainable Development Goals. In: Leal Filho, W., Azul, A., Brandli, L., Özuyar, P., Wall, T. (eds) *Quality Education. Encyclopedia of the UN Sustainable Development Goals.* Springer, Cham. https://doi.org/10.1007/978-3-319-69902-8-106-1

Olesen, B. R., & Nordentoft, H. M. (2018). "On slippery ground"–beyond the innocence of collaborative knowledge production. *Qualitative Research in Organizations and Management: An International Journal, 13*(4), 356–367.

Paytan, A., Halversen, C., Weiss, E., Pedemonte, S., & Mescioglu, E. (2017). Education for climate change adaptation and resilience. *Limnology and Oceanography Bulletin, 26*(3), 71–73.

Pinker, S. (2018). Enlightenment now: The case for reason, science, humanism, and progress. Penguin, UK.

Praskievicz, S. (2022). Ground truth: Finding a "place" for climate change. *Progress in Environmental Geography, 1*(1–4), 137–162.

Project Drawdown. (2022a). *Table of Solutions.* Retrieved 08 Jan 2024, from https://drawdown.org/solutions/table-of-solutions.

Project Drawdown. (2022b). *Clean cooking.* Retrieved 09 Jan 2024, from https://drawdown.org/solutions/clean-cooking.

Pruneau, D., Khattabi, A., & Demers, M. (2010). Challenges and possibilities in climate change education. *Online Submission, 7*(9), 15–24.

Reliefweb, (2023). Africa suffers disproportionately from climate change. Retrieved January 8, 2024, from https://reliefweb.int/report/world/africa-suffers-disproportionately-climate-change.

Reza, M. I. (2016). Sustainability in higher education: Perspectives of Malaysian higher education system. *SAGE Open*, 1–9.

Royster, P. D. (2020). Epistemology of knowledge. In *Decolonizing Arts-Based Methodologies*, pp 150–171. Brill.

Scannell, L., & Gifford, R. (2010). Defining place attachment: A tripartite organizing framework. *Journal of Environmental Psychology, 30*(1), 1–10.

Schlingmann, A., Graham, S., Benyei, P., Corbera, E., Sanesteban, I. M., Marelle, A., & Reyes-García, V. (2021). Global patterns of adaptation to climate change by indigenous peoples and local communities: A systematic review. *Current Opinion in Environmental Sustainability, 51*, 55–64.

Shava, S., & Nkopodi, N. (2020). Indigenising the university curriculum in southern Africa. In *Indigenous studies: Breakthroughs in research and practice*, (pp. 243–254). IGI Global.

Silvestri, S., Bryan, E., Ringler, C., Herrero, M., & Okoba, B. (2012). Climate change perception and adaptation of agro-pastoral communities in Kenya. *Regional Environmental Change, 12*, 791–802.

Simpson, N. P., Andrews, T. M., Krönke, M., Lennard, C., Odoulami, R. C., Ouweneel, B., Steynor, A., & Trisos, C. H. (2021). Climate change literacy in Africa. *Nature Climate Change, 11*(11), 937–944.

Sobel, D. (2005). Place-based education: Connecting classrooms and communities. *Education for Meaning and Social Justice., 17*(3), 63–64.

Ssekamatte, D. (2022). The role of the university and institutional support for climate change education interventions at two African universities. *Higher Education*, 1–15.

Ssekamatte, D. (2020). Towards a theoretical model linking university education to climate change interventions in the African context. *Journal of African Studies and Development, 12*(1), 17–24. https://doi.org/10.5897/jasd2020.0569

Stein, S. (2023). Universities confronting climate change: beyond sustainable development and solutionism. *Higher Education*, 1–19. https://doi.org/10.1007/s10734-023-00999-w

Sterling, S. (2011). Transformative learning and sustainability: sketching the conceptual ground. *Learning and teaching in higher education, 5*(11), 17–33.

Stoner, O., Lewis, J., Martínez, I. L., Gumy, S., Economou, T., & Adair-Rohani, H. (2021). Household cooking fuel estimates at global and country level for 1990 to 2030. *Nature Communications, 12*(1), 5793.

Sultana, F. (2022). The unbearable heaviness of climate coloniality. *Political Geography*. https://doi.org/10.1016/j.polgeo.2022.102638

Thaman, K. H. (2002). Shifting sights: The cultural challenge of sustainability. *Higher Education Policy, 15*(2), 133–142.

Tilbury, D. (2011). Higher education for sustainability: A global overview of commitment and progress. *Higher Education in the World, 4*(1), 18–28.

Ubisi, N. R., Kolanisi, U., & Jiri, O. (2019). Comparative review of indigenous knowledge systems and modern climate science. *Ubuntu: Journal of Conflict and Social Transformation, 8*(2), 53–73.

Udas, E., Wölk, M., & Wilmking, M. (2018). The "carbon-neutral university"–a study from Germany. *International Journal of Sustainability in Higher Education, 19*(1), 130–145.

Ulmer, N., & Wydra, K. (2020). Sustainability in African higher education institutions (HEIs) shifting the focus from researching the gaps to existing activities. *International Journal of Sustainability in Higher Education, 21*(1), 18–33.

UNESCO. (2020). Education for sustainable development: a roadmap. Retrieved Feburary 14, 2022, from https://unesdoc.unesco.org/ark:/48223/pf0000374802.locale=en.

Utter, A., White, A., Méndez, V. E., & Morris, K. (2021). Co-creation of knowledge in agroecology. *Elementa: Science of the Anthropocene, 9*(1), 00026.

Virtenen, A. (2010). Learning for climate responsibility: Via consciousness to action. In W. L. Filho (Ed.), *Universities and climate change* (pp. 231–240). Springer.

Waghid, Z., & Hibbert, L. (2018). Advancing border thinking through defamiliarisation in uncovering the darker side of coloniality and modernity in South African higher education. *South African Journal of Higher Education, 32*(4), 263–283.

Wallace-Wells, D. (2019). The uninhabitable earth: A story of the future. Penguin UK.

White, A., & Utter, A. (2021). Essential attributes in the co-production of knowledge by farmers and agricultural outreach professionals in Vermont (*Unpublished manuscript*). University of Vermont.

Whyte, K. (2017). Indigenous climate change studies: Indigenizing futures, decolonizing the Anthropocene. *English Language Notes, 55*(1), 153–162.

Wiek, A., Withycombe, L., & Redman, C. (2011). Key competencies in sustainability: A reference framework for academic program development. *Sustainability Science, 6*(2), 203–218.

Williams, J., Roth, W.M., Swanson, D., Doig, B., Groves, S., Omuvwie, M., Borromeo Ferri, R., & Mousoulides, N. (2016). *Interdisciplinary mathematics education*. Springer Nature.

Zembylas, M. (2018). Decolonial possibilities in South African higher education: Reconfiguring humanising pedagogies as/with decolonising pedagogies. *South African Journal of Education, 38*(4), 1–11.

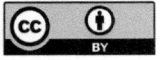

Part II
Pedagogical Insights, Approaches and Capabilities

Chapter 5
ICT-Enabled Climate Change Education and Adaptation in Africa

Evans Olaniyi, Eregha Perekunah, and Osuji Emeka

Abstract Climate change is a global issue that is both inevitable and urgent, and Africa is particularly vulnerable to its impact. This chapter examines how ICT can be used for climate change education and adaptation in Africa. Despite the importance of ICT, there is limited information available in the literature about how it is presently and will be used in Africa's efforts to adapt to climate change. The chapter highlights that ICT is crucial in addressing the significant challenges posed by climate change in Africa, and can be employed to facilitate the dissemination of knowledge required for climate change adaptation at the community level. This can be accomplished by raising awareness, providing access to critical information, and promoting learning and sharing of experiences.

Keywords ICT · Climate change · Climate education · Climate adaptation

Introduction

Recently recorded changes in the climate, such as those attributed to global warming caused by the release of greenhouse gases, indicate catastrophic consequences for natural and human environments worldwide, as reported by the World Meteorological Organization (WMO, 2021). The WMO also found that CO2 and N2O concentrations in the atmosphere were respectively 150%, 262%, and 123% higher than preindustrial levels in 2021. Chatiza (2019) predicts that the world will surpass the threshold for dangerous warming between 2027 and 2042. However, the impacts of human-caused climate change are already evident in Africa. According to the Intergovernmental Panel on Climate Change (IPCC, 2014), the continent has experienced temperature increases of 0.1 to 0.3 degrees every decade, in addition to changing rainfall patterns and more frequent droughts and floods. Trisos et al. (2022) suggest that warming in the South Atlantic and Indian Oceans may have caused a reduction in monsoons, depriving the Sahel region of rainfall. Africa is projected to be more severely affected

E. Olaniyi (✉) · E. Perekunah · O. Emeka
School of Management and Social Sciences, Pan Atlantic University, Lagos, Nigeria
e-mail: oevans@pau.edu.ng

M. F. Mbah et al. (eds.), *Practices, Perceptions and Prospects for Climate Change Education in Africa*, https://doi.org/10.1007/978-3-031-84081-4_5

by climate change than other regions, with warming occurring at a faster rate than the global average. Rising sea levels are expected to impact island nations and the eastern coast of Africa (Trisos et al., 2022).

Information and communication technology (ICT) systems have become a cornerstone of modern civilization (Fettweis and Zimmermann, 2008). The growth of ICT and its integration into daily life has been so significant that it is difficult to imagine a future without smartphones, the internet, and satellite networks (Evans & Mesagan, 2022). Effective use of ICT-enabled tools, such as ICT-enabled climate change education, may aid national climate adaptation efforts in Africa, providing early-warning systems and education to communities about potential consequences and means of adaptation (Trisos et al., 2022). While the extent to which ICT-enabled climate change education is integrated with development goals varies from country to country, there is a need for improvement in utilizing its potential urgently in Africa due to current changes on the continent (Aderemi & Onyekwelu, 2011).

According to Simpson et al., (2022), climate change literacy is essential for intelligent mitigation and adaptation efforts, requiring an understanding of climate change and its anthropogenic causes. As such, international organizations, such as the World Bank and the United Nations, view climate change literacy in the form of ICT expertise as a tool for achieving long-term development, particularly in poorer nations (Baklanov et al., 2016). The broader concept of climate change adaptation, which communities must employ to restore their way of life, includes climate change literacy with ICT tools at its core.

Despite ongoing research, the main barriers to addressing climate change are the complexity and diversity of its impacts (Trisos et al., 2022). This chapter examines climate change education and adaptation in Africa using ICT, emphasizing the diversity within the continent and offering context-specific findings and solutions. Moreover, there is currently a lack of understanding of climate change literacy in Africa. As such, educating the next generation of Africans about climate change is imperative in combating its economic and societal effects (Allianceforscience, 2022).

Methodology

The data collection process involved identifying and reviewing relevant literature on climate change education and adaptation in Africa, as well as the potential of ICT-enabled solutions. Indeed, conceptual papers typically do not require primary data collection, as they are based on a synthesis of existing theories, concepts, and empirical evidence. However, to ensure the accuracy and credibility of the concepts, the synthesis of existing literature was conducted through a review of academic journals, textbooks, as well as online databases, such as Web of Science, Google Scholar, and Scopus. The search terms used were "climate change education," "climate change adaptation," "ICT-enabled solutions," "Africa," and various combinations of these terms. The search items were limited to peer-reviewed publications (published in English) between 2000 and 2022.

Thematic analysis is a commonly used method for analyzing qualitative data in social science research (Braun & Clarke, 2006). We used this method to identify key concepts and themes in the literature on climate change education and adaptation in Africa, as well as the potential of ICT-enabled solutions. Furthermore, the data analysis process in conceptual research involves synthesizing existing concepts and theories to develop a new or refined theoretical model. In this paper, the data analysis process involved the synthesis of existing literature on climate change education and adaptation, as well as the potential of ICT-enabled solutions in Africa. We analyzed and compared various theories and concepts related to climate change education and adaptation, as well as the potential of ICT-enabled solutions.

Climate Change in Africa

Climate change poses a serious threat to the rural livelihoods of Sub-Saharan Africa (Tume et al., 2019). This hazard is increasing agricultural productivity and food security risks, as well as endangering millions of lives worldwide, particularly in Africa (IPCC, 2014). Climate change is having a greater impact on rural African populations due to their increased susceptibility (IPCC, 2014). Hulme (2005) notes that temperatures have increased by approximately 0.5 degrees Celsius per century throughout the twentieth century, with greater warming occurring in the seasons from June to August and September to November than in December to February and March to May. Between 1961 and 2000, there were more warm periods and fewer cold days in southern and western Africa, and the six hottest seasons of this millennium all occurred after 1980, with 1986–1995 being the highest recorded (Kotir, 2011). Scientific evidence indicates an anthropogenic influence in the twentieth century's continental-wide temperature rises (IPCC, 2014; Min et al. 2007; Stott et al., 2010). Unlike the Sahel, where rainfall has decreased over the last two decades, rainfall has recently recovered (WGI AR5). Scientific evidence supports the notion that increased anthropogenic GHG forcing is causing global warming (IPCC 2013).

Developing countries' low-income communities frequently experience the consequences of global warming. One of the most vulnerable regions to climate change is Africa, which is already experiencing the effects of climate change due to soil degradation and loss of agricultural land (Carney et al., 2014; Lambin et al., 2003; Leh et al., 2013). The Fourth Assessment Report of the Intergovernmental Panel on Climate Change (IPCC, 2007) provides more evidence that climate change is happening and poses significant risks to social, political, and economic factors. According to the IPCC, the contemporary "heating of the climate system is unambiguous," as evidenced by increases in global average air and ocean warmth, extensive melting of snow and ice, and an increasing global average sea level (IPCC 2007, p. 5). The Sahara is projected to become the most vulnerable region. These effects will occur concurrently with Africa's rapid population growth, which is expected to rise from 0.9 billion people in 2005 to over 2 billion by 2050 (IPCC, 2007).

African farmers are facing a serious threat from climate change due to increasing temperatures, changing patterns of rainfall and drought, and extreme weather events like storms and heatwaves (Gacheno & Amare, 2021). Africa, with its high population density, has a long history of severe and prolonged disasters, especially droughts. For example, a devastating drought occurred in the Horn of Africa due to four years of below-average rainfall (Weforum, 2022), which affected more than 18 million people in Ethiopia, Somalia, and Kenya who suffered from extreme famine (US government's Humanitarian Information Unit, in Weforum, 2022). Land degradation in Africa has increased the vulnerability of the already precarious situation. Additionally, global economic activities have contributed to recent environmental changes that have led to severe human misery and social instability in Africa (Nyong et al., 2007). Climate change resulting from greenhouse gas (GHG) emissions has caused severe effects on natural and human habitats worldwide (WMO, 2021), with CO2, methane, and N2O emissions in the atmosphere being 150%, 262%, and 123% higher than pre-industrial times in 2021 (WMO, 2021).

Africa has been warming at a rate of about + 0.3 °C per decade between 1991 and 2021, and sea levels are rising faster along African beaches than the global average, putting between 108 and 116 million people at risk from sea level rise by 2030 (WMO, 2022). Droughts in East Africa have been exacerbated by multiple failed wet seasons, as well as increased violence, population displacement, and COVID-19 restrictions. Heatwaves have affected many parts of Northern Africa, including Tunisia, Algeria, Morocco, and Libya, while South Sudan, Nigeria, the Republic of Congo, the Democratic Republic of the Congo (DRC), and Burundi have experienced severe flooding. Sand and dust storms have also been a persistent issue (WMO, 2022).

The main climate-related concerns in Africa are flooding and droughts (WMO, 2022), which have cost the region's economy more than 70 billion USD in the past 50 years. The estimated climate-related costs to African countries could reach $50 billion annually by 2050. Long-term declines in river flows are attributed to rising temperatures, drought, and increasing water demand, and climate-related hazards have been the main cause of displacement in Africa. In 2021, around 14.1 million people in Sub-Saharan Africa were internally displaced. Since 1961, rising temperatures have contributed to a 34% drop in agricultural production in Africa, and a 1.5 °C global warming is expected to result in a 9% decrease in maize output in West Africa (WMO, 2022).

Despite the pressing need to improve climate services across Africa, around 418 million people still lack access to basic drinking water, and 779 million people need basic sanitation facilities (WMO, 2022). Only 27 of the 51 African countries for which data is available have the capacity to undertake Integrated Water Resource Management, and initiatives are often carried out on an ad hoc basis with unstable funding. Only four African countries have end-to-end drought prediction or warning services at full/advanced capability (WMO, 2022).

ICT for Climate Change Adaptation in Africa

The International Panel on Climate Change (IPCC) has predicted that by 2050, agricultural and fodder growing seasons in Western and Southern Africa may be reduced by an average of 20% (Zougmoré et al., 2016). This highlights the need for African communities to become more resilient to climate change and its impact on food security. The use of information and communication technology (ICT) can play a vital role in monitoring, mitigating, and adapting to the effects of climate change. As Pant and Heeks (2012) suggest, ICT can aid in merging current data, providing access to new information and expertise, lowering transaction and service costs, and playing a positive role in ICT-based industry. In order to combat the consequences of climate change, it is necessary to promote extensive climate change measures that utilize ICT as an effective instrument for communicating relevant technology (Adenle et al., 2015).

Africa is especially vulnerable to natural disasters due to its geography, dense population, and poverty (Mbilinyi et al., 2013). Imam et al. (2017) state that tropical cyclones, landslides, and droughts are the primary meteorological catastrophes that impact the environment, ecology, way of life, and socioeconomic development of Africans. In order to promote extensive climate change measures, ICT is expected to be a successful tool for communicating relevant technology to reduce climate change. ICT can be used to share and distribute knowledge that can help in adapting to the effects of climate change, especially in early warning, disaster risk management, early farming, and health care. For instance, in India, locally relevant information provided through ICT has helped residents better adapt to harsh weather disasters and food shortages (Imam et al., 2017). In Burkina Faso, community radio has become a vital resource for raising awareness of the local impacts of climate change. Furthermore, early warning systems, urban and rural planning, health services, and education are all common uses of ICT in Zimbabwe, while participatory videography and digital storytelling are employed in Madagascar (Imam et al., 2017).

The use of ICT can play a crucial role in combating the consequences of climate change in Africa. The UN Climate Change secretariat has recognized this importance by forming a new relationship with the Global e-Sustainability Initiative (GeSI) in 2013. Local-level solutions, such as storm and hurricane shelters, shoreline and river berms, tree plantings, crop diversity, tillage techniques, and diverse agronomic techniques, have also been employed to combat the effects of climate change. As a result, ICT has the potential to transform climate change adaptation strategies in the coming years (Imam et al., 2017).

It is important to acknowledge that there are numerous challenges to the use of ICT in Africa. One major challenge is that the use of ICT is more difficult in rural areas, where communities rely on solar energy for their electricity needs and face annual erosion. This creates obstacles for using ICT tools, as people in these communities often have to go to marketplaces or other individuals with solar home systems to charge their phones. Additionally, poor network access in many African countries affects both mobile phone and internet use for climate change adaptation and mitigation. Illiteracy is also a concern, particularly for using mobile phones and reading dashboards in UDC or CDC (Imam et al., 2017). Moreover, African women are often discouraged from participating in climate change adaptation due to the perception that they are less tech-savvy or disadvantaged.

Studies have highlighted that farmers struggle to adapt to climate change due to a lack of information. In one study, about half of the respondents reported not receiving information about the likely amount and distribution of rainfall in the upcoming season, which reduces their ability to adapt (Chimanga & Kanja, 2020). Therefore, the use of ICT and capacity building can assist small-scale farmers in responding to climate change. Agriculture is a crucial livelihood for many people in Africa, especially in rural areas, and climate change has significantly affected crop seasons. However, most of Africa's rural population lacks the knowledge necessary to make informed decisions about coping with climate change (Morton, 2007). Hence, both developed and developing nations can benefit from ICT services in areas such as agriculture, health, early warning, and catastrophe risk reduction (Morton, 2007).

Other challenges have arisen in implementing ICT technologies, methods, techniques, and policies for adapting to climate-related hazards in Africa. Although some studies have identified these challenges, the difficulty of implementing new technologies arises from the majority of users lacking adequate knowledge about their use and maintenance (Lybbert & Sumner, 2012). Therefore, a regulatory and institutional framework is necessary to promote innovation and the use of existing technology for significant improvement in climate change adaptation, particularly in Africa. This will allow for the possibility of creating and implementing new technologies on a larger scale to move towards a green environment (Lybbert & Sumner, 2012).

The prevalent approach to climate change adaptation has been top-down, planned, and purposeful. However, it is worrying that many climate change adaptation measures are emergent, unplanned, and originate within specific communities, particularly in Africa (Pant & Heeks, 2012). To address this, adaptation projects should include tools for building collective capability (Pant & Heeks, 2012). Fortunately, ICT works towards this direction as most ICT applications, particularly those centered on mobile technology, are designed with a planned strategy in mind (Pant & Heeks, 2012).

Africa's ICT for Education

ICT, a broad term encompassing a wide range of technologies such as radio, telescopic photos, mobile phones, and digital currency exchanges, has become increasingly prevalent in modern society (Salampasis & Theodoridis, 2013). The ability to access, manage, integrate, evaluate, and produce information using digital technology, telecommunication tools, and/or networks in order to function in a knowledge society is defined as ICT literacy by the International ICT Literacy Panel (Educational Testing Service, 2002). This skill set covers a range of abilities, from basic daily living skills to more advanced ICT expertise that can bring about transformational change. The use of computer technologies, which are part of ICT, handles information collection, retrieval, modification, storage, and distribution (Educational Testing Service, 2002). Today, various computer devices, business-focused AI software, communication media, and networking infrastructure are available.

However, the use of ICT in Africa as a transformational tool is hampered by various factors, such as low levels of general literacy and weak abilities in areas like reading, math, and problem-solving (Ukpabi & Karjaluoto, 2017; Zaidan, 2017). Nonetheless, the development of ICT systems is the foundation of modern civilization and innovations in this field are readily embraced. The global economic and social structure have seen significant changes with the emergence of ICT, and it has also transformed the educational process, making it more globally standardized (Evans & Mesagan, 2022; Ukpabi & Karjaluoto, 2017; Zaidan, 2017). Therefore, ICT has become an essential tool in the modern educational landscape, with the capability of providing online education (Ukpabi & Karjaluoto, 2017; Zaidan, 2017).

In Africa, post-secondary education has increased faster than in any other area, with over 77 million students enrolling in institutions in 2020 (Statista, 2022). To take advantage of this potential, many African nations have increased investment in both physical and intellectual capital. The African Union has recognized that innovation, science, and technology are developed on the foundation of education, and many African countries have concentrated on creating ICT policies and National Information and Communication Infrastructure Plans to aid in their socioeconomic development efforts and ICT in education programs. However, obstacles such as lack of comprehensive policies, potential amplification of social, cultural, and economic inequities, and exclusion of underprivileged groups must be acknowledged African Union, 2014).

To seize the opportunities offered by ICT, African countries must continue to develop their physical and human resources. Young people can create jobs out of necessity or free will, and African SMEs are creating regional and customized platforms, apps, content, and solutions that allow Africans to use the internet for communication, commerce, education, and other requirements (African Union, 2014). However, the massive development has had a negative impact on the quality of education and learning because there has been no equivalent increase in funding (African Union, 2014).

Increasing access to and attainment of higher education levels is imperative for more equitable access to healthier living conditions. If education is not provided to enhance human capacity, access to information and ICT will not have the intended developmental effects (UNESCO, 2019). In higher education, one develops the skills necessary for the knowledge economy. Knowledge societies such as in Africa must have the "skills to recognize, produce, process, transform, disseminate, and use data" in order to develop and use knowledge for the advancement of humanity (UNESCO, 2005). Therefore, a mix of education and ICT could be a key development engine for the African region.

ICT for Climate Change Education in Africa

In order to combat the economic and societal effects of climate change, it is imperative to educate the next generation of Africans on this topic, according to the Alliance for Science (2022). Researchers have found that formal education is the most significant contributor to climate literacy (Bhattacharya et al., 2021; Kranz et al., 2022). However, in Africa, a lack of knowledge and information and the absence of a coherent plan that takes climate change into account increases the continent's vulnerability to the impacts of climate change (Bhattacharya et al., 2021; Kranz et al., 2022).

Less developed continents are preferentially affected by climate change challenges, particularly in places where low-tech farming techniques predominate, according to Kates (2000). Africa is one of the continents that suffer the heaviest burden of the effects of climate change, especially on agriculture (Akinyemi et al., 2017). Climate-related phenomena to be concerned about include severe storms, flooding, and their escalating costs. Since the year 2000, there has been an increase in climate-change-induced natural disasters, where climate change literacy through the application of ICT knowledge solutions has been encouraged (Chatiza, 2019).

Climate change education rates differ significantly throughout Africa at both the national and subnational levels, according to Simpson et al., (2022). Moreover, country-level climate change literacy rates are 12.8% lower for women than for males (Simpson et al., 2022). These findings point to the areas where interventions should focus to improve climate change literacy and ensure that responses are based on a better understanding of both the present and future state of the climate.

Studies have also shown that innovative approaches to the difficulties posed by frequent and intense unexpected weather occurrences and stress can be made possible by ICT. By offering an efficient means of communication and warning people about calamities like those caused by climate change, ICT can save lives and preserve the wealth of society (Wakili et al., 2020). The use of social media platforms such as Facebook, Twitter, and Instagram has been crucial to raising individual understanding of climate change (Wakili et al., 2020). Therefore, in order to achieve a sustainable green economy, the transformational use of ICT climate change tools is essential for Africa (Subashini & Fernando, 2017).

However, as several studies have highlighted, including Kremer and Houngbo (2021), digital technology has the potential to alleviate global poverty and hunger, even in rural areas of emerging Africa. Unfortunately, the majority of rural dwellers in Africa have a poor level of ICT education, which makes these data sources either inaccessible or confusing to them. Studies (e.g., Chikaire et al., 2017; Nzonzo & Mogambi, 2016) have shown that the lack of ICT education necessary to integrate ICT tools into climate change mitigation and adaptation processes exacerbates the effects of climate change. Increased awareness of the trends and possible predictors of climate change literacy is critical for empowering more educated responses to anthropogenic climate change, and proper ICT education is a step in the right direction towards a green economy in Africa.

To demonstrate the potential of ICT to increase access to information and support for climate change adaptation, and to build the capacity of local communities to respond to the challenges of a changing climate, a few examples of best practices of ICT-enabled climate change education in Africa include (Bwalya & Mulenga, 2018):

- *The Climate Smart Agriculture Project in Kenya*: This project utilizes ICT to provide smallholder farmers with timely and relevant information on climate change, weather forecasts, and agricultural practices. The project uses mobile phones to disseminate the information to farmers, who can use it to make informed decisions about their farming practices.
- *The Climate Change Adaptation Project in Tanzania*: This project uses ICT to provide information on climate change adaptation strategies to rural communities in Tanzania. The project uses mobile phones and community radio stations to disseminate the information, and also provides training on climate-smart agriculture and sustainable natural resource management.
- *The Adaptation Learning Programme in Mali*: This project uses ICT to provide training and support to farmers and pastoralists in Mali on climate change adaptation. The project uses a combination of radio broadcasts, mobile phones, and community-based training to reach rural communities, and focuses on building the capacity of local organizations to support climate change adaptation efforts.
- *The Rural Climate Exchange in South Africa*: This project uses ICT to facilitate knowledge exchange and collaboration between rural communities, researchers, and policy makers on climate change adaptation. The project provides an online platform for sharing information and best practices, and also facilitates face-to-face meetings and training events.

Discussion

Climate change is a critical issue in sub-Saharan Africa, and rural communities are particularly vulnerable. With agriculture being a crucial source of income for many in these communities, the potential consequences of climate change can have devastating effects on their livelihoods. However, through climate education, it is possible to build resilience against the effects of climate change and promote sustainable practices to mitigate its impact. Education and awareness campaigns can help increase climate literacy and empower communities to take action, making it a crucial step towards creating a sustainable future for the continent. In this regard, ICT and education can work hand in hand to combat the effects of climate change. By utilizing digital technology, climate education can reach a broader audience, making it more accessible to those who may not have had access to traditional forms of education. Additionally, the use of digital technology can enhance the quality and effectiveness of climate education by providing interactive and engaging learning experiences, real-time feedback and assessment, and facilitating collaborative learning.

Moreover, the integration of ICT and climate education can help disseminate information on climate change and its effects. By using digital technology, climate-related data and information can be easily accessed, analyzed, and shared, enabling individuals and communities to make informed decisions and take action to reduce their vulnerability to climate change. For instance, the use of weather monitoring systems and early warning systems can help predict and mitigate the impact of natural disasters and extreme weather events. Investing in the integration of ICT and climate education is crucial in building resilience against climate change in Africa. By providing access to quality education and digital technology, individuals and communities can become more informed, empowered, and capable of taking action towards mitigating the impact of climate change. This can lead to a more sustainable future for the continent.

Climate education can also play a vital role in enabling the effective use of ICT to adapt to the challenges of climate change in Africa. By educating individuals and communities on the potential of ICT and its applications in addressing climate change, they can be empowered to make better use of available technology and access new technologies to mitigate the impact of climate change. Additionally, climate education can help build digital literacy skills, making it easier for rural farmers and other vulnerable communities to utilize ICT tools. By promoting gender equity in education, women in rural areas can also be empowered to utilize digital technology and gain access to vital climate change information.

Furthermore, climate education can create awareness and advocacy for the use of ICT in mitigating and adapting to climate change. By educating policymakers and other stakeholders on the importance of ICT in climate change mitigation and adaptation, they can be encouraged to invest in the necessary infrastructure and resources needed to facilitate the effective use of ICT. In a nutshel, investing in climate education and the integration of ICT is essential in adapting to the challenges of climate change in Africa. By promoting digital literacy, gender equity, and awareness, individuals and communities can be empowered to take advantage of ICT tools to mitigate the impact of climate change and build a more sustainable future. Climate education can be a transformative tool that builds resilience against the effects of climate change and provides a path towards a sustainable future for Africa.

Conclusions and Recommendations

This chapter delves into the topic of climate change education and adaptation in Africa, specifically focusing on how information and communication technologies (ICT) can be used to address these issues. While ICT can be a valuable tool for climate change adaptation, there is currently limited information available on how it is being used in Africa. This chapter highlights the vulnerability of the continent to the impacts of climate change and emphasizes the importance of transformative ICT education to effectively address these issues.

Climate change is a complex challenge that affects many aspects of society, including socio-ecological systems, people, and regions. In Africa, climate change has contributed to a slowdown in economic development and is particularly threatening to rain-fed agricultural production and subsistence farmer livelihoods. Drought and rising temperatures are the most significant climatic factors affecting these regions. Given Africa's over-reliance on sectors that rely on favorable environmental conditions for growth, it is projected to suffer the most from climate change.

Despite the potential benefits of ICT, many African farmers face barriers to accessing and using it due to factors such as illiteracy, lack of technical knowledge, financial constraints, social barriers, and infrastructural concerns. However, ICT has the potential to provide farmers with new information services, allowing them to make more informed decisions regarding their land and water resources. In turn, this could increase their bargaining power with businesses providing agricultural and water services, leading to increased crop yields and income.

Closing the capacity gap in data collection for fundamental hydrometeorological variables, such as drought and flood early warning systems, is crucial to improving climate services in vulnerable LDCs, particularly in Africa. In addition, increased investment in communication networks is necessary to fully realize the potential of ICT in climate change adaptation.

Ultimately, the severity of climate change in Africa underscores the importance of prioritizing ICT-enabled adaptation and mitigation as policy options. However, the lack of ICT literacy necessary to effectively integrate these tools exacerbates the effects of climate change. Therefore, increasing awareness of climate change literacy and proper ICT literacy is critical for empowering more educated responses to anthropogenic climate change in Africa.

References

Adenle, A. A., Azadi, H., & Arbiol, J. (2015). Global assessment of technological innovation for climate change adaptation and mitigation in developing world. *Journal of Environmental Management, 161*, 261–275.

Aderemi, O. O., & Onyekwelu, A. E. (2011). Climate Change Adaptation and ICT-Enabled Education: Case Studies from Africa. In N. Butcher & R. Kozma (Eds.), *ICT in Education in Sub-Saharan Africa: A Comparative Analysis of Basic e-Learning Strategies* (pp. 221–250). World Bank Publications.

African Union. (2014). Agenda 2063: The Africa We Want. https://au.int/en/agenda2063

Akinyemi, O., Alege, P. O., Ajayi, O. O., & Okodua, H. (2017). Energy pricing policy and environmental quality in Nigeria: A dynamic computable general equilibrium approach. *International Journal of Energy Economics and Policy, 7*(1), 268–276.

Allianceforscience 2022: Africa struggles to improve climate change literacy as its effects worsen across the continent. Available online at: https://allianceforscience.cornell.edu/blog/2022/04/africa-struggles-to-improve-literacy-about-climate-change-as-its-effects-worsen-across-the-continent/#:~:text=Researchers%20identified%20formal%20education%20as,to%20education%20on%20the%20continent. Accessed on 15th Oct, 2022.

Baklanov, A., Molina, L. T., & Gauss, M. (2016). Megacities, air quality and climate. *Atmospheric Environment, 126*, 235–249.

Bhattacharya, D., Carroll Steward, K., & Forbes, C. T. (2021). Empirical research on K-16 climate education: A systematic review of the literature. *Journal of Geoscience Education, 69*(3), 223–247.

Braun, V., & Clarke, V. (2006). Using thematic analysis in psychology. *Qualitative Research in Psychology, 3*(2), 77–101.

Bwalya, M. M., & Mulenga, B. P. (2018). Enhancing climate change awareness through information and communication technologies: A review of best practices in Africa. *Journal of Sustainable Development in Africa, 20*(6), 53–68.

Carney, J., Gillespie, T. W., & Rosomoff, R. (2014). Assessing forest change in a priority West African mangrove ecosystem: 1986–2010. *Geoforum, 53*, 126–135.

Chatiza, K. (2019). Cyclone Idai in Zimbabwe: An analysis of policy implications for post-disaster institutional development to strengthen disaster risk management.

Chikaire, J. U., Anaeto, F. C., Emerhirhi, E., & Orusha, J. O. (2017). Effects of use of information and communication technologies (ICTS) on farmers' agricultural practices and welfare in Orlu agricultural zone of Imo state Nigeria. *UDS International Journal of Development, 4*(1), 92–104.

Chimanga, K., & Kanja, K. (2020). The Role of ICTs in Climate Change Adaptation: A Case of Small Scale Farmers in Chinsali District. *Mathematics and Computer Science, 5*(6), 103.

Educational Testing Service. (2002). *Digital transformation: A framework for ICT literacy*. https://www.ets.org/Media/Research/pdf/ICTREPORT.pdf

Evans, O., & Mesagan, E. P. (2022). ICT-trade and pollution in Africa: Do governance and regulation matter? *Journal of Policy Modeling, 44*(3), 511–531.

Fettweis, G., & Zimmermann, E. (2008, September). ICT energy consumption-trends and challenges. In *Proceedings of the 11th International Symposium on Wireless Personal Multimedia Communications* (Vol. 2, No. 4, p. 6). Finland: Lapland.

Gacheno, D., & Amare, G. (2021). *Review of Impact of Climate Change on Ecosystem Services—A Review*.

Hulme, P. E. (2005). Adapting to climate change: Is there scope for ecological management in the face of a global threat? *Journal of Applied Ecology, 42*(5), 784–794.

Imam, N., Hossain, M. K., & Saha, T. R. (2017). Potentials and challenges of using ICT for climate change adaptation: a study of vulnerable community in riverine islands of bangladesh. In *Catalyzing Development through ICT Adoption* (pp. 89–110). Cham: Springer.

Intergovernmental Panel on Climate Change (IPCC). (2007). Climate change 2007: Synthesis report. In R. K. Pachauri & A. Reisinger (Eds.), *Contribution of working groups I, II and III to the fourth assessment report of the intergovernmental panel on climate change* (Core Writing Team). IPCC. https://www.ipcc.ch/report/ar4/syr/

Intergovernmental Panel on Climate Change (IPCC). (2013). Climate change 2013: The physical science basis. In T. F. Stocker, D. Qin, G. K. Plattner, M. Tignor, S. K. Allen, J. Boschung, A. Nauels, Y. Xia, V. Bex & P. M. Midgley (Eds.), *Contribution of working group I to the Fifth assessment report of the intergovernmental panel on climate change*. Cambridge University Press. https://www.ipcc.ch/report/ar5/wg1/

IPCC. (2007). Intergovernmental Panel on Climate Change (IPCC) report 2014. Available online at: https://www.ipcc.ch/report/ar5/syr. Accessed on 15th October 2022.

IPCC. (2014). Intergovernmental Panel on Climate Change (IPCC) report 2019. Available online at: https://www.ipcc.ch/2019/. Accessed on 15th October 2022.

Kates, R. W. (2000). Cautionary tales: Adaptation and the global poor. *Climatic Change, 45*(1), 5–17.

Kotir, J. H. (2011). Climate change and variability in Sub-Saharan Africa: A review of current and future trends and impacts on agriculture and food security. *Environment, Development and Sustainability, 13*(3), 587–605.

Kranz, J., Schwichow, M., Breitenmoser, P., & Niebert, K. (2022). The (Un) political perspective on climate change in education—A systematic review. *Sustainability, 14*(7), 4194.

Kremer, M., & Houngbo, G. F. (2021). Grow back better? Here's how digital agriculture could revolutionise rural communities affected by COVID-19. In *World Economic Forum*. https://www.weforum.org/agenda/2020/07/digital-agriculture-technology.

Lambin, E. F., Geist, H. J., & Lepers, E. (2003). Dynamics of land-use and land-cover change in tropical regions. *Annual Review of Environment and Resources, 28*(1), 205–241.

Leh, M., Bajwa, S., & Chaubey, I. (2013). Impact of land use change on erosion risk: An integrated remote sensing, geographic information system and modeling methodology. *Land Degradation & Development, 24*(5), 409–421.

Lybbert, T. J., & Sumner, D. A. (2012). Agricultural technologies for climate change in developing countries: Policy options for innovation and technology diffusion. *Food Policy, 37*(1), 114–123.

Mbilinyi, A., Saibul, G. O., & Kazi, V. (2013). Impact of climate change to small scale farmers: Voices of farmers in village communities in Tanzania. *Dar es Salaam: Economic and Social Research Foundation, 36*.

Min, S. K., Simonis, D., & Hense, A. (2007). Probabilistic climate change predictions applying Bayesian model averaging. *Philosophical Transactions of the Royal Society a: Mathematical, Physical and Engineering Sciences, 365*(1857), 2103–2116.

Morton, J. F. (2007). The impact of climate change on smallholder and subsistence agriculture. *Proceedings of the National Academy of Sciences, 104*(50), 19680–19685.

Nyong, A., Adesina, F., & Osman Elasha, B. (2007). The value of indigenous knowledge in climate change mitigation and adaptation strategies in the African Sahel. *Mitigation and Adaptation Strategies for Global Change, 12*(5), 787–797.

Nzonzo, D., & Mogambi, H. (2016). An analysis of communication and information communication technologies adoption in irrigated rice production in Kenya. *International Journal of Education and Research, 4*(12), 295–316.

Pant, L. P., & Heeks, R. (2012). ICT-enabled development of capacity for climate change adaptation. *ICTs Clim Change Dev Themes Strateg Actions Univ Manch UK Int Dev Res Cent IDRC Can*, 116–132.

Salampasis, M., & Theodoridis, A. (2013). Information and communication technology in agricultural development. *Procedia Technology, 8*, 1–3.

Simpson, N. P., Clarke, J., Orr, S. A., Cundill, G., Orlove, B., Fatorić, S., Sabour, S., Khalaf, N., Rockman, M., Pinho, P., Maharaj, S. S., Mascarenhas, P. V., Shepherd, N., Sithole, P. M., Ngaruiya, G. W., Roberts, D. C., & Trisos, C. H. (2022). Decolonizing climate change–heritage research. *Nature Climate Change, 12*(3), 210–213.

Statista. (2022). Africa: youths in post-secondary education 2010–2040. Available at: https://www.statista.com/statistics/1286049/number-of-youths-in-post-secondary-education-in-africa/. Accessed 16 October 2022.

Stott, P. A., Gillett, N. P., Hegerl, G. C., Karoly, D. J., Stone, D. A., Zhang, X., & Zwiers, F. (2010). Detection and attribution of climate change: A regional perspective. *Wiley Interdisciplinary Reviews: Climate Change, 1*(2), 192–211.

Subashini, K. P., & Fernando, S. (2017, September). Empowerment of farmers through ICT literacy. In *2017 National Information Technology Conference (NITC)* (pp. 119–124). IEEE.

Trisos, C. H., Adelekan, I. O., Totin, E., Ayanlade, A., Efitre, J., Gemeda, A., Kalaba, F. K., Lennard, C., Masao, C., Mgaya, Y. D., Ngaruiya, G., Olago, D., Simpson, N. P., & Zakieldeen, S. A. (2022). Africa. In H. O. Pörtner, D. C. Roberts, M. Tignor, E. S. Poloczanska, K. Mintenbeck, A. Alegría, M. Craig, S. Langsdorf, S. Löschke, V. Möller, A. Okem, & B. Rama (Eds.), *Climate change 2022: Impacts, adaptation, and vulnerability. Contribution of working group II to the sixth assessment report of the intergovernmental panel on climate change* (pp. 1285–1455). Cambridge University Press.

Tume, S. J. P., Kimengsi, J. N., & Fogwe, Z. N. (2019). Indigenous knowledge and farmer perceptions of climate and ecological changes in the bamenda highlands of Cameroon: Insights from the Bui Plateau. *Climate, 7*(12), 138.

Ukpabi, D. C., & Karjaluoto, H. (2017). Consumers' acceptance of information and communications technology in tourism: A review. *Telematics and Informatics, 34*(5), 618–644.

UNESCO. (2005). He convention on the protection and promotion of the diversity of cultural expressions. Available online at: https://en.unesco.org/creativity/convention. Accessed on 15th October 2022.

UNESCO. (2019). Information and communication technology (ICT) in education in sub-Saharan Africa: A comparative analysis of basic e-readiness. http://uis.unesco.org/sites/default/files/documents/information-and-communication-technology-ict-in-education-in-sub-saharan-africa-a-comparative-analysis-of-basic-e-readiness-2019-en.pdf

Wakili, A. A., Nasiru, L., & Mukhtar, M. I. (2020). ICT innovations for climate change communication and public awareness. *International Journal for Research in Applied Science and Engineering Technology, 8*(5), 1912–1916.

Weforum. (2022). Drought—food—starvation in Africa. Available online at: https://www.weforum.org/agenda/2022/07/africa-drought-food-starvation/. Accessed on 15th October 2022.

World Meteorological Organization (WMO). (2021). *State of the global climate 2021*. Library.wmo.int/records/item/56300-state-of-the-global-climate-2021

WMO. (2022). State of Climate in Africa highlights water stress and hazards. Available online at: https://public.wmo.int/en/media/press-release/state-of-climate-africa-highlights-water-stress-andhazards#:~:text=The%20State%20of%20the%20Climate%20in%20Africa%202021%20report%20has,managed%20water%20resources%20by%202030. Accessed on 15th October 2022.

Zaidan, E. (2017). Analysis of ICT usage patterns, benefits and barriers in tourism SMEs in the Middle Eastern countries: The case of Dubai in UAE. *Journal of Vacation Marketing, 23*(3), 248–263.

Zougmoré, R., Partey, S., Ouédraogo, M., Omitoyin, B., Thomas, T., Ayantunde, A., Ericksen, P., Said, M., & Jalloh, A. (2016). Toward climate-smart agriculture in West Africa: a review of climate change impacts, adaptation strategies and policy developments for the livestock, fishery and crop production sectors. *Agriculture & Food Security*, *5*(1), 1–16.

Chapter 6
Pedagogical Considerations for Climate Change Education in Africa

Marcellus Forh Mbah and Chidi Ezegwu

Abstract Africa is very vulnerable to the adverse effects of climate change despite contributing the least to greenhouse gas emissions. This vulnerability is exacerbated by its limited capacity to predict climate change, mitigate and adapt to its impacts. Education is important for capacity building and mobilising requisite actions against climate change. This paper draws on relevant literature to discuss applicable educational pedagogical perspectives that can foster effective climate change education (CCE) and help the citizenry develop sustainability mindsets in Africa. It highlights the need for decolonised and decentred climate change pedagogies that address existing inequalities and promote context-driven capacity development and indigeneity. It contends that effective pedagogy for climate change education that will support Africa's resilience to climate change should underscore the significance of Indigenous approaches in dealing with localised environmental issues.

Keywords Pedagogy · Africa · Climate change education · Indigeneity

Introduction

Climate change refers to a long-term alteration in the patterns of temperatures and weather resulting from natural causes or human activities (United Nations Framework Convention on Climate Change (UNFCC), 2011). The alteration has been described as a wicked problem because of its enduring complexities and capacity to multiply diverse human and general ecosystem health challenges. The associated complications threaten human and animal health, the survival of various species, natural assets, agriculture, water resources and coastal areas (Schneider & Lane, 2006; Anthony, 2012; Black & Butler, 2014; Incropera, 2016; Maron et al., 2016; Harrison, 2023; Kallbekken, 2023). Education is one of the major instruments for

M. F. Mbah (✉) · C. Ezegwu
School of Environment, Education and Development, Manchester Institute of Education, University of Manchester, Manchester, UK
e-mail: marcellus.mbah@manchester.ac.uk

M. F. Mbah et al. (eds.), *Practices, Perceptions and Prospects for Climate Change Education in Africa*, https://doi.org/10.1007/978-3-031-84081-4_6

raising awareness and preparing people for impactful climate change action. Unfortunately, information from the literature indicates that it has not been adequately tapped (UNFCCC, 2022a; UNESCO, 2020, 2021; Mbah & Ezegwu, 2024).

A UNESCO (2021) study that examined education sector plans and curriculum frameworks of 46 countries reveals that the depth of inclusion of climate change and environmental issues in education policies and curriculums was very low on average. It also notes that limited attention has been given to developing socio-emotional and action competencies relevant to environmental and climate actions. Hence, holistic pedagogy that engages students socially and emotionally, integrates Local and Indigenous Knowledge (LIK) through broad consultation with Indigenous knowledge holders and inspires learners to action is required (UNESCO, 2021). These findings further draw attention to existing scholarly observations on the need to promote meaningful environmental and climate change education that promotes learning about (a) the environment through the people, (b) learning about people in and through their environment—for the health of the environment and (c) appropriate and context-relevant pedagogy of climate and environment (Kimaryo, 2011; Ajaps & Forh Mbah, 2022; Mbah & Ezegwu, 2024). These raise questions about how schools have planned and delivered climate education.

This chapter discusses pedagogical approaches for climate change education (CCE) in Africa. Africa needs functionally effective climate CCE pedagogies to build local capacities and help its citizens develop sustainability and adaptation mindsets. Mindsets represent an individual or group's internal lens from which they see, relate to and navigate the vicissitude of life (Wamsler et al., 2020). In this paper, we use climate change sustainability and adaptability mindsets to represent positive dispositions, attitudes, perspectives, philosophies, beliefs, frames of mind, values and outlooks toward climate sustainability and adaptation. Falkner and Sheard (2019, p. 446) define pedagogy as "the science of how we promote learning, and it consists of the learning activities, strategies, and techniques that provide the environment where learning may take place". Siraj-Blatchford et al. (2002) add that pedagogy encapsulates instructional techniques and strategies required for learning to happen. Others note that appropriate pedagogies (including the suitable use of instructive strategies and interactive processes between teachers and learners) is a very critical step in helping learners learn and develop relevant competencies needed to mitigate and adapt to climate change (Davidson & Lyth, 2012; Falkner & Sheard, 2019; Moyles et al., 2002; Sainz & Khoo, 2020). In the following section, we summarise Africa's climate conundrums that necessitate a rethink on the role of education.

Climate Change and Its Impact on Africa

Climate change poses a critical challenge to the achievement of the Sustainable Development Goals (SDGs) in Africa due to its huge impacts on essential socioeconomic structures and people's well-being (United Nations Environment Programme, 2021; African Development Bank (AfDB), 2022a, 2022b). The continent is believed to be warming quicker than the global land and ocean average. The World Meteorological Organization's (WMO) (2023) State of the Climate Africa 2022 report shows that the mean near-surface air temperature anomaly in Africa in 2022 was 0.16 °C higher than the average long-term of 1991–2020 and was also 0.88 °C higher than the 1961–1990 average. As summarised in Table 6.1, between 1991 and 2022, the continent's average rate of warming stood at + 0.3 °C/decade. This exceeded the reported +0.2 °C/decade for 1961–1990 (World Meteorological Organizations (WMO), 2023). Among the 16 countries that are most affected by different elements of the climate crisis are 12 African countries: Somalia, Burkina Faso, DR Congo, Ethiopia, Mali, Mozambique, Cameroon, Nigeria, Central African Republic, Niger, South Sudan and Sudan (International Organisation for Migration (IOM), 2022; International Rescue Committee, 2023). The observed warming has been intensifying rapidly in the northern region of Africa, which experienced the worst drought in 40 years (as observed in Ethiopia, Kenya and Somalia), as well as extreme heat and wildfires, particularly noted in Algeria and Tunisia (African Development Bank (AfDB), 2022a, 2022b; World Meteorological Organization, 2023). The average rainfall was below normal in many locations, with East Africa experiencing five successive failures of rainfall season. In contrast, countries like Chad, Niger and Nigeria experienced extreme rainfall with consequential disastrous flooding (World Meteorological Organization (WMO), 2023).

Existing information shows that Africa accounts for only between 2 and 3% of the world's carbon dioxide emissions. However, it suffers disproportionately from global climate change impact and lacks the requisite capacity to effectively adapt to

Table 6.1 Near-surface air temperature anomalies in °C for 2022 relative to the 1991–2020 and 1961–1990 reference periods

	1991–2020	1961–1990
North Africa	0.50 °C [0.41–0.65 °C]	1.40 °C [1.24–1.64 °C]
West Africa	0.03 °C [–0.18–0.14 °C]	0.71 °C [0.39–0.87 °C]
Central Africa	0.13 °C [–0.04–0.37 °C]	0.80 °C [0.60–1.11 °C]
East Africa	0.14 °C [–0.02–0.28 °C]	0.90 °C [0.69–1.12 °C]
Southern Africa	0.01 °C [–0.17–0.12 °C]	0.61 °C [0.42–0.75 °C]
Indian Ocean Island Countries	0.03 °C [–0.04–0.10 °C]	0.60 °C [0.49–0.70 °C]
Africa	0.16 °C [0.06–0.28 °C]	0.88 °C [0.74–1.07 °C]

Source World Meteorological Organization (WMO) (2023, p. 9). The WMO calculation (reproduced in the table) is based on six different data sets, including observational data sets such as HadCRUT5, NOAAGlobalTemp, GISTEMP, and Berkeley Earth and mitigate the impacts (World Meteorological Organization, 2023; African Development Bank, 2022a, b). Information in academic and grey literature indicates that the continent's limited weather and climate observation capability contributes to complicating its vulnerability: persistent limitations of its internally sourced knowledge tend to weaken African countries' ability to predict and adapt to impending weather changes (Mbah et al., 2021; UNDP, 2023; World Meteorological Organization, 2023). The UNFCCC (2022b) identified climate education as one of the key strategies for strengthening capacity and promoting sustainable climate actions. The opportunities education offers are briefly discussed below.

The Existing and Potential Roles of Climate Change Education in Africa

The important roles of education in climate change mitigation and adaptation are increasingly recognised, which has led to the promotion of climate change education (CCE) by various international development agencies to improve the understanding of it and actions needed to be taken by diverse stakeholders (UNESCO, 2018, 2020, 2021; UNFCC, 2022a, b). Climate change education is expected to inform learners about climate-induced risks, their impact on the human environment, socioeconomic activities and well-being, and the diverse kinds of adaptation and mitigation strategies required (Greer & Glackin, 2021; Thew et al., 2021; Molthan-Hill et al., 2022; Mbah & Ezegwu, 2024). It is useful for the promotion of behaviour change, raising citizens that will hold their governments accountable by demanding proactive and responsive policies, opening opportunities for increased research collaboration and building a workforce with green skills (Thew et al., 2021; Molthan-Hill et al., 2022).

The prevailing CCE is dominated by Western approaches with very limited incorporation of Indigenous knowledge systems (Tanyanyiwa, 2019). However, the potential benefits of decolonising CCE and adopting more relatable and effective approaches in non-Western contexts are documented (see Mbah et al., 2021). Within Africa, different Indigenous groups have maintained rich environmental practices and stocks of knowledge that can be adapted and integrated into the mainstream CCE. Tanyanyiwa (2019) and Mbah et al. (2021) note the uniqueness of Indigenous and traditional knowledge and its connection to places upon which climate change impacts. They describe how such impacts inspire natural responses towards mitigation or adaptation. Indigenous people continuously evolve and improve their interaction with their environment through farming activities, soil improvement and water conservation (Mbah et al., 2021). For example, Chianese (2016) documents how Gamo people in Ethiopia have continued to use Indigenous knowledge to assess the impacts of environmental changes on their livelihoods, water sources, soil fertility and crop yields. Similarly, various groups, such as Bahima in Uganda and Samburu in East Africa, use local knowledge for seasonal weather prediction. Such categories of knowledge are continuously refined through years of practical experience and learning (Chianese, 2016; Mbah et al., 2021).

Some recent interdisciplinary initiatives have sought to blend Local and Indigenous Knowledge (LIK) with the Western education framework for teaching and learning climate change issues in many African countries. However, many known initiatives are limited in scope (see Zimu-Biyela, 2019; Magagula, 2020; Singh-Pillay, 2020; Mbah & Ezegwu, 2024). For example, de Sousa et al. (2021) reported a climate change project in South Africa that brought community members and Grade 7 teachers together to learn about a local community-based climate change problem. The participatory research project empowered learners and community members to contribute to finding solutions to relevant local environmental issues. A similar initiative was a Farming for Change project, which brought together academics, farmers and community development experts to work with farming households in Malawi and Tanzania. Farming for Change was designed to create an innovative curriculum that synthesises the knowledge of smallholder farmers and crystallises existing nutrition, agroecology, social equity and climate change best practices in ways that relay their interconnectedness (Kerr et al., 2022). Duggan et al. (2020) also shared experiences of middle school modules in South Africa that employed integrated curriculum design with social and situated learning framework, which drew practical exercises from local and community environments while incorporating data from their regional marine science study. A major limitation of these projects is that they have not been integrated into mainstream curricula or broadly adopted in schools. Hence, their enduring efficacies are yet to be tested. In the following section, we discuss selected philosophical and pedagogical approaches and highlight how they provide or fail to provide opportunities to embed and adapt climate change and environmental knowledge to the peculiarities and realities of different African societies.

Mainstream Pedagogical Approaches in Climate Change Education

Pedagogical approaches play important roles in learning and knowledge acquisition and influence learners' understanding of subjects. A crucial element of pedagogy is centred on how teaching is organised and delivered rather than the actual content, and different perspectives have influenced this (Wall et al., 2015; Falkner & Sheard, 2019). We cautiously summarise some of these perspectives and associated pedagogical approaches for brevity. Multiple perspectives have also influenced some pedagogical approaches.

Behavioural Perspective

The behavioural perspective gives attention to observable and measurable behaviours. Pedagogical approaches associated with behaviourism tend towards objectivism, which believes in the existence of a reality with a certain structure that needs to be assimilated by learners. From a behavioural perspective, the role of educators is to present, explain and interpret what is seen as the reality to learners (Zoellner, 1969; Diekelmann, 1993; Child Australia, 2017). An instructivist approach is linked to the behavioural perspective and assumes that sanctioned information should be effectively transmitted to learners who passively receive and assimilate them through memorisation and recall (Porcaro, 2011). From this perspective, educators tend to reinforce what is considered appropriate behaviour through various forms of rewards and affirmative actions and discourage deviant behaviours by drawing learners' attention to the potential implications of non-conforming behaviours (Child Australia, 2017; Falkner & Sheard, 2019).

Behavioural-instructivist approaches pervade many African countries' education systems, influencing how climate change-related topics are taught and learned. Various scholars document how colonial forces sought to erase other forms of knowledge, displaced Indigenous knowledge systems and replaced them with hegemonic forms of knowledge (Kwauk, 2020; Mather, 2017). According to Kwauk (2020, p. 12), the traditional Western education approach that dominates the African education landscape "tends to be individualistic, one-directional, and transmissive rather than collective, interactive, and transformative. It tends to view children as receivers rather than co-creators of knowledge", This approach has contributed to weakening the transformative capacity of education and activation of learners' agency, voice, and creativity through education. Thus, the traditional Western education approach does not adequately recognise the important roles learners could play in building climate resilience in their communities (Kwauk, 2020). Consequently, it tends to promote epistemic exclusion[1] of learners' Indigenous knowledge in CCE. This makes it inadequate for inclusive teaching and learning of climate change issues in the African context.

Cognitive Perspective

The cognitive perspective gives attention to the learners' cognitive activity, processes, and structure of knowledge. Constructivist pedagogy is linked to the cognitive perspective and believes that learners are not merely receiving and storing transmitted knowledge but are constructing knowledge (Piaget & Cook, 1952; Ben-Ari,

[1] Epistemic exclusion refers to the delegitimisation policies and practices that are rooted in disciplinary, racial and sociocultural biases about what is accepted and valued as valid knowledge. It constitutes an infringement on the epistemic agency of knowers in ways that reduce their ability to contribute and participate in their epistemic communities (Dotson, 2012, 2014).

1998; Falkner & Sheard, 2019). Duffy and Cunningham (1996) contend that there is an active construction process during learning that is supported by classroom instructions. Constructivist pedagogy, therefore, promotes critical thinking and ownership of the learning processes. There are two schools of constructivism: cognitive constructivism and social constructivism. The first considers individuals' construction of knowledge through an interplay of their new experiences and conventional ideas, while the second recognises individuals' construction through teachers' and learners' interactions (see Piaget & Cook, 1952; Kalina & Powell, 2009; Vygotsky, 2012; Falkner & Sheard2019). Constructive pedagogy enables learners to reflect and develop their knowledge by articulating their prior knowledge and giving attention to local values (Hughes, 2011).

Some studies in Africa suggest that, when pertinently arranged and adapted, the constructivist approach could provide useful procedures for knowledge co-creation because of its potential to help learners become self-motivated and responsible for their development and learning (Booyse & Chetty, 2016). Some African game and project-based learning initiatives have drawn from this approach to enhance climate and environmental education. Project-based learning, for example, is a learner-centred constructivist education approach that engages learners through carefully designed projects to improve their knowledge (Singh-Pillay, 2020). The game-based learning approach has been used to teach various climate change issues, including food security, analysis and utilisation of climate information, disaster preparedness and response, and urban waste management (Suarez & Bachofen, 2013).

Sociocultural Perspective

Closely related to the constructive perspective is the sociocultural perspective, which promotes relationships and participation as important frames for learning. It recognises learners' differences and multiple ways of learning. Socio-culturalists believe that learners acquire information on how to become interdependent through their participation in daily activities as members of the community and different social groups (Murphy & Ivinson, 2003; Gee, 2008; van Compernolle & Williams, 2013; Child Australia, 2017). They also believe that development results from learning, and children have agency, participating and contributing actively to their social groups (Child Australia, 2017). Pedagogical approaches associated with it encourage policymakers and educators to take into consideration the context of children's lives and lived experiences in curriculum development and delivery, promoting students' independent understandings and respecting their own cognitive experiences (Child Australia, 2017).

Place-based education (also called a pedagogy of place) is an example. It is centred on the conceptualisation of abstract theories using local and environmental issues, outdoor learning and interactions with natural and human communities where learners live, and using community values and local resources in the learning processes in ways that promote partnerships between schools and local communities (Velempini et al., 2018; Ferreira, 2020; Ajaps & Forh Mbah, 2022). Ajaps and Forh Mbah (2022, p. 512) explain place-based pedagogy as "education that is based on the people and environment engaged with it. Place, land, and environment are synonymous for many Indigenous peoples". It is a pedagogy that influences social and ecological well-being and, in particular, provides an opportunity for learners to bond with the natural world. It originates from a particular place's characteristics and tends to be inherently multidisciplinary and experiential. Philosophically, it is linked to inclusive lifelong learning that connects with people's self and community and looks beyond learning for earning (Ontong & Grange, 2015). A place-based education experiment in Okavango Delta, Botswana, got teachers to integrate environmental education, local environmental knowledge, and traditional ways of knowing into the delivery of the school curriculum. This was implemented in consultation with educational authorities, teachers, students and community members (Velempini et al., 2018).

From its definition, characterisation and experimentation (see Semken et al., 2017; Ajaps & Forh Mbah, 2022; Yemini et al., 2023; Mbah & Ezegwu, 2024), place-based education looks like a promising pedagogy for climate change education in Africa. As a pedagogical approach, its emphasis is on bringing together the learning process and the geophysical place where teachers and learners are located. It provides an opportunity to incorporate local concepts, meanings and lived experiences of people in a place into learning processes, which makes it appealing for the promotion of Africa-centred climate change education. However, caution needs to be exercised to ensure learning is topically relevant and focused because there is a possibility that educators might hide behind place-based education to introduce diverse kinds of irrelevant topics and issues in the learning process. We also note that Africa is very diverse—Nigeria alone has over 500 Indigenous languages (see Ezegwu, 2020). Hence, there is a need to give attention to respective Indigenous contexts and avoid the hegemonic imposition and influence of one culture over another.

Critical Perspective

The Critical perspective holds that learners have natural rights as well as agency that bequeath them with some capacity to make choices and decisions about their learning. Pedagogical approaches associated with it, therefore, see learners as co-creators of knowledge and co-participants with educators in charting the course of learning (Blake & Masschelein, 2003; Child Australia, 2017). From this perspective, learning is perceived as a partnership that necessitates learners' liberation, putting them at the centre of learning, enhancing cognitive acts and eliminating transferals of information methods (Freire, 1994, 2000; Peters & Mathias, 2018). It also helps to abolish the "teachers-of-the-students and the students-of-the-teacher" standpoint learning method and enrich "teacher-student with students-teachers" standpoints (Freire, 2000, pp. 79–80). This creates a milieu in which the traditional teachers and students learn in classroom dialogue. Thus, they become jointly responsible for the learning process and their own development (Freire, 2000). This is particularly relevant in Africa because it can potentially create space for communities and learners to bring their local experiences and knowledge to school.

It is important to note that critical pedagogy is broad (and somewhat integrates elements of sociocultural, constructive and post-structural perspectives) and encompasses various methods that embrace criticality and interrogate the status quo (see Coemans & Hannes, 2022; Malka & Lotan, 2023). Hughes (2011) explains that an important part of a critical perspective view on climate change relates to the ability to assimilate, summarise and evaluate enormously different categories of information from different sources. Photovoice is one of the critical pedagogical tools used to study climate change and environmental issues in Africa. It is a visual, dialogical and participatory approach to learning and knowledge co-creation that allows participants to use cameras to document, communicate and reflect on a particular issue of concern (Budig et al., 2018; Haffejee, 2021). The approach was used in South Africa to engage groups of undergraduate students who took photographs of various environmental factors that cause diseases and engaged in a group dialogue, reflecting and discussing how visible environmental factors affect people's health (Haffejee, 2021). It promoted a higher learning order that includes critical thinking and problem-solving skills that help to transform students from rote learners to critical thinkers. It empowered students to reflect on what they learned and how it related to their experiences and community realities (Haffejee, 2021).

Post-Structuralist Perspective

The post-structuralist perspective conveys that learners have complex and shifting identities, and there are both multiple and contested approaches to knowledge acquisition (Clemitshaw, 2013; Martusewicz, 2001; Whitson, 1995). Its linked pedagogical approaches focus on power relations and issues around equality, such as how power plays out between learners and educators and how to promote democratic participation and inclusion in the learning processes, including the accommodation of different ways of knowing and Indigenous knowledge systems (Child Australia, 2017). Dewey (1938) highlights the importance of people's experience in learning, which, according to Mbah and Che (2019), can be activated in the process of an individual's reflective and critical thinking to generate ideas, new meanings and insights. Participation and collaboration tend to provide opportunities for people to draw on their Indigenous and local knowledge and experiences to advance their learning and development.

A review of the literature revealed several collaborative and partnership CCE projects in various African countries, which may have purposefully or unintentionally drawn on critical and post-structural theories to promote inclusion and partnership. These include CCE investigations that recognised the complex and shifting identities and contested spheres of knowledge production, exchange and transfer and sought to ensure integration of local participation and perspectives (e.g., Govender, 2019; Kerr et al., 2022; Zimu-Biyela, 2019; Magagula, 2020; Singh-Pillay, 2020; Mbah et al, 2021; de Sousa et al., 2021; Opoku & James, 2021; Wilson et al., 2021). An example is a project in South Africa that sought to integrate Zulu Indigenous and local knowledge with existing scientific knowledge in the teaching of environment (Opoku and James, 2021). Also, de Sousa et al. (2021) report a participatory climate change research project that brought community members and teachers together to learn about local community-based climate change problems and engaged teachers to mobilise students as citizen scientists.

Connectivism

Connectivism is a learning perspective that believes students learn best as a clustered community of people with similar areas of interest, which fosters interaction, dialogue, thinking together and knowledge exchange. It promotes networking and information exchange via digital platforms (Siemens, 2004, 2008; Downes, 2008; Kop & Hill, 2008). The philosophy of distributive knowledge underpins connectivism, and thus, it perceives learning as a network phenomenon that inclines towards technology and socialisation (Goldie, 2016). Connectivism views traditional learning frameworks as limited due to their "failure to address the learning that is located within technology and organisations and their lack of contribution to the value judgments that need to be made in knowledge-rich environments" (Goldie, 2016, p. 1064). Collaborative pedagogy may be associated with connectivism. From this perspective, the individualist learning approach is considered ineffective for the new generation of non-traditional learners, giving rise to collaborative pedagogy that encompasses a wide range of practices that decentre the classroom and promote collaboration in the forms of peer tutoring and group discussion (Howard, 2000). Collaborative pedagogy can take a blended form (such as physical and technological) or non-blended learning approaches. Solís et al. (2019) shared a report on the Global Connections and Exchange Youth TechCamps initiative that centred on students and peer-engaged project-based learning exchange, which involved selected high school students in Bolivia, Panama, South Africa and the USA. The participants collaborated online and in person during the training events in the participating countries that focused on various themes around GeoTechnologies for Climate Change and Environment and how they could adapt and apply them in their communities.

These varying perspectives and associated pedagogies may be useful in different contexts and times, depending on the topic, participants and available resources. In the following section, we discuss further considerations for Africa-centred pedagogies.

Towards African-Centred Pedagogies for Climate Change Education

Given the considerations above, a key question will be: What should a socially just and context-relevant climate change education in Africa look like? While it has been noted that no country's education sector meets up with climate action responsibility, current approaches to education for sustainable development (ESD) remain at odds with transformational learning (Kwauk, 2020). The transformational learning perspective encourages holistic and whole-person learning, communities of practice, critical dialogue, autonomous thinking, inclusive outlook and increased collaboration (Winchester-Seeto et al., 2017). African countries specifically need radical adjustment of their CCE pedagogies to meet the continent's transformational ESD needs. Wall et al. (2015) explain that mainstream pedagogy in each country is influenced by various factors, including national development objectives and existing knowledge. Information in the literature (see Mbah & Ezegwu, 2024) points to four contextualised issues that Africa-centred CCE needs to give attention to. These are inequality, capacity needs, youthful population and Indigeneity.

First, there exists some form of global inequality and disproportional experiences of climate change in terms of its causes and effects that have serious implications for Africa. Africa contributes less to global warming but is most impacted by it, as noted in the introductory section of this chapter. Second, there is low adaptive capacity in Africa, which differs across countries on the continent. African countries have differential economic, technological and manpower vis-a-vis countries in other global regions. Hence, there is a need to build local capacity in relation to local needs to maximise local resources in respective African countries. Third, African capacity development strategies also need to recognise that the continent has the youngest population globally (Favier et al., 2021; Makina & Mudungwe, 2023), which will suffer the intensified and devastating effects of climate change more than the older generations. Finally, the complexities of climate change mitigation and adaptation, Africa's dependence on climate-sensitive natural resources and limited hi-tech capacity make it necessary for Africa to evolve and improve its Indigenous knowledge and initiatives for dealing with climate change and its localised challenges.

The Centrality of Indigeneity

Considering that local initiatives are very effective in addressing localised problems (Ezegwu, 2015, 2020), there is a need to document Indigenous environmental and climate knowledge to support pedagogical adjustments. Sustainable development literature emphasises how indigenous people are best placed to fight local problems and how local interventions are most suitable for combating local challenges, but such knowledge and information on such initiatives need to be available to educators (Ménard, 2013; Whitham, 2012; Ezegwu, 2015, 2020). For example, UNESCO (2018) documents how many African pastoralists who depend on natural resources have become particularly sensitive to weather and climate variations and have evolved ways to monitor and predict weather and climate changes to sustain their lifestyles. For generations, they have interacted with their environment and accumulated context-specific and bespoke knowledge of their climate and its variations, which they have used to decide and facilitate local adaptations.

Based on a comparative study in Ghana, Nigeria, South Africa and Kenya that analysed Indigenous and homegrown development strategies and programmes, Okereke and Agupusi (2015) contend that externally driven development programmes have partial success in the promotion of Africa's development. Thus, to create an enabling environment for the unravelling of appropriate pedagogical approaches for Africa, Indigenous education researchers need to document the existing Indigenous and local knowledge and solutions, making them available to education policymakers, curriculum developers and teacher educators to enhance their adoption in the education system. Most of the existing knowledge about the local environment resides with local and indigenous communities. Therefore, the task before African researchers is to "understand how the best available knowledge about the climate and environment may be mobilised and communicated in relation to the capacity of meteorological sciences to meet developmental challenges" (UNESCO, 2018, p. 11). This will contribute to enhancing understanding of the evolving climate change context and its impacts, as well as how to support and manage the existing resources to build the needed resilience and lower African societies' vulnerabilities.

Similar calls have been made for African scholars and higher educational institutions to lead the process of synchronising local knowledge systems with the existing formal education system. For example, ecologically driven education institutions have been recommended to enhance a shift from the present dominant education model and promote a context-specific, community-engaged and decolonised approach that recognises the history, culture and local environment of the learner, including their past, present and future identity as equally and critically important to their development (see Agozino & Anyanike, 2007; Barnett, 2011; Dei, 2014; Emeagwali & Dei, 2014; Rufai et al., 2019; Johnson & Mbah, 2021). This implies adopting a pedagogy and apprenticeship system that recognises, includes and accepts the local Indigenous communities (it seeks to serve) and their knowledge system.

Leveraging on Africa's Indigenous Knowledges for Relevant Apprenticeship or Pedagogy

Further to the documentation of African knowledge and initiatives, it is also critical to consider the Indigenous African approach to teaching and learning. Many traditional African systems of education were largely diffused and less formalised (with some aspects of the training focusing on spiritual, economic and arts education), taking the forms of apprenticeship and learning through role plays, oral literature, instruction, observation and direct participation, with an emphasis on practical knowledge (Fafunwa, 1974; Ohadike, 1994; Amaele, 2006; Ezegwu, 2012, 2020). Apprenticeship is one of the more formalised approaches and involves learners working alongside their teachers and elders (Ohadike, 1994; Ezegwu, 2012). Walther (2008) shares examples of some restructuring of the apprenticeship system in post-primary schools in Benin, Mali and Togo. These are based on a dual apprenticeship model, which necessitates some form of partnership and division of labour between colleges that provide a theoretical dimension of the training and craftsman's workshops that offer opportunities for hands-on skills. While these may be considered limited in terms of their separation of theory and practice (rather than blending), they highlight some potential that may be adapted for climate change education. Our literature search suggests a critical gap in research and adoption of apprenticeships in African countries' climate change education.

Additionally, effective climate change education pedagogy that will work for Africa may need to draw on a combination of relevant external and Indigenous knowledge and initiatives that reflect Africa's history, local realities, and development needs. Both Indigenous and relevant external philosophy and pedagogical approaches may be blended as long as they are adapted to mirror the peculiarity of the African environment and experiences. Mbah (2014, 2016) shares examples of how African education systems can midwife an Africa-centred climate education. Using universities as examples, he argues that African education institutions need to shift from dumping externally constructed learning agendas on local communities. Instead, they need to work with respective communities to articulate learning agendas (and, by extension, learning pedagogies) by listening to community voices and co-creating knowledge and development choices with them. This does not construe an anti-Western system. It rather demands collaborations to develop and adapt what is relevant to each society into its education system. This may somewhat tilt towards some of the pedagogical approaches discussed above that promote learners' and communities' participation, which, according to Mbah (2014), underlie Africa's Ubuntu spirit, which values caring, sharing, respect and solidarity among community members, and advances a framework of interconnections between education institutions and the communities.

Pointer to the Way Forward

It is noteworthy that there may not be a single answer to the question of what socially just and context-relevant climate change education would look like in Africa because Africa is not a single country, and a unified solution may lead to the displacement of a far more multi-layered reality that is ethnoculturally dissimilar, linguistically diverse and ecologically dispersed African geography—which can lead to costly mistakes and missed opportunities (Chakravorti, 2015). According to UNESCO (2018), the knowledges of various Indigenous groups in Africa are highly diverse and diverge across communities and Indigenous institutional systems. This necessitates a shift from one-size-fits-all strategies to accommodate varying approaches that work for people in different locations. As Moyles et al. (2002, p.3) explain, "Effective pedagogy not only produces outcome results in relation to input but also represents a common core of values and objectives to which all those involved can subscribe". An effective pedagogical approach to climate change education in Africa should not be externally recommended or influenced. Instead, it should be internally evolved in respective sociocultural contexts and directed towards helping African learners understand climate change issues, including present inequalities, capacity gaps and needs, and the pre-existing Indigenous approaches for dealing with environmental challenges vis-à-vis the globally acclaimed frameworks. Its targeted results will include helping learners understand both localised and globalised challenges, values, and objectives and inspiring and empowering them to take meaningful actions towards their immediate environmental sustainability.

Notwithstanding, it appears that pedagogical approaches that are linked to constructivists, as well as critical and post-structural perspectives, provide a comparable adaptable CCE pedagogical framework for Africa, as could be observed above. Besides, historical and cultural studies suggest that many groups in Africa were originally democratic, leading cooperative and communal lifestyles (Adeyemi & Adeyinka, 2002; Ominde et al., 2020; Obodoegbulam et al., 2021). A more egalitarian pedagogical approach is required to recognise, integrate, co-create and share climate change-relevant information in the classroom. Constructivists, as well as critical and post-structural-linked pedagogies, will create space to reflect the communal education system that historically existed in many parts of Africa. Communalism is at the centre of the African Indigenous education system, and integrating communalist principles in the school curriculum will contribute to advancing African ethics of responsibility, fairness, citizenship and participation (Adeyemi & Adeyinka, 2002; Ominde et al., 2020).

Conclusion

There have been ongoing discussions (e.g., Mbah et al., 2021; Ajaps & Forh Mbah, 2022; Mbah & Ezegwu, 2024) on the need to promote decolonised and Afrocentric education. Such discussions include demands for the adoption of Indigenous and local knowledge systems, which without being anti-Western, are grounded in Indigenous African culture, history, experience and approaches and also have the potential to address some observed decades of marginalisation and disenfranchisement of African ways of knowing and knowledge transfer. In this paper, we contend that climate change education that will work for Africa needs to go beyond the integration of African Indigenous knowledge and history. It needs to draw on relevant philosophy and pedagogical approaches that create, embed and adapt climate change and environmental knowledge to the peculiarity of the African environment and realities. This demands an unbiased and non-discriminatory engagement and collaboration with respective communities and their knowledge holders to create space for their participation, contribution and partnership in the creation of a climate change education framework that works for the people. Such collaboration will also provide space for Africans to draw on their experiences to co-create climate change knowledge based on their historical and social-cultural realities. Another benefit of such collaboration is that while it works towards effective climate change education that promotes local climate actions, it will also help preserve the positive local and Indigenous climate change and traditional environmental knowledge systems and transfer them to the coming generation.

References

Adeyemi, M. B., & Adeyinka, A. A. (2002). Some key issues in African traditional education. *McGill Journal of Education/Revue des sciences de l'éducation de McGill, 37*(002).

African Development Bank. (2022a). *Climate change and green growth department–2021 annual report: Improving access to financing for green growth.* African Development Bank. https://www.afdb.org/en/documents/climate-change-and-green-growth-2021-annual-report

African Development Bank. (2022b). North Africa economic outlook 2022: Supporting climate resilience and a just energy transition. African Development Bank.

Agozino, B., & Anyanike, I. (2007). IMU AHIA: Traditional Igbo business school and global commerce culture. *Dialectical Anthropology, 31*, 233–252.

Ajaps, S., & Forh Mbah, M. (2022). Towards a critical pedagogy of place for environmental conservation. *Environmental Education Research, 28*(4), 508–523.

Amaele, S. (2006). Nigerian Traditional Education. In G. Akanbi (ed.), *History of education in Nigeria.* National Open University of Nigeria.

Anthony, R. (2012). Taming the unruly side of ethics: Overcoming challenges of a bottom-up approach to ethics in the areas of food policy and climate change. *Journal of Agricultural and Environmental Ethics, 25*, 813–841.

Barnett, R. (2011). The coming of the ecological university. *Oxford Review of Education, 37*(4), 439–455.

Ben-Ari, M. (1998). Constructivism in computer science education. *ACM SIGCSE Bulletin, 30*(1), 257–261.

Black, P. F., & Butler, C. D. (2014). One health in a world with climate change. *Rev Sci Tech, 33*(2), 465–473.

Blake, N., & Masschelein, J. (2003). Critical theory and critical pedagogy. *The Blackwell guide to the philosophy of education*, 38–56.

Booyse, C., & Chetty, R. (2016). The significance of constructivist classroom practice in national curricular design. *Africa Education Review, 13*(1), 135–149.

Budig, K., Diez, J., Conde, P., Sastre, M., Hernán, M., & Franco, M. (2018). Photovoice and empowerment: Evaluating the transformative potential of a participatory action research project. *BMC Public Health, 18*(1), 1–9.

Chakravorti, B. (2015). It is time to get past the "single story" about Africa. Indian and Africa Forging a Strategic Partnership. Brookings Institute. https://www.brookings.edu/wp-content/uploads/2015/10/indiaafrica_bhaskar.pdf

Chianese, F. (2016). *The traditional knowledge advantage: Indigenous peoples' knowledge in climate change adaptation and mitigation strategies. IFAD Advantage Series.* IFAD.

Child Australia. (2017). What is pedagogy? How does it influence our practice? https://www.childaustralia.org.au/wp-content/uploads/2017/02/CA-Statement-Pedagogy.pdf

Clemitshaw, G. (2013). Critical pedagogy as educational resistance: A post-structuralist reflection. *Power and Education, 5*(3), 268–279.

Coemans, S., & Hannes, K. (2022). Photovoice as critical pedagogy. In: *30 years of photovoice: past, present, and future, Date: 2022/10/20–2022/05/22*. Photovoiceworldwide.

Davidson, J., & Lyth, A. (2012). Education for climate change adaptation—enhancing the contemporary relevance of planning education for a range of wicked problems. *Journal for Education in the Built Environment, 7*(2), 63–83.

de Sousa, L. O., Hay, E. A., Raath, S. P., Fransman, A. A., & Richter, B. W. (2021). Shifting gears: Lessons learnt from critical, collaborative, self-reflection on community-based research. *Educational Research for Social Change, 10*(1), 70–82.

Dei, G. J. S. (2014). Indigenising the school curriculum. In G. Emeagwali & G. J. S. Dei (Eds.), *African Indigenous knowledge and the disciplines* (pp. 165–180). Sense Publishers.

Dewey, J. (1938). *Experience and Education.* In Later Works, 1935–1953, (vol. 13, pp. 1–63). Southern Illinois University Press.

Diekelmann, N. L. (1993). Behavioral pedagogy: A Heideggerian hermeneutical analysis of the lived experiences of students and teachers in baccalaureate nursing education. *Journal ofNursing Education, 32*(6), 245–250.

Dotson, K. (2014). Conceptualising epistemic oppression. *Social Epistemology, 28*(2), 115–138.

Dotson, K. (2012). A cautionary tale: On limiting epistemic oppression. *Frontiers: A Journal of Women Studies, 33*(1), 24–47.

Downes, S. (2008). An introduction to connective knowledge. In T. Hug (Ed.), *Media, knowledge & education: Exploring new spaces, relations and dynamics in digital media ecologies.* Innsbruck University Press.

Duffy, T. M., & Cunningham, D. J. (1996). Constructivism: Implications for the design and delivery of instruction. In D. H. Jonassen (Ed.), *Handbook of research for educational communications and technology.* Macmillan Library Reference USA.

Duggan, G. L., Jarre, A., & Murray, G. (2020). Learning for change: Integrated teaching modules and situated learning for marine social-ecological systems change. *The Journal of Environmental Education, 52*(2), 118–132.

Emeagwali, G., & Dei, G. J. S. (2014). *African indigenous knowledge and the disciplines: Anti-colonial educational perspectives for transformative change.* Sense Publishing.

Ezegwu, C. (2012). *Masculinity and low male secondary school enrolment in Anambra State, Nigeria* [MA Dissertation]. The University of London, Institute of Education.

Ezegwu, C. (2015). *Home-grown initiatives for local challenges on gender inequality in basic education in Nigeria.* http://www.edorennigeria.wordpress.com/2015/03/04/home-grown-initiatives-for-local-challenges-on-gender-inequality-in-basic-education-nigeria

Ezegwu, C. (2020). *Masculinity and access to basic education in Nigeria.* Lancaster University. https://books.google.com/books/about/Masculinity_and_Access_to_Basic_Educatio.html?id=y5QTzgEACAAJ

Fafunwa, A. (1974). *History of education in Nigeria.* George Allen and Unwin.

Falkner, K., & Sheard, J. (2019). Pedagogic approaches. In S. Fincher, & A. Robins (Eds.), The *Cambridge handbook of computing education research* (Cambridge Handbooks in Psychology, pp. 445–480). Cambridge University Press. https://doi.org/10.1017/9781108654555.016

Favier, T., Van Gorp, B., Cyvin, J. B., & Cyvin, J. (2021). Learning to teach climate change: Students in teacher training and their progression in pedagogical content knowledge. *Journal of Geography in Higher Education, 45*(4), 594–620.

Ferreira, J. G. (2020). Student perceptions of a place-based outdoor environmental education initiative: A case study of the "Kids in Parks" program. *Applied Environmental Education & Communication, 19*(1), 19–28.

Freire, P. (1994). *Pedagogy of hope.* Continuum.

Freire, P. (2000). *Pedagogy of the oppressed* (30th anniversary ed.). Continuum.

Gee, J. P. (2008). A sociocultural perspective on opportunity to learn. *Assessment, Equity, and Opportunity to Learn,* 76–108.

Goldie, J. G. S. (2016). Connectivism: A knowledge learning theory for the digital age? *Medical Teacher, 38*(10), 1064–1069.
Govender, N. (2019). Subsistence farmers' knowledge in developing integrated critical pedagogy education curricula. *Education as Change, 23*(1), 1–23.
Greer, K., & Glackin, M. (2021). 'What counts' as climate change education? Perspectives from policy influencers. *School Science Review, 103*(383), 16–22.
Haffejee, F. (2021). The use of photovoice to transform health science students into critical thinkers. *BMC Medical Education, 21*, 1–10.
Harrison, R. T. (2023). W (h) ither entrepreneurship? Discipline, legitimacy and super-wicked problems on the road to nowhere. *Journal of Business Venturing Insights, 19*, e00363.
Howard, R. (2000). Collaborative Pedagogy. In G. Tate et al. (Eds.), *A guide to composition pedagogies*. Oxford University Press.
Hughes, P. (2011). Climate pedagogy: A critical, collaborative, constructivist approach for the social sciences. In S. Haslett, & S. Gedye (Eds.), *Pedagogy of climate change*. Plymouth: Higher Education Academy, pp 51–61.
Incropera, F. P. (2016). *Climate change: A wicked problem: Complexity and uncertainty at the intersection of science, economics, politics, and human behavior*. Cambridge University Press.
International Organisation for Migration (IOM). (2022). *Climate change and migration in vulnerable countries: A snapshot of least developed countries, landlocked developing countries and small island developing States*. IOM
International Rescue Committee. (2023). *Climate action for the epicenter of crisis: How COP28 can address the injustices facing conflict-affected communities*. International Rescue Committee. https://www.rescue.org/sites/default/files/2023-11/IRC-ClimateAction-CrisisEpicenter112023.pdf
Johnson, A. T., & Mbah, M. F. (2021). (Un) subjugating indigenous knowledge for sustainable development: Considerations for community-based research in African higher education. *Journal of Comparative and International Higher Education, 13*(3), 43–64.
Kalina, C., & Powell, K. C. (2009). Cognitive and social constructivism: Developing tools for an effective classroom. *Education, 130*(2), 241–250.
Kallbekken, S. (2023). Research on public support for climate policy instruments must broaden its scope. *Nature Climate Change, 13*(3), 206–208.
Kerr, R. B., Young, S. L., Young, C., Santoso, M. V., Magalasi, M., Entz, M., & Snapp, S. S. (2022). Farming for change: developing a participatory curriculum on agroecology, nutrition, climate change and social equity in Malawi and Tanzania. In *Critical adult education in food movements* (pp. 29–46). Springer Nature Switzerland.
Kimaryo, L. (2011). *Integrating environmental education in primary school education in Tanzania: Teachers' perceptions and teaching practices*. Doctoral dissertation, Åbo Akademis förlag-Åbo Akademi University Press.
Kop, R., & Hill, A. (2008). Connectivism: Learning theory of the future or vestige of the past? *International Review of Research in Open and Distributed Learning, 9*(3), 1–13.
Kwauk, C. (2020). *Roadblocks to quality education in a time of climate change*. Center for Universal Education at the Brookings Institution.
Magagula, H. B. (2020). Military integrated environmental management programme of the South African National Defence Force. *South African Geographical Journal= Suid-Afrikaanse Geografiese Tydskrif, 102*(2), 170–189.
Makina, D., & Mudungwe, P. (2023). Patterns and Trends of International Migration within and Out of Africa. In *Routledge handbook of contemporary African migration* (pp. 79–98). Routledge.
Malka, M., & Lotan, S. (2023). Beyond the risk discourse: Photovoice as critical-pedagogical tool of sexuality education for adolescents. *Sexuality Research and Social Policy, 20*(1), 103–119.
Maron, M., Ives, C. D., Kujala, H., Bull, J. W., Maseyk, F. J., Bekessy, S., ... & Evans, M. C. (2016). Taming awicked problem: Resolving controversies in biodiversity offsetting. *BioScience, 66*(6), 489–498.

Martusewicz, R. A. (2001). *Seeking passage: Post-structuralism, pedagogy, ethics.* Teachers College Press.

Mather, E. (2017). Do contemporary practices of schooling reinforce colonial relations of power? https://www.e-ir.info/2017/10/03/do-contemporary-practices-of-schooling-reinforce-colonial-relations-of-power/

Mbah, M. F. (2014). *Rethinking university engagement to address local priority needs within the context of community development: a case study.* Canterbury Christ Church University (United Kingdom).

Mbah, M. F. (2016). Towards the idea of the interconnected university for sustainable community development. *Higher Education Research & Development, 35*(6), 1228–1241.

Mbah, M., & Che, C. (2019). University's catalytic effect in engendering local development drives: Insight into the instrumentality of community-based service learning. *Journal of Sustainable Development, 12*(3), 22–34.

Mbah, M., Ajaps, S., & Molthan-Hill, P. (2021). A systematic review of the deployment of indigenous knowledge systems towards climate change adaptation in developing world contexts: Implications for climate change education. *Sustainability, 13*(9), 4811.

Mbah, M. F., & Ezegwu, C. (2024). The decolonisation of climate change and environmental education in Africa. *Sustainability, 16*(9), 3744. https://doi.org/10.3390/su16093744

Ménard, G. (2013). Environmental non-governmental organisations: Key players in development in a changing climate—a case study of Mali. *Environment, Development and Sustainability, 15*, 117–131.

Molthan-Hill, P., Blaj-Ward, L., Mbah, M. F., & Ledley, T. S. (2022). Climate change education at universities: Relevance and strategies for every discipline. *Handbook of climate change mitigation and adaptation* (pp. 3395–3457). Springer International Publishing.

Moyles, J., Adams, S., & Musgrove, A. (2002). *Study of pedagogical effectiveness in early learning. Brief No: RB363.* Anglia Polytechnic University.

Murphy, P., & Ivinson, G. (2003). *Pedagogy and cultural knowledge: a sociocultural perspective.*

Obodoegbulam, A. O., Adeyini, J., & Amadi, S. E. (2021). Table and trust: The African culture of communalism in Ogba, Egenni and Ikwerre Traditions. *Journal of African Studies and Sustainable Development*, 155–170.

Ohadike, D. (1994). *Anioma: A social history of the Western Igbo People.* Ohio University Press.

Okereke, C., & Agupusi, P. (2015). *Homegrown development in Africa: Reality or illusion?* Routledge.

Ominde, E. S., K'Odhiambo, A. K., Gunga, S. O. (2020). Analysis of praxis of African communalism: A model of ethical values in primary school curriculum in Kenya. *Journal of Educational Research in Developing Areas, 1*(3), 278–288. https://doi.org/10.47434/JEREDA.1.3.2020.278

Ontong, K., & Le Grange, L. (2015). The need for place-based education in South African schools: The case of Greenfields Primary. *Perspectives in Education, 33*(3), 42–57.

Opoku, M. J., & James, A. (2021). Pedagogical model for decolonising, indigenising and transforming science education curricula: A case of South Africa. *Journal of Baltic Science Education,* 20(1), 93–107.

Peters, J., & Mathias, L. (2018). Enacting student partnership as though we really mean it: Some Freirean principles for a pedagogy of partnership. *International Journal for Students as Partners, 2*(2), 53–70.

Piaget, J., & Cook, M. (1952). *The origins of intelligence in children* (vol. 8(5), pp. 18–1952). International Universities Press.

Porcaro, D. (2011). Applying constructivism in instructivist learning cultures. *Multicultural Education & Technology Journal, 5*(1), 39–54.

Rufai, A., Assim, V., & Iroh, E. (2019). Improving the survival rate of SMEs: Modernising the Igbo apprenticeship system (Imu Ahia). *The International Journal of Humanities & Social Studies.*

Sainz, G. M., & Khoo, S. M. (2020). Development education and climate change. *A Development Education Review, 30*, 1–7.

Schneider, S. H., & Lane, J. (2006). An overview of 'dangerous' climate change. *Avoiding Dangerous Climate Change, 7*(11).

Semken, S., Ward, E. G., Moosavi, S., & Chinn, P. W. (2017). Place-based education in geoscience: Theory, research, practice, and assessment. *Journal of Geoscience Education, 65*(4), 542–562.

Siemens, G. (2004). Elearnspace. Connectivism: A learning theory for the digital age. *Elearnspace.org*, 14–16.

Siemens, G. (2008). Learning and knowing in networks: Changing roles for educators and designers. Paper 105: University of Georgia IT Forum.

Singh-Pillay, A. (2020). Pre-service technology teachers' experiences of project based learning as pedagogy for education for sustainable development. *Universal Journal of Educational Research, 8*(5), 1935–1943.

Siraj-Blatchford, I., Sylva, K., Muttock, S., Gilden, R., & Bell, D. (2002). *Brief No: 356 researching effective pedagogy in the early years.* www.ioe.ac.uk/REPEY_research_brief.pdf

Solís, P., Huynh, N. T., Huot, P., Zeballos, M., Ng, A., & Menkiti, N. (2019). Towards an overdetermined design for informal high school girls' learning in geospatial technologies for climate change. *International Research in Geographical and Environmental Education, 28*(2), 151–174.

Suarez, P., & Bachofen, C. (2013). Using games to experience climate risk: Empowering Africa's decision makers. *CDKN Action Lab Innovation Grant Final Rep., Climate and Development Knowledge Network.*

Tanyanyiwa, V. I. (2019). Indigenous knowledge systems and the teaching of climate change in Zimbabwean secondary schools. *SAGE Open, 9*(4), 2158244019885149.

Thew, H., Graves, C., Reay, D., Smith, S., Petersen, K., Bomberg, E., Boxley, S., Causley, J., Congreve, A., Cross, I., Dunk, R., Dunlop, L., Facer, K., Gamage, K. A. A., Greenhalgh, C., Greig, A., Kiamba, L., Kinakh, V., Kioupi, V., Lee, M., Klapper, R., Kurul, E., Marshall-Cook, J., McGivern, A., Mörk, J., Nijman, V., O'Brien, J., Preist, C., Price, E., Samangooei, M., Schrodt, F., Sharmina, M., Toney, J., Walsh, C., Walsh, T., Wood, R. Wood, P., & Worsfold, N. T. (2021). *Mainstreaming climate education in higher education institutions.* COP26 Universities Network Working Paper.

UNDP. (2023). *Gender-responsive climate change actions in Africa.* UNDP. https://climatepromise.undp.org/research-and-reports/gender-responsive-climate-change-actions-africa

UNESCO. (2018). *The report of the UNESCO expert meeting on indigenous knowledge and climate change, Nairobi, Kenya.* Retrieved June 27–28, 2018, United Nations Educational, Scientific and Cultural Organization.

UNESCO. (2020). Global education monitoring report: Inclusion and education. *All means all.* UNESCO. Available at: https://unesdoc.unesco.org/ark:/48223/pf0000373718

UNESCO. (2021). *Learn for our planet: A global review of how environmental issues are integrated in education.* UNESCO. Available at: https://unesdoc.unesco.org/ark:/48223/pf0000377362

UNFCCC. (2022a). *Climate change education.* Retrieved April 25, 2022, https://unfccc.int/blog/climate-change-education

UNFCCC (2022b). *Africa climate week 2022: Youth affiliated event.* UNFCC. https://unfccc.int/sites/default/files/resource/Youth%20Affiliated%20Event%20ACW22%20Summary%20Report.pdf

United Nations Environment Programme. (2021). *Adaptation gap report 2021: The gathering storm—adapting to climate change in a post-pandemic world.*

United Nations Framework Convention on Climate Change (UNFCC). (2011). Fact sheet: Climate change science–the status of climate change science today. https://unfccc.int/files/press/backgrounders/application/pdf/press_factsh_science.pdf

van Compernolle, R. A., & Williams, L. (2013). Sociocultural theory and second language pedagogy. *Language Teaching Research, 17*(3), 277–281.

Velempini, K., Martin, B., Smucker, T., Ward Randolph, A., & Henning, J. E. (2018). Environmental education in southern Africa: A case study of a secondary school in the Okavango Delta of Botswana. *Environmental Education Research, 24*(7), 1000–1016.

Vygotsky, L. S. (2012). *Thought and language (revised and expanded edition).* MIT Press.

Wall, S., Litjens, I., & Taguma, M. (2015). *Early childhood education and care pedagogy review.* OECD.

Walther, R. (2008). Towards a renewal of apprenticeship in West Africa. Agence Française de Développement (AFD).

Wamsler, C., Schäpke, N., Fraude, C., Stasiak, D., Bruhn, T., Lawrence, M., Schroeder, H., & Mundaca, L. (2020). Enabling new mindsets and transformative skills for negotiating and activating climate action: Lessons from UNFCCC conferences of the parties. *Environmental Science & Policy, 112*, 227–235. https://doi.org/10.1016/j.envsci.2020.06.005

Whitham, M. P. (2012). *It's about community-finding local solutions to local problems: Exploring responses to the social problem of youth unemployment* [Doctoral dissertation]. University of Waikato.

Whitson, J. A. (1995). Post-structuralist pedagogy as counter-hegemonic praxis: Can we find the baby in the bath water?. *Postmodernism, Post-colonialism and Pedagogy*, 121–143.

Wilson, L., Vedel, K. A., Samuel, G. M., & Nielsen, C. S. (2021). Example of best practice: Getting to the core of Red Apples-Green Apples: A dance and visual arts learning project between South Africa and Denmark. *Intercultural Education, 32*(6), 682–690.

Winchester-Seeto, T., McLachlan, K., Rowe, A., Solomonides, I., & Williamson, K. (2017). Transformational learning–possibilities, theories, questions and challenges. In *Learning through community engagement: Vision and practice in higher education* (pp. 99–114).

World Meteorological Organization. (2023). *The state of the climate in Africa 2022.* World Meteorological Organization. Retrieved June 27, 2024, https://library.wmo.int/viewer/67761/?offset=#page=11&viewer=picture&o=bookmark&n=0&q=

Yemini, M., Engel, L., & Ben Simon, A. (2023). Place-based education–a systematic review of literature. *Educational Review*, 1–21.

Zimu-Biyela, N. (2019). Using the school environmental education programme (SEEP) to decolonise the curriculum: Lessons from Ufasimba Primary School in South Africa. *International Journal of African Renaissance Studies-Multi-, Inter-and Transdisciplinarity, 14*(1), 42–66.

Zoellner, R. (1969). Talk-write: A behavioral pedagogy for composition. *College English, 30*(4), 267–320. https://doi.org/10.2307/374179

Chapter 7
Valuation of High-Level Climate Change Education Discourse in Africa: Evidence-Based Insights for Mainstreaming Disaster Risk Management

Henry Ngenyam Bang

Abstract With scary predictions of frequent, diverse, and severe impacts of disaster and climate change (CC)-induced risks in Africa, discourses on viable CC and disaster mitigation and adaptation measures have intensified in recent decades. Endorsed by the Sendai Framework for Disaster Risk Reduction (SFDRR) and the United Nations Educational, Scientific and Cultural Organization (UNESCO), education is perceived as a powerful disaster risk reduction (DRR) tool. Underpinned by a novel "*Determinants of DRR Education*" conceptual framework, this chapter employs a qualitative research approach to explore the pedagogic space for CC/DRR education in Africa. Aimed at assessing whether CC/DRR education has been comprehensively inculcated into discourses at the highest level in Africa, this chapter interrogates the depth and scope of assimilation of CC/DRR education during the African Regional Platforms (ARP) and High-Level ministerial meetings for DRR. By applying a Likert Scale grading to the analysis, assessments were done on two main objectives: (1) the consolidation of CC/DRR education into the various Platforms and (2) how immersed the various determinants of DRR/CC education are in the Platforms. Informed by basic statistical analysis (percentages) the first objective was assessed as "*poor-mediocre*" while the second objective was "*fair*". Overall, the inclusion of DRR/CC education in the Platforms is considered "*Mediocre-Fair*". From the result, we argue that the High-Level discussions have minimal consideration for education as a potent disaster risk mitigation strategy. Given the crucial role of education in mitigating climate risks, this chapter argues for inculcating DRR/CC education into the ARPs to expedite uptake by African countries. The chapter has policy implications for disaster/climate risk education, contributes to the literature in the field, and is also beneficial to CC/DRM practitioners, professionals and students/teachers.

H. N. Bang (✉)
Department of Disaster and Emergency Management, School of the Environment, Coventry University, Coventry, UK
e-mail: henry.bang@coventry.ac.uk

M. F. Mbah et al. (eds.), *Practices, Perceptions and Prospects for Climate Change Education in Africa*, https://doi.org/10.1007/978-3-031-84081-4_7

Keywords Climate change education · Africa · Africa regional platforms for disaster risk reduction · Disaster risk management · Determinants of disaster risk reduction education

Introduction

The climate crisis is continuously impacting many countries with more devastating consequences in Africa which has low organisational resilience to tackle the ramifications of climate change (CC) (UN, 2020). This was emphasised in Sharm el-Sheikh (Egypt) during the 2022 United Nations Climate Change Conference (UNCCC; COP 27). Informed by contemporary CC research which indicates increasing intensity/frequency and duration of climate risks over the next decades (IPCC, 2020), governments were criticized for limited action on disaster/climate risk reduction actions and urged to be more proactive in reducing disaster/climate risks (UNCCC, 2022). Given the potential ramifications of climate change, the need for urgent climate action has never been so pressing! Before the Sharm el-Sheikh conference, very little progress had been made to reduce CC effects although during COP 21 in France a few years earlier, more ambitious actions at reducing the effects of CC were agreed including through disaster/CC education.

As emphasised by priority three of the Hyogo Framework for Action (HFA) (2005–2015), countries should embed innovative measures for safety and resilience at all levels (local, regional/state, national), including the use of knowledge/education (UNISDR, 2005). DRR education provides knowledge and a better understanding of the causes of disasters and measures to reduce/mitigate disaster risks (Adiyoso and Kanegae, 2010). Hence, the reason education is perceived as a central apparatus for enhancing CC/DRR resilience (UNESCO/UNICEF, 2012). Since the disaster risk reduction (DRR) concept encapsulates efforts at minimising disaster/climate risks, DRR education also aims to educate about CC-induced risk reduction. Hence CC education is discussed in this chapter under the auspices of DRR education.

As a discipline, DRR education is new (UNESCO, 2011) and perceived as cost-effective with enhanced operational and functional attributes for effective DRM (Asian Disaster Preparedness Centre, 2008). As a novel area, there is great potential for DRR pedagogy beyond formal school/curriculum teaching (UNESCO/UNICEF, 2012; Patel, 2008). A study that was carried out by the Asian Disaster Preparedness Centre (2008) found that communities that have been educated on disaster preparation are more proactive in responding to disasters/crises than those that have not received any tuition/learning. Similarly, low-risk awareness, consciousness and/or knowledge of climate/disaster hazards and vulnerabilities can adversely affect DRR preparation and recovery (UNESCO, 2011). With Africa having a relatively low resilience to climate/disaster risks (UN, 2020), enhancing risk reduction through education is, therefore, vital.

Africa is susceptible to CC due to many factors including a projected increase in population growth. Coastal towns/cities with very high population densities are relatively more vulnerable to climate risks such as storms, rise in sea level, floods, hurricanes etc. (Bang & Burton, 2021). Arguably, the future of Africans is jeopardized if urgent action is not taken to mitigate the effects of CC on the continent (UN, 2020, 2022). Consequently, there is an urgent requirement for enhanced climate action (UNCCC, 2022) in Africa including mainstreaming DRR/CC education in DRR strategies from the community to national levels.

The motivation of this research is twofold: First, to assess the depth to which the African Regional Platforms (ARP) and High-Level Ministerial Meetings for DRR (hereinafter referred to as Platforms) have inculcated education as a vital instrument for mitigating disaster/climate risks. Second, this research seeks to use the determinants of DRR education to gauge the degree of embeddedness of DRR/CC education into the various Platforms. This study is innovative and arguably, pioneers research analysing the depth with which CC/DRR education has been immersed into the Platforms.

This chapter fosters understanding and knowledge of the importance of integrating CC/DRR education in the Platforms to facilitate the governance of disaster/climate risks and enriches the discourse on CC/DRR education at the regional Africa level. Literature on the subject will be enhanced, including accelerating climate action as emphasised in Sustainable Development Goal (SDG) 13. By informing regional and national policies, including knowledge on CC/DRR education, this chapter argues that CC/DRR education is an invaluable risk aversion tool and should be inculcated into the African educational pedagogic system at all levels from primary school to University.

Succinct Appraisal of Underlying Concepts

Climate Change (CC)

The concept of CC discussed in this chapter is about the impact of climate risks (severe storms, hurricanes/cyclones, floods, etc.), with one of the worst being recurrent flash flooding (Bang and Burton, 2021). Of the various conceptualisations of CC, perhaps that of the United Nations is more succinct and considers CC as any change of climate observed over a comparable extended period which could be linked to natural climate variability (directly or indirectly) or resulting from anthropogenic activities (UNFCCC, 2011). Humanity is wary of CC because of the effects of climate risks that could have devastating consequences on various infrastructures, including ecosystems and public services (ibid). Its potential impact on human lives, livelihoods, well-being, health, socioeconomic, and cultural assets (IPCC, 2020) has intensified discussions on risk reduction strategies together with education on CC/DRR, which is the motivation of this chapter. In light of evidence indicating a rise in the severity, frequency and effects of climate risks (IPCC, 2020; Wing et al., 2018), debates on the most potent measures to mitigate the consequences are now a topical issue. As will be analysed later in the chapter, High-Level DRR discourses in Africa have addressed these concerns albeit with limitations.

Disaster Risks and DRR

The United Nations conceptualises disaster risk as "*the potential loss of life, injury, or destroyed or damaged assets which could occur to a system, society or a community in a specific period of time, determined probabilistically as a function of hazard, exposure, vulnerability, and capacity*" (UNISDR, 2017, p. 14). This definition aligns with the view of disaster risk in this chapter with CC perceived as a critical disaster risk posing a grave threat to humanity and economic growth worldwide (IPCC, 2020). Since climate risks constitute a core disaster risk, they can be managed through DRR, hence the discourse in this chapter also encapsulates climate risk mitigation.

DRR is usually perceived as a concept and practice of systematic risk reduction (including climate risks) and is aimed at reducing vulnerabilities and enhancing resilience to contemporary risk, averting emerging and/or new threats/risks, and ensuring residual risks are appropriately controlled (UNISDR, 2017). As underscored by UNESCO (2011), DRR/CC education is the best tool for mitigating these vulnerabilities/hazards and preventing them from becoming disasters. This is aptly discussed in the next section.

DRR/CC Education

The popular analogy "*What people know is more important than what they have*" denotes the relevance of education in enhancing resilience to CC/DRR. The conceptualisations of DRR education relevant to this chapter embrace that of UNESCO (2011) which perceives DRR education as fostering disaster management learning/teaching to alleviate lives/livelihoods and increase the resilience of communities/populations to disaster risks. In like manner, Nakano and Yamori (2021) consider disaster education as a knowledge acquisition activity involving teachers/instructors and students or learners where information, skills and expertise in preparing for, responding to, or proactively mitigating disaster/climate risks are shared. DRR education does not only involve formal teaching, the concept also encapsulates DRR/CC communication and/or information dissemination and raising awareness that encourages preemptive measures to increase resilience to disaster/climate risks (Delicado et al., 2017; Shaw & Krishnamurthy, 2010). Likewise, DRR education has been associated with disaster management planning (macro, meso, and micro levels), that involves assessment, protection, and response (Petal, 2008). As highlighted before, DRR/CC education is the central theme of discussion in this chapter, to assess its relevance in DRR discourses and practices at the highest regional level in Africa.

Contemporary Actions and Practices for DRR/CC Education: Succinct Perusal

Embedding DRR Education into Disaster Management Platforms/Meetings

International disaster management frameworks such as the Sendai Framework for Disaster Risk Reduction (SFDRR) (2015–2050) and the HFA (2005–2015) endorsed education in disaster risk consciousness and risk reduction at all levels (national, regional/provincial/state and local/community) as well as strengthening the culture of resilience/safety (UNISDR, 2005, 2015). In alignment with international frameworks, the African and Asian regional frameworks for DRR have urged their member countries to collect and dispense DRR/CC knowledge to foster risk reduction (Bang, 2023; Torani et al., 2019). In compliance, some governments have been designing novel DRR educational approaches to train the public, especially communities that are more prone to disaster/climate risks (Bosschaart et al., 2016; Kagawa and Selby, 2012). Nevertheless, as analysed later in the chapter, the degree of assimilation of DRR/CC education within the Platforms leaves much to be desired.

Benefits of CC/DRR Education

The literature mentions several advantages of CC/DRR education (traditional/formal): enhances measures for CC adaption; informs on awareness, knowledge and the appropriate skills for climate and disaster risks; offers insights on how the populace can reduce the effects (economic, cultural, social, impact) of climate/disasters risks; raises awareness on the appropriate actions, attitudes and/or behaviours required to confront, withstand, prepare for or bounce back after disasters/crises, as well as the desired risk reduction actions for lessening vulnerabilities/hazards plaguing individuals/communities (Patel, 2008; Rohrmann, 2008; UNESCO, 2011; UNISDR, 2015). O'Brien et al. (2006) and Cherniack (2008) opine that for effective DRR, it is essential to adopt a bespoke approach that targets individuals/groups with unique roles in DRR education and/or "*at risk*" communities. Examples include policymakers, educators (academics, teachers, professors), administrators (governors, directors, ministers etc.) service providers (emergency responders/rescuers, firefighters, volunteers, and relief/rescue teams etc.) and communities prone to/vulnerable to disaster risks (Dolatabadi et al., 2016; Torani et al., 2019). Torani et al. (2019) and Muttarak and Lutz (2014) analysed the effectiveness of DRR education and recommended that to maximise impact, the educational approach utilised (method of teaching and training) for the target communities or specific groups should be informed by their knowledge/experience, authority or leadership level, and level of educational attainment or qualification. The most vulnerable communities should be the first target for reasons discussed in the next section.

Prioritising the Most Vulnerable People

Vulnerability is conceptualised as situations caused by physical, socio-economic, and environmental elements or activities that make societies, countries, people, communities, or their various resources susceptible to hazard impacts. Vulnerable people/communities face unique challenges and therefore, need specialist help which could be in the form of education, training and/or sensitisation on risk management measures, which should be offered by disaster management professionals, specialists or educators (Muttarak & Pothisiri, 2013). The most vulnerable people such as the elderly, children and women, (Dolatabadi et al., 2016; Torani et al., 2019; UNESCO/UNICEF, 2012), require tailored disaster education, as discussed in the following paragraphs.

Specific gender issues have been associated with women who face distinct challenges during disasters (Eisenman et al., 2007; Wisner, 2006). Research has shown that after receiving basic training in disaster education, women became very active in the community to address risk issues. (Muzenda-Mudavanhu et al., 2016). Furthermore, it has been argued that when women are educated in DRR, the entire family will have a greater awareness of the issues since women are more proactive in disaster education at home (Muttarak and Pothisir, 2013; O'Brien et al., 2006).

Research-informed suggestions oblige disaster education for elderly and disabled people to focus mainly on their physical health and cognitive/mental ability in coping with, preparing for and responding to disaster risks (Thomas et al., 2015). For instance, there is evidence that training or informing/educating people with disabilities can help them survive disasters since the ability to save themselves with minimal assistance would have been enhanced (Morrow, 1999; Torani et al., 2019). This is similar for children.

Children are considered vulnerable to disasters, partly because they are emotionally and physically dependent on their parents/adults for protection against natural hazard-induced disasters, crises, and emergencies (UNESCO/UNICEF, 2012; Bosschaart et al., 2016; Tuladhar et al., 2015). The advantages of teaching children about disasters at an early age have been considered an innovative approach to DRR since children will grow with the knowledge in them and studies have shown that children hardly forget what they learned while they were younger (Bray, 2000). Disaster education in children will inform their knowledge of preventive and preparedness measures with an increased tendency to act for their safety during disasters (Bosschaart et al., 2016). Furthermore, there is a greater potential to educate children in their homes, which easily enhances their disaster risk perception (Sawada, 2007; Torani et al., 2019; Faber et al., 2014).

Despite the benefits and good practices for DRR education, it has been acknowledged that although DRR has the potential to foster knowledge on risk aversion awareness/consciousness, the overarching motivation for anticipatory risk reduction actions is scarcely achieved. Contrarily, Wisner (2006) reiterated that limited knowledge and low awareness/understanding of disaster risks unfavourably affect community resilience and preparation for any forthcoming disaster risks. That is why education on disaster/climate risks is the focus of this chapter with the discourse and application in Africa.

Assessing CC/DRR Education Discourse Within the Various ARPs

During the ARPs, national and international DRR stakeholders—regional stakeholders (economic communities/entities, directors/ministers from governments) and international development organisations converge to discuss/share experiences, lessons learned, skills and insights on any advancement in DRR efforts in Africa. Delivering Member States' pledges to the African Union's (AU) Programme of Action (PoA) of implementing the guidelines of the SFDRR/HFA and adopting the outcomes of the Platform is the key motivation for these High-Level meetings (UNDRR, 2021). Eight regional Platforms have occurred in Africa between 2005 and 2021. The next section elaborates on DRR education during the meetings including the AU's PoA for implementing the SFDRR.

The 1st ARP for DRR was formed in Nairobi Kenya between the 26th and 27th of April 2007. Organised as a consultative meeting, the motive of this maiden gathering of High-Level officials was to assess improvements made in the application of DRR measures promised during the 2005–2015 HFA and the Africa PoA for DRR. This first meeting underscored the need to combat CC through the strengthening of institutional mechanisms to create what the organizers called "*Ministers for Climate Change*". Discussions on education as a DRR tool were sparse albeit education together with knowledge, and innovation featured as a prominent theme (ARC, 2009).

The 2nd Ministerial Conference on DRR which took place from the 14th to 16th of April 2010 in Kenya's capital (Nairobi) briefly discussed DRR/CC education. The African Development Bank and the AU urged African governments to mainstream knowledge and education on risk reduction in developing a resilient and safety culture. African countries were requested to boost formal and informal education systems for DRR, integrate DRR into educational institutions at the various educational levels (tertiary, elementary/primary and secondary/colleges) and assess the education sectors' resilience and capacities (AU, 2010).

From the 9th to the 13th of May 2011, Addis Ababa (Ethiopia) hosted the 3rd ARP for DRR. The Platform recommended that DRR and CC adaptation should be a national education priority and be merged into the educational systems of countries. Emphasis was placed on utilising education and acquired knowledge in building a culture of resilience and safety in the multiple sectors/levels in countries. Worthy of note, was an emphasis on developing the disaster resilience of vulnerability education through public awareness and information dissemination campaigns, including the need to inculcate disaster management training in the tertiary education sector including in postgraduate degree programmes (AU & UNISDR, 2011).

Arusha, in Tanzania, hosted the 4th ARP in 2013, between the 13th and 15th of February. This meeting elaborated more on DRR education with more in-depth discussions on the relevance of incentivising the education sector to incorporate DRR into the curriculum/programme of educational institutions through more technical and financial support; strengthening collaboration/cooperation with tertiary educational institutions of different countries, including in urban risk management skills, and taking risk reduction decisions; allowing women and children to easily access DRR education; amalgamate community reduction on DRR with awareness/sensitisation and foster its coordination across the public and private sectors; and expedite agreement made during the second Ministerial Conference on DRR to create a network of education institutions at the sub-regional/regional levels (UNISDR, 2013).

The 5th ARP for DRR was combined with the third Ministerial Meeting for DRR and held in May 2014 from the 12th to the 16th in Abuja (Nigeria). This High-Level gathering also stressed the assimilation of DRR education within the curriculum of schools/universities and that education should be used as an instrument to enhance safety and resilience. The delegates also discussed how DRR/CC education can be used in the private sector for training/skills development. In recognition of the limited exposure of DRR in Africa, the Platform mandated African countries to rapidly develop human capital in risk reduction through more engagement with schools, universities and other institutions delivering DRR education. Given the ongoing conflict in the Democratic Republic of Congo at the time, the meeting discussed the potential of disaster education in mitigating complex emergencies and conflicts through education programs on law and order, peace and nonviolence (AFRP Bulletin, 2014).

Between the 22nd and 25th of November 2016, Mauritius hosted the 5th High-Level Meeting on DRR together with the 6th ARP for DRR. Education was highlighted as critical in reducing vulnerabilities to disasters and in overcoming challenges in risk reduction. The host nation reported a forceful DRR and sensitisation/awareness-raising approach in its schools and communities. There was an indication that other African countries were gradually embracing DRR education when Zimbabwe delegates mentioned that plans were underway to develop novel policies on CC and embed DRR into the curriculum and education programs in Zimbabwe. The meeting reiterated the desire for higher education institutions in Africa to deliver training and qualifications in disaster management education and the delegates were urged to take action on this issue in their respective countries (AFRP Bulletin, 2016).

The triple 4th Arab Conference on DRR, the 6th High-Level Meeting on DRR and the 7th Session of the ARP for DRR were held in the capital of Tunisia (Tunis) from the 9th to 13th October 2018. The main DRR-related education issues discussed were the relevance of member states to institute DRR education programs; for countries to apply SDG 4 on having quality CC/DRR education; develop risk management courses/programs in Universities across Africa; use of DRR education in building the skills/capacity of youths; develop education programs that provide insights on multiple hazards; foster gender inclusion through educational programs on risk; and to further build in-country abilities for DRR professional practice (UNISDR Bulletin, 2018).

From the 16th to the 19th of November of 2021, the 8th Platform was hosted in Kenya. A session focused on local and indigenous knowledge systems and practices/activities associated with DRR. How indigenous knowledge systems enhance community participation in policy decisions was underlined. Indigenous women's knowledge of CC and weather forecasting was highlighted. While assessing the inclusion of disability in DRR across Africa, a delegate mentioned education as a means used to promote disability inclusiveness in Zimbabwe. The meeting developed a PoA matrix for the period 2021–2025 to implement the SFDRR in which the incorporation of climate risk in educational systems across the continent was considered as one of the strategic areas. Mauritius, a country that is prone to CC specifically indicated its intention to assimilate and forster risk management education in its primary/elementary school programs (DRR Bulletin, 2021).

The AU's PoA for the implementation of the SFDRR highlighted education as one of the local-level structures to be used by local government and sub-national agencies via participatory approaches through the mobilisation, sensitisation and empowering of people at the community level (AU, 2017). The PoA encouraged higher education institutions to engage in increasing the capacity of vulnerable communities in building collaborative DRR and management teaching and research. In addition, DRM stakeholders were requested to work with the public sector (government officials/institutions) to develop and support a culture of DRR education and awareness. The role of the media was recognised as fundamental to fostering DRR education during disaster mitigation, preparedness, response, and recovery including via accurate, fair, widespread, timely, and comprehensive reporting of disaster risk information. The PoA recommended that academic institutions should inform DRR education at the scientific, and technical higher-level education institutions. Cooperation and partnership with stakeholders in both the private sector and public agencies via educational training were also mandated. Furthermore, education was discussed in connection to enlightening risks, and the integration of DRR with adequate resourcing, advocacy, awareness raising, and mobilization for DRR. The PoA also recognised the relevance of education in strengthening its activities—and requested the UN Africa Office for Disaster Reduction to use education and training to strengthen the monitoring and implementation of activities related to the SFDRR.

Discussion: Critical Analysis of the Determinants of DRR/CC Education in the Platforms

The extent to which key CC/DRR education determinants are recognised as DRM instruments during the Platforms is discussed in this section. Key DRR/CC education determinants identified in the literature have been used to assess the depth of the education discourse during the Platforms. The eight most effective determinants of DRR/CC education identified and used in the analysis are an assessment of educational needs, approaches to education, educational planning, educational tools, knowledge-raising strategies, the content of educational delivery, educational challenges, and education organizations. This assessment informs the uptake of DRR/CC education in Africa since it arguably depends on how the Platforms are explicit in selling the concept to African countries.

Educational Needs Assessment

An evaluation of community educational needs is vital in fostering adequate programs on DRR/CC education. Aghaei et al. (2018) discuss the need for critical scrutiny of the expertise, skills, capabilities, knowledge and interests of people and their communities to assess their educational requirements. Other researchers have acknowledged many other attributes such as checking the strengths, weaknesses, ages and the most convenient educational tools for people/communities requiring DRR education (Meesters et al., 2009; Perry and Lindell, 2003; Prashar et al., 2013); an assessment of community resilience and the inherent risks-hazards, vulnerability, exposure (Apronti et al., 2015; Siripong, 2010) and how they can be mapped (Apronti et al., 2015; Perry and Lindell, 2003); and operational research and DRR plans/programs (Aghaei et al., 2018; UNISDR, 2005). At least two of these attributes are mentioned in all the Platforms (see Table 7.1). The most articulated education needs are DRR target groups and the need to integrate DRR education into schools. Both are also mentioned in five of the nine Platforms assessed followed by themes related to the requirement for building a resilience and safety culture including CC as a priority for DRR—both articulated in five of all the frameworks. While the former is mentioned thrice, the latter is mentioned twice. Bang (2023) analysed the most prominent disaster educational needs assessment as shown in Fig. 7.1.

Strategies for Raising Knowledge

DRR scholarship and knowledge/capacity building are invaluable in ensuring effective policy, operational practice, decision making and collaboration in disaster management activities (Weichselgartner & Pigeon, 2015). Such knowledge, as

Table 7.1 Summary of how the various platforms have articulated the CC/DRR determinants

Effective determinants of DRR/CC education	Regional platforms and high level ministerial meetings for disaster risk reduction in Africa									
	1st ARP for DRR	2nd ARP for DRR	3rd ARP for DRR	4th ARP for DRR	5th ARP for DRR	6th ARP for DRR	7th ARP for DRR	8th ARP for DRR	African Union's PoA for implementing the SFDRR	*Overall assessment of the CC/DRR determinants analysed
Educational needs assessment	• Need to fight CC or mitigate its impacts • CC education is also conceptual-ized as knowledge and innovation in CC	• Build a resilience and safety culture • Assess the resilience and capacity of educational systems	• Ensure DRR education and CC adaptation are national priorities • Apply knowledge of DRR education in developing a resilience and safety culture • Vulnerable communities should be targeted for DRR education • Mainstream DRR education into the tertiary/higher education sector	• The target for DRR education is women and children • Provide funding and resources as incentives for the education sector	• Developing a resilience and safey culture • Mainstream DRR education into school programs		• Target groups were building youth capacity and gender inclusiveness in DRR education in consideration of women • Develop DRR education courses in higher education institutions	• The target groups were local and indigenous people/women and those with disabilities	• Significantly reduce disaster risks in the continent by 2030 • Mobilization, sensitization, and empowering of people at the community level • Vulnerable communities are identified as a target group • Developing and supporting a culture of education and risk awareness • Support a culture of education and risk awareness • Stakeholder cooperation in the public and private sectors	Good

(continued)

Table 7.1 (continued)

Effective determinants of DRR/CC education	Regional platforms and high level ministerial meetings for disaster risk reduction in Africa									
	1st ARP for DRR	2nd ARP for DRR	3rd ARP for DRR	4th ARP for DRR	5th ARP for DRR	6th ARP for DRR	7th ARP for DRR	8th ARP for DRR	African Union's PoA for implementing the SFDRR	*Overall assessment of the CC/DRR determinants analysed
Strategies for raising knowledge	Not identified	Not identified	Not identified	Not identified	Not identified	Not identified	• Public awareness campaigns	Not identified	Not identified	Poor
Educational approaches	Not identified	Not identified	Not identified	Not identified	Not identified	• Some countries like mauritius indicated they have undertaken an aggressive DRR education and awareness campaign, including simulations	Not identified	• Local community participation	• Participatory approaches through the mobilization, sensitization and empowering of people at the community level • Collaborative teaching • The use of media for disaster education is recognized • Accurate, fair, widespread, timely, and comprehensive reporting of disaster risk information • Advocacy, awareness raising and mobilization for DRR	Mediocre

(continued)

Table 7.1 (continued)

Effective determinants of DRR/CC education	Regional platforms and high level ministerial meetings for disaster risk reduction in Africa									
	1st ARP for DRR	2nd ARP for DRR	3rd ARP for DRR	4th ARP for DRR	5th ARP for DRR	6th ARP for DRR	7th ARP for DRR	8th ARP for DRR	African Union's PoA for implementing the SFDRR	*Overall assessment of the CC/DRR determinants analysed
Educational planning	Not identified	• Use DRR education to build a culture of safety and resilience • DRR education should be formal and informal • DRR education is to be integrated into educational institutions at the primary, secondary, and tertiary levels	• Ensure DRR education and CC adaptation are a national priority • Integrate CC adaption and DRR education into the educational system • Develop school curricula and training of teachers in DRR education • Resilience and safety culture should be established at all levels • A target group for DRR education is vulnerable communities • DRR education should be mainstreamed into tertiary education	• Coordination of community DRR education across the public sector/ government and private sector • The funding of DRR education including human resources like technical support • Integrate DRR education into school programs	• Risk prevention and building resilience in disaster risks • DRR education should be inculcated into school curricula	• Disaster prevention, preparedness and risk reduction • Some countries like Zimbabwe plan to mainstream DRR education into their educational system and develop a new CC policy and response measures	• Expediting DRR plans through education programs • Establishing DRR education programs in member states of the AU • Countries should create avenues to build capacity in DRR education and professional practice • African countries plan to address SDG 4 on quality education • Zimbabwe reported that education was used to promote disability inclusiveness in Zimbabwe	• Local community participation in policy-making • Incorporate climate risks into the educational system • Mauritius indicated the intention to integrate DRR education into the curriculum of primary schools	• The higher education sector is to be used as the main structure to implement DRR education in the continent • Government educational agencies and subregional agencies • Collaboration in the public and the private sector on DRR • Academic institutions like scientific, academic, and technical higher-level education institutions should be used for DRR education • The Africa UN Office for DRR has been mandated to strengthen the monitoring and implementation of the SFDRR through the provision of DRR education and training • Collaboration/cooperation of DRR stakeholders in the public and private sectors • DRR education resourcing	Good

(continued)

Table 7.1 (continued)

Effective determinants of DRR/CC education	Regional platforms and high level ministerial meetings for disaster risk reduction in Africa									
	1st ARP for DRR	2nd ARP for DRR	3rd ARP for DRR	4th ARP for DRR	5th ARP for DRR	6th ARP for DRR	7th ARP for DRR	8th ARP for DRR	African Union's PoA for implementing the SFDRR	*Overall assessment of the CC/DRR determinants analysed
Educational tools	Not identified	Not identified	Not identified	Not identified	Not identified	Not identified	Not identified	Not identified	Not identified	Poor
Educational content	Not identified	Culture of safety and resilience in DRR education	• Disaster safety and resilience • CC adaptation • Training of teachers for DRR education	• Build knowledge in urban risk management	• Relationship between DRR education and conflict • Education about non-violence especially in violence-induced crisis-prone countries and territories	• Disaster preparedness and prevention	• DRR education on the various risk types	• Local and Indigenous knowledge systems' contribution to DRR • DRR practices • Indigenous women's knowledge of CC and weather forecasting • Climate risks	• Identifying disaster risk • Disaster mitigation, preparedness and response • Enlightening risks and risk management products	Satisfaactory
Educational organizations	• Strengthening institutional mechanisms for CC education	Not identified	Not identified	Not identified	Not identified	Not identified	Not identified	Not identified	Not identified	Poor

(continued)

Table 7.1 (continued)

Effective determinants of DRR/CC education	Regional platforms and high level ministerial meetings for disaster risk reduction in Africa									
	1st ARP for DRR	2nd ARP for DRR	3rd ARP for DRR	4th ARP for DRR	5th ARP for DRR	6th ARP for DRR	7th ARP for DRR	8th ARP for DRR	African Union's PoA for implementing the SFDRR	*Overall assessment of the CC/DRR determinants analysed
Educational Challenges	Not identified	Not identified	• Disaster safety and resilience • CC adaptation • Trainiing in DRR education	Not identified	• Limited DRR education and public awareness in Africa • Need for more engagement with higher education institutions • The requirement to enhance human resources in DRR education	Not identified	Not identified	Not identified	Not identified	Mediocre
*Overall assessment of DRR/CC Education inclusion in the various Platforms	Mediocre	Fair	Satisfactory	Fair	Satisfactory	Fair	Fair	Fair	Satisfactory	General assessment of both criteria/ objectives is Mediocre-Fair

Source Table developed by the author
*Generally, assessment of the CC/DRR determinants is subjective and informed by the grading of a Likert scale comprising "*Poor, Mediocre, Fair, Satisfactory, Good, and Excellent*" from to the high end

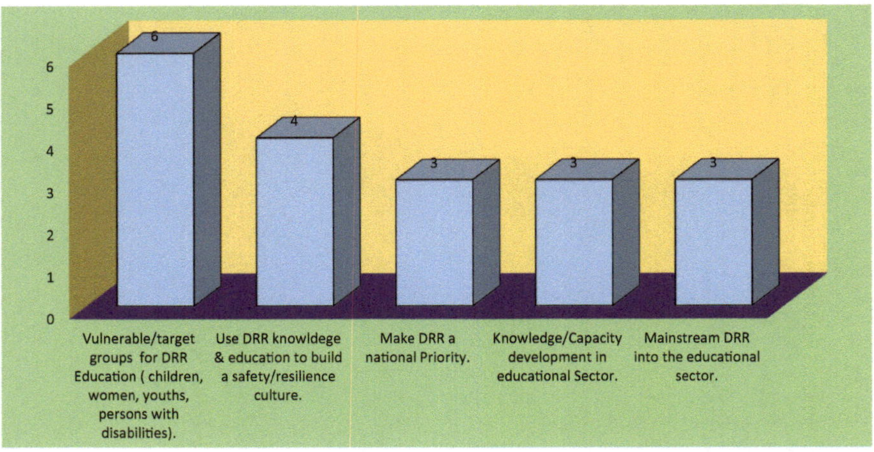

Fig. 7.1 The most prominent educational needs assessment. *Source* Adapted from Bang (2023)

emphasised by the SFDRR (UNISDR, 2015), can be shaped by people/community perceptions, culture and experiences. Viable knowledge-raising and transfer measures include information dissemination through various means (Lee et al., 2007) and joined community or team action in resolving disaster issues (Efthymis, 2014). Indeed, educational methods for DRR–inquiry/problem-based, require participation, collaboration and sharing of opinions, experiences and ideas amongst DRR stakeholders. Nanaka and Takeuchi (1995) endorse this method of intangible knowledge acquisition and emphasise that people's emotions, values, judgements, beliefs and perceptions are valuable considerations for DRR to be effective.

From the analysis, it can be deduced that community involvement and public awareness in DRR/CC educational issues are sparsely discussed during the ARP. Considering Weichselgartner and Pigeon (2015) assertion of limited assimilation of knowledge systems in Africa, the relevance of more engagement on DRR/CC knowledge development during the Platforms cannot be overemphasised. Notwithstanding, it should be noted that if DRR/CC knowledge is improperly implemented, it will not lead to risk aversion/mitigation (Weichselgartner and Obersteiner, 2002), irrespective of the scale and scope of dissemination. On a similar note, Briceno (2015) affirmed an uneven DRR knowledge development with implications for practical and effective risk reduction. Furthermore, Gall et al. (2015) noticed implementation challenges in DRR professional practice and research. Further investigations have revealed that DRR/CC research and knowledge development has a shallow evidence-based application, DRR/CC education has limited cross-discipline application and the co-production of DRR/CC research and knowledge dissemination is led by Western scholars, compared to their peers in developing countries (ibid). Bang (2023) analysed the most prominent strategies for raising knowledge as shown in Fig. 7.2.

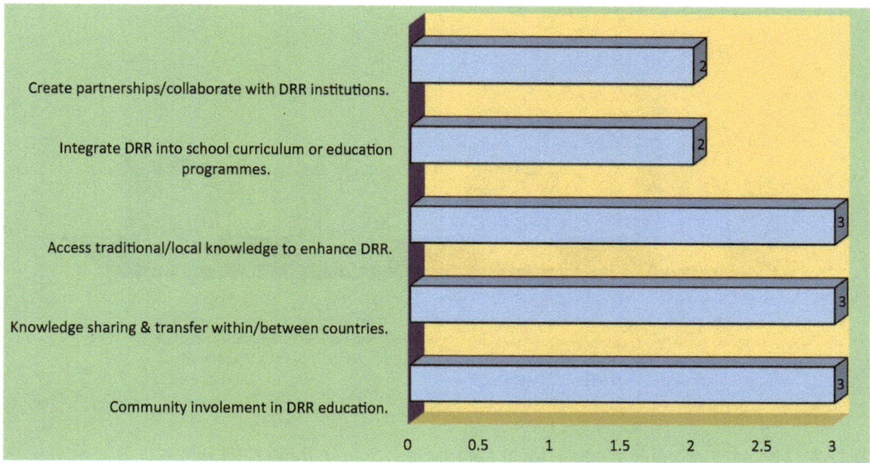

Fig. 7.2 The most prominent strategies for raising knowledge. *Source* Adapted from Bang (2023)

Educational Approaches

A core approach to fostering DRR education is inculcating it within the programs of schools, universities and/or educational establishments (Apronti et al., 2015; Efthymis, 2014). The literature has a variety of educational methods/approaches employed in DRR. A joint study by UNESCO and UNICEF on educational approaches in the field of DRR in 30 countries revealed a broad range of DRR approaches. The main ones are methods that focus on environmental education and sustainable development viewed as a symbiosis approach; one that requires expertise and proficiency in developing educational programs (centralised competency-based approach); an inquiry, interactive, action-learning, and experiential approach (albeit limited) and the more traditional textbook-driven method (UNESCO/UNICEF, 2012). Patel (2008) notes that having critical analytic and problem-solving skills is essential to DRR education. Of the different approaches, the most common is knowledge transmission from teachers to learners or students, perceived as the traditional teaching style (Arvai, 2014). A common critique of the traditional teaching approach is that it creates knowledge disparity/gap between students and other learners and their teachers/tutors (Nakano & Yamori, 2021). Contrarily, there is abundant research evidence demonstrating that more effective DRR educational programs are driven by community engagements, programs and activities in risk awareness and mitigation strategies (Aghaei et al., 2018; Chen & Lee, 2012; Perry & Lindell, 2003). In this era of digital communication, community education can be facilitated through different outlets including the media (TV, radio etc.) and many other online Platforms (Aghaei et al., 2018) such as Facebook, Instagram, Twitter, etc. Nevertheless, social media can be used for deception, telling lies, misinformation, and promoting fraudulent activities, hence, the need for caution (Patel, 2008). A recent study (Bang, 2023)

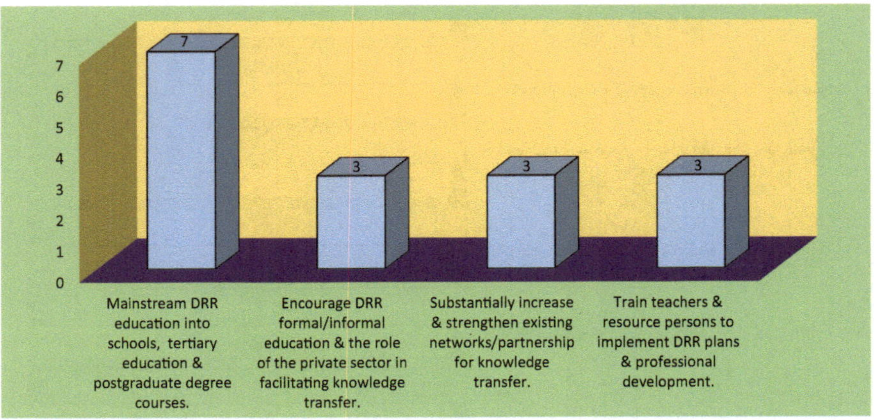

Fig. 7.3 The most prominent educational approaches. *Source* Bang (2023)

revealed the most common educational approaches as shown in Fig. 7.3. The analysis of the approaches in the various Platforms reveals community engagement and media use are the most popular educational approaches discussed, albeit mentioned just a few times.

Educational Planning

Authors (Chen & Lee, 2012) have noted that education stakeholders such as the appropriate government officials; DRR professionals, scholars and experts (people with specialist knowledge on various aspects of DRR/CC); community leaders, and the populace are instrumental in designing/developing and administering DRR educational programs at the local, regional and national levels. Differentiating the community groups/people requiring DRR/CC education while also noting their educational needs is essential (Martin et al., 1984; Lopez et al., 2016). The literature has also identified other essential peculiarities required for planning such as experiences, skills, age, gender, political, religious and socio-economic factors (Apronti et al., 2015; Perry and Lindell, 2003; Aghaei et al., 2018). Patel (2008) underscored the role of synergies between indigenous/local, and scientific/technical knowledge in informing the planning process. An analysis of the Platforms shows that educational planning manifests most and the most prominent educational determinants are integrating DRR education into the school curriculum; the private, informal, and formal sectors; and establishing a resilience/safety culture. Discussion on assimilating DRR education at all levels and sectors of society features in almost all the Platforms, including the recent meeting where Mauritius stipulated the country's intention to inculcate DRR education into the educational system at the primary level. Reiterating this point in successive Platforms could imply its application remains minimal

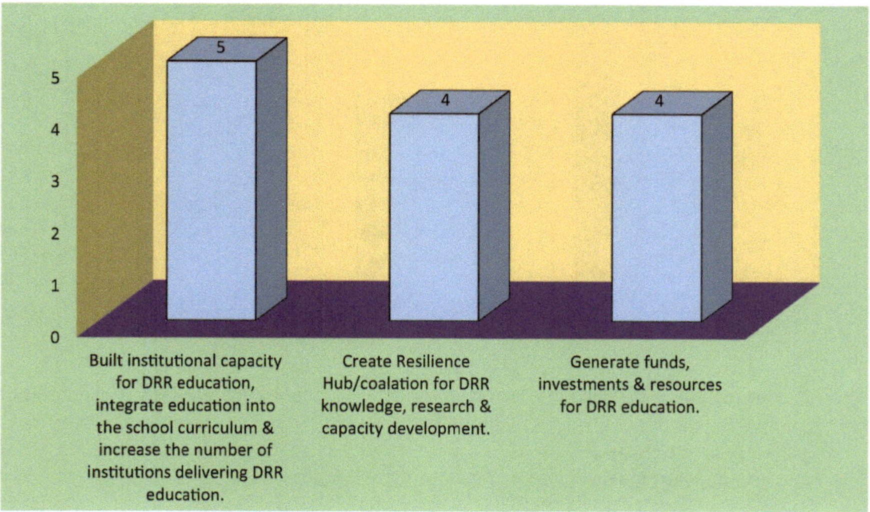

Fig. 7.4 The most prominent educational planning. *Source* Adapted from Bang (2023)

in most countries. An analysis of the most prominent educational planning ideas is presented in Fig. 7.4.

Educational Tools

Various DRR educational tools are suitable for different groups and sectors of society. Effective training for students utilises multimedia tools such as graphics, video, audio, and animations; and other learning materials such as textbooks, and manuals including multimedia and online Platforms. These tools have also been found to be appropriate for educating children, including posters, painting, 3D simulations, animations, and exhibitions (Apronti et al., 2015; Grunwald and Corsbie-Massayand, 2006; Efthymis, 2014). Educative DRR/CC interactive learning programs designed for children with different age groups (Aghaei et al., 2018) and educative DRR media adverts/programs run over the radio and TV increase DRR awareness and knowledge (Fisch, 2000). Statistics from Fletcher's (1990) investigation revealed using different education tools in children facilitates learning—children retain 20% of what they learn through hearing, and 40% of what they learn visually and studies that combinations of interactive, auditory and visual activities constitute 75% of what stays in their memory. Hence, the recommendation for bespoke educational tools for children is aligned with their age, skills or any special requirements (Aghaei et al., 2018). Research evidence suggests viable educational tools for women should consider participatory approaches including regular community campaigns/meetings including through the use of online social media tools and Platforms (Apronti et al.,

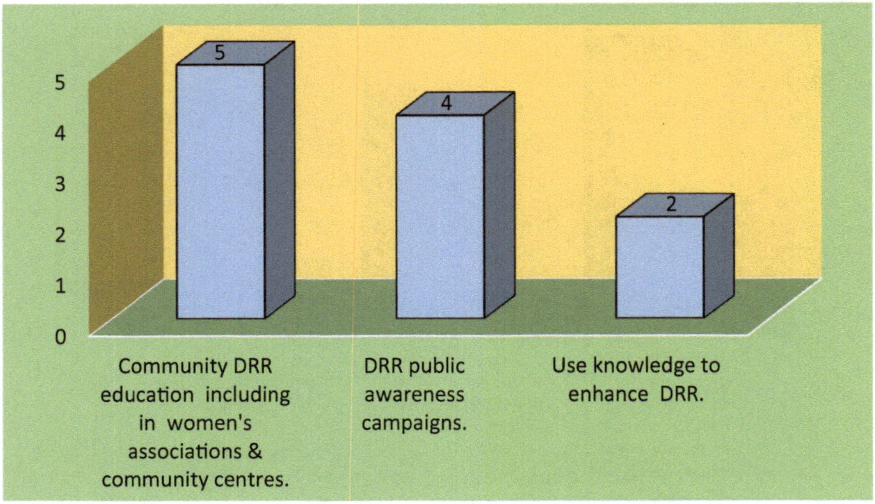

Fig. 7.5 The most prominent educational tools. *Source* Adapted from Bang (2023)

2015; Redmond and Radjak, 2014; Meesters et al., 2009). UNESCO/UNICEF (2012) emphasized participatory methodologies on the premise that they enhance active/experiential learning such as doing risk (hazard, vulnerability, exposure and capacity) assessment. Formal teaching in schools is often informed through academic textbooks, manuals and other scholarly tools such as publications on the subject (ibid). To galvanise DRR knowledge, ProventionWeb, in collaboration with the UNISDR developed a universal easily accessible library in DRR awareness and education that informs DRR/CC technical/scientific, and Indigenous and other knowledge systems (UNDRR, 2019). An in-depth analysis of the Platforms (Bang, 2023) reveals minimal and inadequate discussions on DRR/CC educational tools. Indigenous knowledge, community and public sensitisation/awareness campaigns were identified as the most common instruments used (ibid; see Fig. 7.5).

Educational Content

The goal of DRR education is to transmit knowledge on socio-environmental conditions and human actions/decisions that may enhance or limit hazards/disasters. Hence, the principal aim is to ensure the educational content can articulate disaster risks to different societal groups (families, agencies, departments, communities, policymakers, individuals, etc.), non-action and action to mitigate vulnerabilities and to recognize society's role in creating a resilience culture (Patel, 2008; UNISDR, 2005, 2015). Just like educational tools, it has been determined that bespoke DRR education content is best for different community groups. The appropriate educational needs

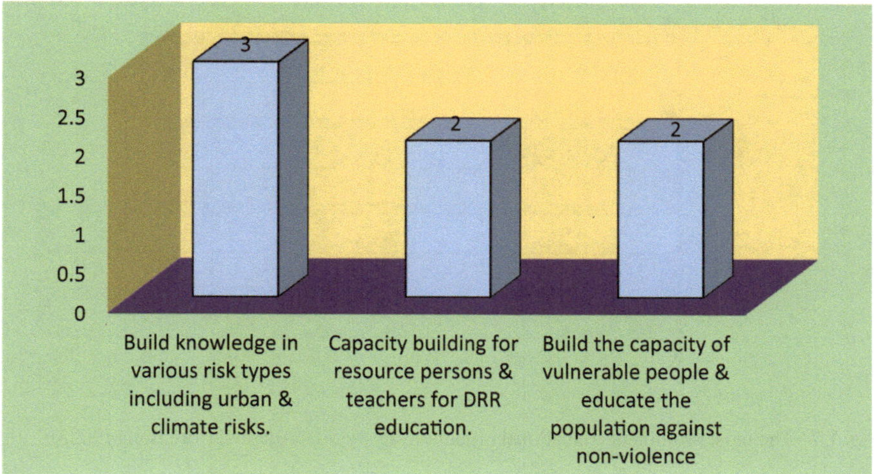

Fig. 7.6 The most prominent educational content. *Source* Adapted from Bang (2023)

of the various groups can be known from educational advisors, DRR education-related textbooks, catalogues, online resources, and community leaders conversant with the profile and interests of target communities (Hogan, 1997). The content can be based on the demographic structure of the population, socioeconomic and cultural status, risk perception, gender, or employment status (Apronti et al., 2015). This is vital to ensure that DRR education meets the requirements of the appropriate groups since the needs might evolve. Integrating disaster-related topics/themes into specific school subjects such as the natural and physical sciences is popular in many countries (UNESCO/UNICEF, 2012). Risk assessment, including hazard, vulnerability, and capacity identification at all levels is a core DRR subject which according to Patel (2008) should go hand in hand with DRR planning. The analysis shows that articulation of DRR educational content is generally sparse in the ARPs with the most mentioned relating to themes on DRR, resilience, safety, disaster preparedness, and response. Figure 7.6 is an analysis of the most prominent education content.

Educational Organisations

Educational organisations, agencies, or bodies are vital in developing, crafting, and administering DRR education. DRR/CC education can be under the authority of various ministries with tailored responsibilities such as education, health and research (Chen and Lee, 2012; FitzGerald et al., 2010) but the Ministry of Education in most countries is in charge of disaster/CC education (Apronti et al., 2015; Chen & Lee, 2012). Disaster and emergency management agencies such as rescue and safety agencies (firefighters, police, military), disaster service providers such as NGOs

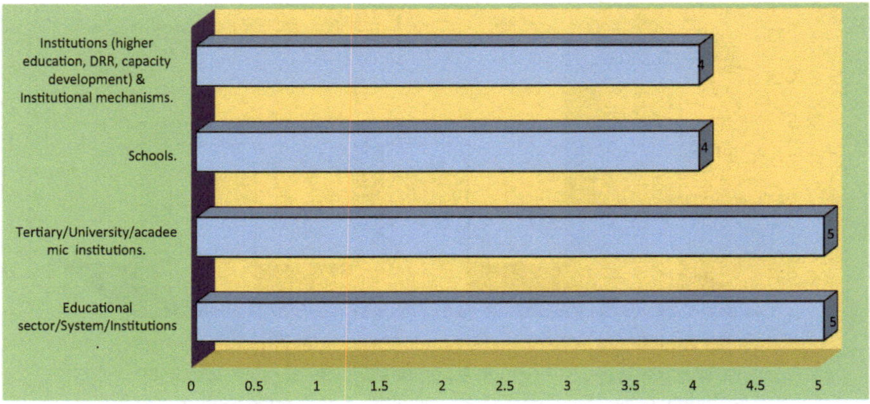

Fig. 7.7 The most prominent educational organisations. *Source* Adapted from Bang (2023)

and community/civil organisations can also be instrumental in fostering educational activities for risk reduction (Aghaei et al., 2018; Perry and Lindell, 2003; Siripong, 2010). In addition, Aitsi-Selmi et al. (2015) underlined the importance of financial and legal frameworks in boosting DRR education. Themes on educational organisations and frameworks were not identified in the ARP, nevertheless, the most prominent educational organisations are presented in Fig. 7.7.

Educational Challenges

Without diagnosing the issues plaguing DRR education in Africa, it will be challenging to mitigate them. The main challenges are limited assimilation of disaster mitigation/response in education programs; lack of impetus and resources for DRR education; divided views amongst stakeholders on how to approach DRR/CC education; non-application of DRR policies and plans; the inherent inabilities and limitations preventing a majority of the population from engaging in DRR educational activities; lack of DRR/CC skills, awareness and knowledge amongst key decision makers and sometimes in DRR educators, tutors, teachers, lecturers etc. (Aghaei et al., 2018; Bang, 2023; Chen & Lee, 2012; Porter & Graham, 2016). Although the training of teachers has been perceived as invaluable in enhancing the quality of education (Bonifacio et al., 2010), professional training of disaster educators remains limited (Chan and Lee, 2012). Language, rigid institutional/legislative frameworks, socio-economic challenges and inadequate collaboration and coordination amongst stakeholders are other barriers that need attention (Apronti et al., 2015; Gilbert, 2005). The UNFCC had noted the minimal inclusion of CC mitigation/adaptation in education programs, a view affirmed in this study since the analysis of the Platforms revealed limited discourse on DRR education challenges. The 5th ARP for DRR highlights meagre DRR educational resources limited engagements with higher

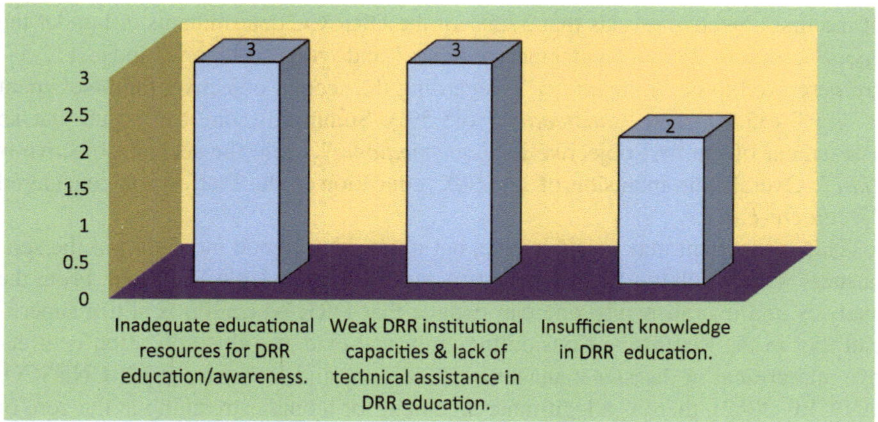

Fig. 7.8 The most prominent educational challenges. *Source* Adapted from Bang (2023)

education institutions and public awareness as impediments. According to a recent analysis, inadequate educational resources for DRR education/awareness, weak DRR institutional capacities lack of technical assistance in DRR education and insufficient knowledge in DRR education have been identified as the most prominent educational challenges (Bang, 2023; Fig. 7.8). If African countries had embedded DRR education into their educational frameworks, one would expect them to regularly review progress made during the Platforms, hence discussing challenges encountered and possible solutions.

Overall Assessment of DRR Education as a Critical Tool in the Platforms

A summary of the assessment and analysis made from the key enquiries in this chapter is presented in Table 7.1. All the 9 regional Platforms in the Table were assessed for the two main objectives of this study. First, the extent of assimilation of the DRR/CC education determinants during the Platforms (check the last row of the table) and second, the depth to which each of the determinants was discussed during the various Platforms (check the last column of the table). The depth to which each determinant was discussed during the Platforms was the criteria used for the first objective (assessed horizontally in Table 7.1). The second objective was informed by the total number of determinants discussed in each Platform, and how comprehensively the specific determinant was discussed (assessed vertically in Table 7.1).

In assessing both objectives, a Likert scale grading having attributes of "*excellent*", "*good*", "*satisfactory*", "*fair*", "*mediocre*" and "*poor*", was employed in the analysis with "*excellent*" at the highest end and "*poor*" at the lowest end. Statistical analysis of the first objective reveals that 37.5% of the DRR/CC determinants fall under the "*poor*" category, 25% fall under the "mediocre" and "good" categories and just 12.5% are perceived to be "*satisfactory*". Regarding the second objective, the assessment is "*fair*" (55.5%) and "*satisfactory*" (45.5%). Summarily, one can argue that an assessment of the first objective is "*poor-mediocre*" while the second objective is "*fair*". Overall, the inclusion of DRR/CC education in the Platform is considered "*Mediocre-Fair*".

This assessment may be subjective, but it provides a good indication of the seriousness with which the ARPs have taken education as a viable DRR tool. From the analysis and overall result, one can deduce that DRR education is still a superficial tool in the Platforms. Considering empirical evidence that education is effective and critical in disaster management (Adiyoso and Kanegae, 2012; UNESCO/UNICEF, 2012), there is a legitimate argument for its mainstreaming at the core of the Platforms' risk mitigation strategy.

Conclusion: Prospects for Integrating DRR Education into the Platforms

The discourse in this chapter has focused on how DRR/CC education is perceived in the ARP. The central role of risk reduction education in CC and disaster reduction is underscored by the relevant education actions and practices required for enhanced climate and risk reduction actions within African countries. Using the determinants of disaster/climate education, the chapter analysed and assessed the extent and/or depth/scope to which the ARPs have assimilated DRR education. By applying a Likert scale analysis, an assessment of the level of assimilation of climate/disaster education in the various Platforms has been determined "*mediocre- fair*".

It can be deduced from the findings that the impact of education on DRR/CC risk mitigation is not systematically evaluated during High-Level ministerial and regional gatherings. Arguably, limited feedback on the implementation of CC/DRR educational efforts from the delegates indicates that either education in risk reduction is not considered a key tool for DRM, or the relevant stakeholders do not collect education-related data, information, or facts. This has implications for effectively mitigating disaster risks in Africa, especially mainstreaming CC/DRR education into school/academic programs.

Consequently, this study recommends prioritisation of CC/DRR education during the meetings and explicit discourse on the subject to provide more insights to the participating countries. This chapter contributes to literature and knowledge on measures to enhance climate/disaster education in Africa with implications for other developing countries. In particular, the findings inform country-level CC/DRR educational policy and professional practice at the local, state (provincial) or national levels. It provides a roadmap for strategic discussions on the essence of assimilating risk reduction education as an essential and inevitable DRM instrument in Africa.

References

Adiyoso, W., & Kanegae H. (2010). The effect of different disaster education programs on tsunami. *Disaster Mitigation of Cultural Heritage and Historic Cities, 6*, 165–172. file:///C:/Users/hbang/Downloads/dmuch6_23%20(1).pdf.

AFRP Bulletin. (2014). A summary report of the fifth Africa regional Platform for disaster risk reduction and third ministerial meeting for disaster risk reduction. https://www.unisdr.org/files/35308_5thafricaregionalPlatform.pdf

AFRP Bulletin. (2016). 6th session of the Africa regional Platform and 5th High-Level meeting on disaster risk reduction. Retrieved November 22–25, 2016. enbplus141num8e.pdf (amazonaws.com)

Aghaei, N., Seyedin, H., & Sanaeinasab, H. (2018). Strategies for disaster risk reduction education: A systematic review. *Journal of Education and Health Promotion, 7*, 98.

Aitsi-Selmi, A., Egawa, S., Sasaki, H., Wannous, C., & Murray, V. (2015). The Sendai framework for disaster risk reduction: Renewing the global commitment to people's resilience, health, and well-being. *International Journal of Disaster Risk Science, 6*(2), 164–176. https://doi.org/10.1007/s13753-015-0050-9

Apronti, P. T., Osamu, S., Otsuki, K., & Kranjac-Berisavljevic, G. (2015). Education for disaster risk reduction (DRR): Linking theory with practice in Ghana's basic schools. *Sustainability, 7*(7), 9160–9186. https://doi.org/10.3390/su7079160

ARC. (2009). ARC Briefing Note on the second Africa regional Platform for disaster risk reduction consultative meeting. https://enb.iisd.org/africa/brief/briefing1901e.html

Arvai, J. (2014). The end of risk communication as we know it. *Journal of Risk Research, 17*(10). https://doi.org/10.1080/13669877.2014.919519

Asian Disaster Preparedness Center. (2008). *A study on the impact of disasters on the education sector in Cambodia.* Asian Disaster Preparedness Center.

AU and UNISDR. (2011). Africa report to the third global Platform for disaster risk reduction. Retrieved May 9–13, 2011. https://www.preventionweb.net/files/globalPlatform/entry_bg_paper~finaldraftafricareporttothegpeng4may2010pb.pdf

AU. (2010). Report of the second ministerial conference on disaster risk reduction. Retrieved April 14–16, 2010. https://www.preventionweb.net/files/18733_englishreport.pdf

AU. (2017). Programme of action for the implementation of the sendai framework for disaster risk reduction 2015–2030 in Africa. https://www.preventionweb.net/files/67054_PoAimplementationofthesendaiframewo%5B1%5D.pdf

Bang, H., Miles, L., & Gordon, R. (2019). Disaster risk reduction in Cameroon: Are contemporary disaster management frameworks accommodating the Sendai Framework Agenda 2030? *International Journal of Disaster Risk Sciences, 10*, 462–477. https://doi.org/10.1007/s13753-019-00238-w

Bang, H., & Burton, N. (2021). Contemporary flood risk perceptions in England: Implications for flood risk management foresight. *Climate Risk Management, 32*. https://doi.org/10.1016/j.crm.2021.100317

Bang, H. (2023). Strengthening disaster management discourse: An evaluation of disaster education in Africa's regional Platforms for disaster risk reduction. *African Journal of Education and Practice, 9*(3), 1–29. https://doi.org/10.47604/ajep.2030

Bonifacio, A. C., Takeuchi, Y., & Shaw, R. (2010). Mainstreaming climate change adaptation and disaster risk reduction through school education: Perspectives and challenges. In R. Shaw, J. M. Pulhin, & J. Jacqueline Pereira (Eds), *Climate change adaptation and disaster risk reduction: issues and challenges* (vol 4, pp 143–169). https://doi.org/10.1108/S2040-7262(2010)0000004013

Bosschaart, A., van der Schee, J., Kuiper, W., & Schoonenboom, J. (2016). Evaluating a flood-risk education program in the Netherlands. *Studies in Educational Evaluation, 50*, 53–61. https://doi.org/10.1016/j.stueduc.2016.07.002

Bray, T. (2000). *Community partnerships in education: Dimensions, variations and implications.* http://web.worldbank.org/archive/website00238I/WEB/PDF/COMMUNIT.PDF

Briceño, S. (2015). Looking back and beyond Sendai: 25 years of international policy experience on disaster risk reduction. *International Journal of Disaster Risk Science, 6*(1), 1–7. https://doi.org/10.1007/s13753-015-0040-y

Chen, C. Y., & Lee, W. C. (2012). Damages to school infrastructure and development to disaster prevention education strategy after Typhoon Morakot in Taiwan. *Disaster Prevention and Management: An International Journal., 21*(5), 541–555. https://doi.org/10.1108/09653561211278680

Cherniack, E. P. (2008). The impact of natural disasters on the elderly. *American Journal of Disaster Medicine, 3*(3), 133–139. https://doi.org/10.5055/ajdm.2008.0018

Delicado, A., Rowland, J., Fonseca, S., et al. (2017). Children in disaster risk reduction in Portugal: Policies, education, and (Non) participation. *International Journal of Disaster Risk Science, 8*, 246–257. https://doi.org/10.1007/s13753-017-0138-5

Dolatabadi, Z. A., Seyedin, H., & Aryankhesal, A. (2016). Policies on protecting vulnerable people during disasters in Iran: A document analysis. *Trauma monthly, 21*(3). https://doi.org/10.5812/traumamon.31341

DRR Bulletin. (2021). Eighth Africa regional Platform for disaster risk reduction. Retrieved November, 16–19, 2021. https://afrp.undrr.org/sites/default/files/2021-12/8th_africa_regional_drr_0.pdf

Efthymis, L. (2014). Disaster data centre—an innovative educational tool for disaster reduction through education in schools. *Journal of Power and Energy Engineering, 2*(9), 35–40. https://doi.org/10.4236/jpee.2014.29006

Eisenman, D. P., Cordasco, K. M., Asch, S., Golden, J. F., & Glik, D. (2007). Disaster planning and risk communication with vulnerable communities: lessons from Hurricane Katrina. *American Journal of Public Health, 97*(Supplement_1), S109-S115. 10.2105%2FAJPH.2005.084335

Faber, M. H., Giuliani, L., Revez, A., Jayasena, S., Sparf, J., Mendez, J. M. (2014). Interdisciplinary approach to disaster resilience education and research. *Procedia Economics and Finance, 18*, 601–609. https://doi.org/10.1016/S2212-5671(14)00981-2

Fisch, S. M. (2000). A capacity model of children's comprehension of educational content on television. *Media Psychology, 2*(1), 63–91. https://doi.org/10.1207/S1532785XMEP0201_4

FitzGerald, G. J., Aitken, P., Arbon, P., Archer, F., Cooper, D., Leggat, P., Myers, C., Robertson, A., Tarrant, M., & Davis, E. R. (2010). A national framework for disaster health education in Australia. *Prehospital and Disaster Medicine, 25*(1), 4–11. https://doi.org/10.1017/s1049023x00007585

Fletcher, J. D. (1990). Effectiveness and cost of interactive videodisc instruction in defense training and education. Institute for Defense Analyses Alexandria VA. https://apps.dtic.mil/sti/citations/ADA228387

Gall, M., Nguyen, K. H., & Cutter, S. L. (2015). Integrated research on disaster risk: Is it really integrated? *International Journal of Disaster Risk Reduction, 12*, 255–267. https://doi.org/10.1016/j.ijdrr.2015.01.010

Gilbert, J. H. (2005). Interprofessional learning and higher education structural barriers. *Journal of Interprofessional Care, 19*(sup1), 87–106. https://doi.org/10.1080/13561820500067132

Grunwald, T., & Corsbie-Massay, C. (2006). Guidelines for cognitively efficient multimedia learning tools: Educational strategies, cognitive load, and interface design. *Academic Medicine, 81*(3), 213–223. https://doi.org/10.1097/00001888-200603000-00003

Hogan, L. (1997). Planning responsibly for adult education. A guide to negotiating power and interests. *The Journal of Higher Education, 68*(3), 359–361. https://doi.org/10.1080/00221546.1997.11778988

IPCC. (2020). *The concept of risk in the IPCC Sixth Assessment Report: a summary of cross-Working Group discussions.* https://www.ipcc.ch/site/assets/uploads/2021/02/Risk-guidance-FINAL_15Feb2021.pdf

Kagawa, F., & Selby, D. (2012). Ready for the storm: Education for disaster risk reduction and climate change adaptation and mitigation1. *Journal of Education for Sustainable Development, 6*(2), 207–217. https://doi.org/10.1177/0973408212475200

Lee, J., Lee, Y., Ryu, Y., & Kang, T. H. (2007). Information quality drivers of KMS. In *2007 International conference on convergence information technology (ICCIT 2007)* (pp. 1494–1499). IEEE. https://doi.org/10.1109/ICCIT.2007.271

Lopez, L. M., Grey, T. W., Tolley, E. E., & Chen, M. (2016). Brief educational strategies for improving contraception use in young people. *Cochrane Database of Systematic Reviews*, (3). https://doi.org/10.1002/14651858.cd012025.pub2l

Martin, B. G., Rolen, H. B., & Goodman, D. C. (1984). Educational strategies for prospective payment. *Laboratory Medicine, 15*(8), 551–553. https://doi.org/10.1093/labmed/15.8.551

Meesters, J. J., Vliet Vlieland, T. P., Hill, J., & Ndosi, M. E. (2009). Measuring educational needs among patients with rheumatoid arthritis using the Dutch version of the Educational Needs Assessment Tool (DENAT). *Clinical Rheumatology, 28*(9), 1073–1077. https://doi.org/10.1007/s10067-009-1190-3

Morrow, B. H. (1999). Identifying and mapping community vulnerability. *Disasters, 23*(1), 1–18. https://doi.org/10.1111/1467-7717.00102

Muttarak, R., & Lutz, W. (2014). Is education a key to reducing vulnerability to natural disasters and hence unavoidable climate change? *Ecology and society, 19*(1). https://doi.org/10.5751/ES-06476-190142

Muttarak, R., & Pothisiri, W. (2013). The role of education on disaster preparedness: case study of 2012 Indian Ocean earthquakes on Thailand's Andaman Coast. *Ecology and Society*, 18(4). https://doi.org/10.5751/ES-06101-180451

Muzenda-Mudavanhu, C., Manyena, B., & Collins, A. E. (2016). Disaster risk reduction knowledge among children in Muzarabani District, Zimbabwe. *Natural Hazards, 84*(2), 911–931. https://doi.org/10.1007/s11069-016-2465-z

Nakano, G., & Yamori, K. (2021). Disaster risk reduction education that enhances the proactive attitudes of learners: A bridge between knowledge and behaviour. *International Journal of Disaster Risk Sciences, 66*. https://doi.org/10.1016/j.ijdrr.2021.102620

Nonaka, I., & Takeuchi, H. (1995). *The knowledge creating company*. Oxford University Press.

O'Brien, G., O'Keefe, P., Rose, J., & Wisner, B. (2006). Climate change and disaster management. *Disasters, 30*, 64–80. https://doi.org/10.1111/j.1467-9523.2006.00307.x

Patel, M. (2008). Disaster risk reduction education. In R. Shaw, R. Krishnamurty (Eds.), *Disaster management: global challenges and local solutions*. University Press. https://www.researchgate.net/publication/277821689_Disaster_Risk_Reduction_Education

Perry, R., & Lindell, M. (2003). Preparedness for emergency response: Guidelines for the emergency planning process. *Disasters, 27*(4), 336–350. https://doi.org/10.1111/j.0361-3666.2003.00237.x

Porter, W. W., & Graham, C. R. (2016). Institutional drivers and barriers to faculty adoption of blended learning in higher education. *British Journal of Educational Technology, 47*(4), 748–762. https://doi.org/10.1111/bjet.12269

Prashar, S., Shaw, R., & Takeuchi, Y. (2013). Community action planning in East Delhi: A participatory approach to build urban disaster resilience. *Mitigation and Adaptation Strategies for Global Change, 18*(4), 429–448. https://doi.org/10.1007/s11027-012-9368-4

Preston, J. (2012). What is disaster education? In J. Preston (Ed.), *Disaster education*. SensePublishers. https://doi.org/10.1007/978-94-6091-873-5_1

Redmond, A. D., & Radjak, A. (2014). Development of evidence-based technical guidance and evidence-based technical guidance and education/training programs for the advancement of health and disaster risk management capabilities. *Disaster Medicine and Public Health Preparedness, 8*(4), 369–371. https://doi.org/10.1017/dmp.2014.73

Rohrmann, B. (2008). Risk perception, risk attitude, risk communication, risk management: A conceptual appraisal. In *15th International Emergency Management Society (TIEMS) Annual Conference* (vol. 2008).

Sawada, Y. (2007). The impact of natural and manmade disasters on household welfare. *Agricultural Economists, 37*(1), 59–72. https://doi.org/10.1111/j.1574-0862.2007.00235.x

Shaw, R., & Krishnamurthy, R. (2010). Disaster management: global challenges and local solutions. *Disaster Prevention and Management, 19*(4). https://doi.org/10.1108/dpm.2010.19.4.518.7

Siripong, A. (2010). Education for disaster risk reduction in Thailand. *Journal of Earthquake and Tsunami, 4*(02), 61–72. https://doi.org/10.1142/S1793431110000716

Tarazona, M., & Gallegos, J. (2011). Recent trends in disaster impacts on child welfare and development 1999–2009. In *Global assessment report on disaster risk reduction.* Children in a Changing Climate. https://www.preventionweb.net/english/hyogo/gar/2011/en/bgdocs/Tarazona_&_Gallegos_2010.pdf

Thomas, N., Leander-Griffith, M., Harp, V., & Cioffi, J. (2015). Influences of preparedness knowledge and beliefs on household disaster preparedness. *Morbidity and Mortality Weekly Report, 64*(35), 965–971. https://doi.org/10.15585/mmwr.mm6435a2

Torani, S., Majd, P. M., Maroufi, S. S., Dowlati, M., & Sheikhi, R. A. (2019). The importance of education on disasters and emergencies: A review article. *Journal of Education and Health Promotion., 64*(35), 965–971. https://doi.org/10.4103/jehp.jehp_262_18

Tuladhar, G., Yatabe, R., Dahal, R. K., & Bhandary, N. P. (2015). Assessment of disaster risk reduction knowledge of schoolteachers in Nepal. *International Journal of Health System and Disaster Management, 3*(1), 20. https://doi.org/10.4103/2347-9019.147142

UN. (2020). *Climate change is an increasing threat to Africa.* https://unfccc.int/news/climate-change-is-an-increasing-threat-to-africa

UN. (2022). *Introduction to climate action.* https://unfccc.int/climate-action/introduction-climate-action

UNCCC. (2022). *Joint work on implementation of climate action on agriculture and food security.* file:///C:/Users/hbang/Downloads/cop27_auv_3ab_Koronivia.pdf

UNDRR. (2019). *Preventionweb: the knowledge platform for disaster risk reduction: Learning about disaster risk reduction.* https://sdgintegration.undp.org/preventionweb-knowledge-Platform-disaster-risk-reduction

UNDRR. (2021). *Eighth Africa regional Platform for disaster risk reduction: Towards disaster risk-informed development for a resilient Africa in a COVID-19 transformed world.* https://afrp.undrr.org/#:~:text=The%20biennial%20Africa%20Regional%20Platform,learned%20on%20disaster%20risk%20reduction

UNESCO. (2011). Disaster risk reduction in education. an imperative for education policymakers. https://unesdoc.unesco.org/ark:/48223/pf0000213925

UNESCO/UNICEF. (2012). *Disaster risk reduction in school curricula: Case studies from thirty countries.* https://www.preventionweb.net/files/26470_drrincurriculamapping30countriesfin.pdf

UNFCCC. (2011). *Fact sheet: Climate change science–the status of climate change science today.* https://unfccc.int/files/press/backgrounders/application/pdf/press_factsh_science.pdf

UNICEF. (2019). *It is getting hot. Call for education systems to respond to the climate crisis.* https://www.unicef.org/eap/media/4596/file/It%20is%20getting%20hot:%20Call%20for%20education%20systems%20to%20respond%20to%20the%20climate%20crisis.pdf

UNISDR Bulletin. (2018). *The Africa-Arab platform on disaster risk reduction.* Retrieved October 9–13, 2018. enbplus141num12e.pdf (amazonaws.com).

UNISDR. (2005). Hyogo framework for action 2005–2015: building the resilience of nations and communities to disasters. https://www.unisdr.org/2005/wcdr/intergover/official-doc/L-docs/Hyogo-framework-for-action-english.pdf

UNISDR. (2013). Report of the 4th Africa regional Platform on disaster risk reduction. https://www.preventionweb.net/files/30143_4thafrpproceedingsreport.pdf

UNISDR. (2015). *Sendai framework for disaster risk reduction.* https://www.preventionweb.net/files/43291_sendaiframeworkfordrren.pdf

UNISDR. (2017). *Sustainable development: Risk reduction.* Report of the open-ended intergovernmental expert working group on indicators and terminology relating to disaster risk reduction, UN A/71/644,16–21184 (E). file:///H:/50683_oiewgreportenglish%20(1).pdf

Weichselgartner and Obersteiner, 2002 Weichselgartner, J., & Obersteiner, M. (2002). Knowing sufficient and applying more: Challenges in hazards management *Global Environmental Change Part b: Environmental Hazards, 4*(2–3), 73–77. https://doi.org/10.3763/ehaz.2002.0407

Weichselgartner, J., & Pigeon, P. (2015). The role of knowledge in disaster risk reduction. *International Journal of Disaster Risk Science, 6*, 107–116. https://doi.org/10.1007/s13753-015-0052-7

Wing, E., Bates, D., Smith, M., Sampson, C., Johnson, A., Fargione, J., & Morefield, P. (2018). Estimates of present and future flood risk in the conterminous United States. *Environmental Research Letters, 13*(3), 1–7. https://doi.org/10.1088/1748-9326/aaac65

Wisner, B. (2006). A Review of the role of education and knowledge in disaster risk reduction. https://www.researchgate.net/publication/44836374_Let_our_children_teach_us_a_review_of_the_role_of_education_and_knowledge_in_disaster_risk_reduction

Chapter 8
Gender and Climate Change Education in sub-Saharan Africa as the Missing Component in Climate Change Adaptation for an Effective Management of Natural Resources

Henry Bikwibili Tantoh, Suiven John Paul Tume, Nyong Princely Awazi, and Tracey J. M. McKay

Abstract Natural resources remain fundamental to rural livelihoods and wellbeing in Africa, where they serve as foundation of economic development. However, climate change continues to exert pressure on natural resources particularly, water resources which are central to socio-economic development. Hence, the effects of climate change affect different demographic groups differently. Rural women, for example, who are amongst the poorest are one of the most marginalized groups and vulnerable in terms of education about climate change. Given the fast pace at which the world is changing, rural women have to contend with the shortages and difficulties of water, while these problems are exacerbated by their limited level of education and challenges posed by a rising population. Most significant among

H. B. Tantoh (✉)
Department of Environmental Sciences, College of Agriculture and Environmental Sciences, University of South Africa, Pretoria, South Africa
e-mail: bikwibilith@gmail.com; tantohb@unisa.ac.za

H. B. Tantoh · S. J. P. Tume
Department of Geography and Planning, Faculty of Arts, The University of Bamenda, Bamenda, Cameroon
e-mail: suiven.john@uniba.cm

N. P. Awazi
Department of Forestry and Wildlife Technology, The University of Bamenda, Bamenda, Cameroon

FOKABS, Ottawa, Canada

FOKABS, Yaounde, Cameroon

N. P. Awazi
e-mail: awazinyong@uniba.cm; pnyong@fokabs.com

T. J. M. McKay
Olympus Online Education, Gauteng, South Africa
e-mail: tracey@olympus.org.za

M. F. Mbah et al. (eds.), *Practices, Perceptions and Prospects for Climate Change Education in Africa*, https://doi.org/10.1007/978-3-031-84081-4_8

these is how to effectively improve the education of rural women who are vital in natural resource management to meet their needs and those of their communities. Inequalities in land tenure and access to natural resources, is everywhere governed by power dynamics that often contribute to the marginalization of rural women in poor communities. In addition, studies on adaptive capacity and differentiated-gender vulnerability are limited. This study seeks to investigate the role of gender education in climate change adaptation and effective natural resources management. A systematic review of academic literature consisted in the methodology adopted for this study. The results indicated that a gendered approach that values the capacities, limits and vulnerabilities of rural women is required for effective natural resource management and climate change adaptation.

Keywords Climate change · Climate change adaptation · Gender · Natural resource management · Rural · Women

Introduction

Climate change is a global phenomenon and represents one of the fundamental ultimate challenges of the twenty-first century (Malala Fund, 2021; Vincent, 2007). Several academic studies have documented that climate change is real, as demonstrated by rising temperatures, fluctuating patterns of rainfall, increasing incidences of drought episodes, inundation, and hurricanes among others (IPCC, 2014; Tantoh et al., 2022a). It is a fact that most of the warming occurring over the past five decades is a human-induced phenomenon. Thus, global temperatures have risen to about 0.6 °C over the past century resulting in significant changes in precipitation patterns (IPCC, 2021). These shifts have an adverse impact on the primary, secondary and tertiary sectors, especially in the economies of sub-Saharan Africa (SSA). This state of affairs has intensified unending difficulties related to the degradation of natural resources and ecosystems, spurring food insecurity, increasing the numbers of environmental refugees, while boosting the emergence of new zoonotic diseases (Malala Fund, 2021). Consequently, rural communities and particularly women are mostly affected. Increasing temperatures have further deteriorated rural livelihood and wellbeing intensifying inequalities, including gender-based inequality (Newman and Smith, 2021). This has worsened by an inadequate educational system in several low and middle-income countries that remain trapped in a state of continuous failure, with the majority of students attending school but not learning (Newman and Smith, 2021; IPCC, 2021).

Studies show that climate change affects different demographic groups, social classes, men and women disproportionately (Carvajal-Escobar et al., 2008; Mbah et al., 2021; Tantoh et al., 2021). Rural women, for example, who are amongst the poorest and the most marginalized and vulnerable groups in terms of being recipients for a climate change education. Although they have been seldom acknowledged, women in the developing world play a pivotal role in climate change adaptation and in the management of environmental resources, by collecting non-timber forest products, wood for heating and cooking, or water for domestic use. Yet, existing evidence attests the significance of recognizing gendered-related disparities in developmental projects (Lambrou & Piana, 2006; Pentlow, 2020). Also, the patriarchal structure and ideologies that shape the culture in developing countries continue unabated to maintain the invisibility of women (Carvajal-Escobar et al., 2008; Tantoh & Mckay, 2020). On the same token, the inaccessibility to education hampers a sustainable management of natural resources, particularly which they depend upon for their livelihood and wellbeing and the fact that men's jobs often pay more than that women's increase women's exposure to climate-related risks and impacts their capacity to effectively adapt, impede or improve from it (Jordan, 2019). It is argued that girls can learn critical thinking skills required to counter climate-related disasters as well as anyone else. Thus, an investment in girls' education was predicated as essential by Muttarak and Lutz (2014), to reduce the number of human deaths from droughts and inundation episodes. Further evidence suggests that the effects of education on decreasing vulnerability to climate change will translate to a reduction of the negative effects of weather-related disasters According to Streissnig et al. (2013), girls' secondary education was identified as the most important social and economic factor in reducing their vulnerability to extreme weather events, and weather-related disasters. Therefore, the impacts of extreme weather and weather-related disasters can be alleviated through efficient associated applicable and child-centered disaster risk reduction education (Sellabos et al., 2011).

Recently, researchers have shown increased interest in the **link between education, gender and climate change** (Malala Fund, 2021; Newman and Smith, 2021). It is estimated that at least 200 million adolescent girls living in the poorest communities in the developing world face a heightened risk from the effects of climate change. Evidence on the role of education and girls' to address climate change through adaptation, resilience and mitigation is limited, albeit growing necessity (Jordan, 2019; Newman and Smith, 2021). Increasing research on **gender and climate change education has been the subject of growing attention in recent years, but this approach must be considered with attention to nuances** (Sims, 2021). For example, it may be tempting to portray global education, as a means to tackle climate change and improve natural resource management and governance. While there are empirical and moral reasons to be cautious of this approach, education which builds foundational skills among women particularly can play a crucial role in building resilience to the negative impacts of global climate change in poor countries (Malala Fund, 2021).

The evidence-based on the role of education in gender mainstreaming to mitigate climate change remains limited to effectively improve natural resource management, despite a growing body of literature in these areas, especially in low and middle-income countries (Carvajal-Escobar et al., 2008; Jordan, 2019). Additional research investigates the impacts of climate change on health with inadequate analysis of the possible effects on education and how resilience can be built (Muttarak et al., 2014; Peek et al., 2018; Smith, 2021), **Indigenous Knowledge System and climate change adaptation in the developing world context** (Blankespoor et al., 2010;

Nguyen & Ross, 2017; Nalau et al., 2018; Mbah et al., 2021) **amongst others**. The overlapping inequalities faced by women (income inequality, sexual orientation, social background etc.) further compound their vulnerability for accessing education and managing natural resources (Lambrou & Piana, 2006; Sims, 2021). Given the fast pace at which the world is changing, rural women have to contend with the challenges faced in the management of natural resources which are exacerbated by the rising population and their limited level of education. A major question is how to effectively improve the education of rural women so that they may be able to adapt to the effects of climate change in natural resource management? This study, therefore, focuses on the significant role of gender education in climate change adaptation and effective natural resources management.

Gender and Climate Change Education

Climate change is a sad reality and everyone everywhere is experiencing the effects. National governments, International Development Organisations, Non-Governmental Organisations (NGOs), climate groups and scholars have attempted to reduce global warming and its effects through a series of technical solutions (improvements to energy efficiency) without success (IPCC, 2014; World Economic Forum (WEF), 2022). This is because global temperatures continue to rise (about 0.6 °C) over the past century (IPCC, 2021). These measures have not been able to adequately solve the climate emergency crisis. With the passage of time, rising population, rising urbanization, extension and intensification of agriculture and changes in consumption patterns have exerted enormous pressures on natural resources resulting in their over-exploitation and environmental degradation (Tantoh & Mckay, 2020). Natural resource extraction, for example, accounts for about 90% of global biodiversity loss and water stress and half the global greenhouse gas emission (Malala Fund, 2021). These activities have pushed both the natural and human systems beyond their boundaries, while destroying ecosystems and livelihoods (World Economic Forum (WEF), 2022).

Recent evidence suggests that gender and other drivers of vulnerability and exclusion is a fundamental determinant, and influences how climate change is experienced (Kwauk et al., 2019). For example, about 200 million teenage girls live in the poorest communities in the developing world and encounter increased risk from the effects of climate change (Atkinson & Bruce, 2015). Despite the fact that a greater proportion of women and girls live in rural poor communities, they are seldom seen as silent victims of climate change (Ravera et al., 2016). This is likely to cause misunderstanding of the causes of vulnerability and can conceal their role as influential agents of change. For example, there is growing evidence of the important role of education to support climate resilience, adaptation and mitigation. The direct and indirect impacts of education on the girl child has the probability to decrease vulnerability to climate change and environmental degradation by decreasing the impacts of weather-related disasters (Muttarak & Lutz, 2014). Recent academic studies, however, illustrate that women are increasingly been involved in more climate change learning activities than what is documented and appreciated by the community. It has been reported that women have fundamental knowledge, and skills and play significant roles in the management of natural resources while adapting to the effects of climate change (Sims, 2021; Newman & Smith, 2021).

Linking Education and Climate Change in Natural Resources Management

A long-term efficient and effective approach to preventing climate change and its effects should go past scientific innovation to address the differences and manipulative arrangements at the root of the crisis and promote instead greater social justice (Malala Fund, 2021). Given the dynamics and risks related to climate change, it has been widely acknowledged that local management and participation is the only possible option to sensitize, educate and motivate the public against risk reduction and control (Lavell et al., 2004). Therefore, vulnerable women have been recognized to play fundamental roles such as mobilizing the community in different aspects of natural resource management in Africa (Kwauk et al., 2019; Tantoh & McKay, 2020). It has been argued that rural women in sub-Saharan Africa, for example, are responsible for the provision of potable water and fuel wood for heating and cooking (Newman and Smith, 2021). A number of researchers have reported that women are crucial resource consumers and managers and their involvement in natural resource planning and management endeavours can possibly create a platform for confidence-building and improve their roles and responsibilities in decision-making and benefit sharing (Lendelvo et al., 2012; Tantoh, 2021b). They also seek solutions to access to health services and education, reducing the tendency of the vulnerability of their communities in the event of hydrometeorological episodes related to climate change coupled with establishing networks with other women and development associations to boost their social capital (Carvajal-Escobar et al., 2008).

In their domestic activities and professional development, women are habitually in a better position to identify the lark of basic resources in the household and the depreciation of natural resources, which they depend upon for their livelihood and wellbeing. For example, Harding (1998) noted that women are knowledgeable about a number of environmental hazards and the prevalence and patterns of sicknesses among children in the communities, and can immediately detect irregularities in the water during laundry (see also Peek et al., 2018). According to Fonjong (2008), this assertion is based on the idea that the traditional roles of women in rural communities of most developing countries are linked to the domestic management of water and other natural resources. In the Northwestern part of Cameroon for example, women play an essential role in participating through manual work in community-based developmental projects, managing, and maintaining community-based water supply systems (Tantoh, 2021b, 2021a). These socially constructed gender roles, which are evident and common in most rural communities across sub-Saharan Africa, are intricately linked with their socio-economic status of the community.

It known that the roles and duties of women and men vary across Africa. This is because they are shaped by diverse social and cultural norms and their association with different use and management of environmental resources (Fonjong, 2008; Lendelvo et al., 2012). Furthermore, gender-specific duties and responsibilities in food crop production besides the generation of cash income, regularly result in diverse desires, priorities, opportunities, and concerns for women and men (Sunderland et al., 2014). Conversely, women are believed to be largely responsible for collecting wild resources for household use and concentrate more on subsistence agriculture, with a specific focus on products that provide instant household-level food security (Cavendish, 2000). However, despite the fact that women seem to commodify resources such as firewood, medicinal plants, and non-forest timber products less often than men, the sale of the non-forest timber products seems to be a fundamental source of revenue for women, who do not have many of the opportunities for generating revenue that is more generally obtainable by men (Sunderland et al., 2014). Instead, the men control resources with monetary value that are seldom used but are able to generate considerable income such as wildlife and timber (Sunderland et al., 2014). Thus, women, have greater wisdom than men in managing natural resources because of their constant use of these, for their daily activities. They tend to develop a broad knowledge and understanding of nature and its processes, through their day-to-day experiences with the environment (Fonjong, 2008). Also, women tend to be disciplined when they participate in activities that are not part of their defined roles and this can prevent them from participating in future conservation ventures. This usually affects future decision-making abilities and further limits livelihood options, especially in situations where their participation is conditioned by unequal power dynamics. Furthermore, where the vulnerable and marginalized are powerless to express their views due to rigid, traditional leadership, participation and management of natural resources become compromised in most cases (Tantoh & McKay, 2020).

Climate Change as a Threat to Gender Equality in Natural Resources Management

Studies on climate change are widely documented for its effects occurring all around the world. The mitigation strategies recommended by climate-related institutions in several climate change fora on the global environment are not the same (Intergovernmental Panel on Climate Change (IPCC), 2018). However, despite the fact that climate change is universal, the effects on different sectors affect different social and demographic groups differently (Tantoh et al., 2022b). Consequently, the threats alter responsibilities and vulnerabilities and these are also distributed differently, particularly among men and women. For example, rural communities in sub-Saharan Africa share a complex socio-cultural relationship with their surroundings. These peoples rely on their environment for their livelihoods yet, such a reliance makes them vulnerable to climate change. In addition, the differences in gender roles render people vulnerable to the effects of climate change (Tantoh et al., 2021). Tiondi (2000) substantiated the veracity of this condition because gender roles are regularly underrated. This discontentment makes women invisible thus, reducing their involvement as active development agents. Still, cultural norms, and traditional practices rarely consider the important contributions of women in the management of natural resources such as water, forests, and food production (Tantoh & McKay, 2020).

Further evidence highlights future threats to gender equality and the vulnerability to climate crisis urging for investments and advances in gender education for climate change (Chigwanda, 2016). Similarly, disparities in societies and drivers of vulnerability, continue to affect women, particularly after weather-related disasters. This makes women even more vulnerable to the effects of climate change and exposes them more directly to the suffering from environmental degradation (Villamor et al., 2018). Still, in the domain of natural resources, water scarcity caused by drought, can negatively impact adolescent girls' education because of challenges in managing menstrual hygiene, which can prevent them from going to school (Chigwanda, 2016). In addition to this, impacts on livelihoods and household finances can result in sanitary products not being purchased. This is amplified by their natural weakness and the socio-cultural roles assigned to women in Africa (Tantoh & McKay, 2020). The rational of the connection between rural women to the natural environment and the effects of climate change in sub-Saharan Africa depicts the significance of appropriate and comprehensive feminist policy frameworks (Tiondi, 2000).

Gender Education, Climate Change Adaptation and Resilience in Natural Resources Management

The risks associated with environmental degradation, climate change and natural resource management can only be attenuated with an inclusion of women in local participation, which is essential to create awareness, knowledge and incentives for effective management (Sims, 2021). At this level, women have been identified to play exceptional roles and several studies have reported about their capabilities to initiate, mobilize and participate in different stages of the developmental process (Galvin, 2011; Lendelvo et al., 2012; Newman and Lane Smith, 2021; Tantoh & McKay, 2020). For example, women have taken leading roles in contributing and participating in the realization of rural water supply in Northwest Cameroon (Tantoh, 2021a, 2021b). It is argued that women adopt and play multidimensional roles in natural resource management and are essential in environmental conservation (Carvajal-Escobar et al., 2008; Harding, 1998; Lendelvo et al., 2012). Thus, they are sensitive to any harm that might affect environmental resources such as climate crisis. For example, women seek solutions to lack of access to potable water supply, health and education, reducing elements of climate vulnerability in rural communities in the event of hydro meteorological episodes associated with the effects of climate change (Carvajal-Escobar et al., 2008).

Despite the strength and ability of women to participate in several developments and natural resource management activities, some have not been able to actively participate. It is argued that some women do not possess appropriate skills in relation to educational achievements and specific skills required for certain activities (Lendelvo et al., 2012). Furthermore, gender roles are usually undervalued and overlooked (Tiondi, 2000). This lack of recognition renders them invisible and diminishes their contribution as active agents in development. There is also gender bias between males and females in most communities in sub-Saharan Africa and this is directly related to patriarchal cultural tendencies. Thus, women are the most vulnerable to climate crisis despite their unique capabilities as community managers of natural resources and their misjudged potential and so, underused in other community development and environmental management activities.

Methodology

Description of the Study Site

This study examined relevant literature on gender and climate change education, climate change adaptation and natural resource management in a sub-Saharan African context. The region is one of the most diverse regions with 10% of the world's population. This includes the African countries in the southern part of the Sahara Desert. sub-Saharan Africa is divided into regional components including, West Africa, Central Africa, East Africa and Southern Africa. The Horn of Africa often included in the Eastern Zone of Africa is another important section at the Eastern end of the continent. This region is home to about 750 million people and is labelled as the world's most impoverished region (Stockdale, 2017). This is due to the socio-economic and environmental problems despite the abundant natural resources and relatively pleasant climatic zones. Despite the vastness of the Sahara Desert characterized by weather extremes, between day time temperature and night time temperatures, the greater part of Africa enjoys hospitable, distinctive seasonal climates. The SSA region is a treasure of mineral deposits with half of the world's diamond and gold, timber, uranium, and wildlife petroleum among others. It is reported that SSA has the highest population growth rate projected to double in the next 20 years (Stockdale, 2017). This high population growth, rising urbanization and changing consumption patterns exerts enormous pressures on natural resources, resulting in over-exploitation and environmental degradation. This has been exacerbated by the climate crisis, which has rendered women and children more vulnerable. The vulnerability of women and girls has been intensified by the high incidence of school dropouts. Girls steadily lag behind in terms of educational attainment and school completion and with <15% in the Central Africa Republic, Chad and South Sudan and <30% in Equatorial Guinea, Malawi, Mali Niger Uganda.[1] Another notable feature among the cultures is the patriarchal nature of society. The voices of women are seldom heard and this has impacted limited them from actively participating in decision-making processes that concern them (Lendelvo et al., 2012).

Research Methods

The study was based on a systematic review of related literature on gender, education and natural resources management. The articles reviewed were derived from several databases—Google Scholar, Scopus, Science Direct and Web of Science (See Fig. 8.1). These databases.

were chosen because they are open and comprehensive. Furthermore, they are amongst the world's leading scientific citation search with topical issues on climate change education, climate change adaptation, natural resources management among others. A three-staged approach was used to select relevant literature for the study (Fig. 8.1). Initially, key words such as Gender, Climate Change Education, Climate Change Adaptation and Natural Resources Management were entered in the search engines. A total of 847 articles were generated and systematically reviewed. The titles and abstracts of the articles were later screened and 336 duplicated articles were excluded from the study. Another 426 articles were later excluded due to inexplicit methods—211, no intent to review the literature-119 and irrelevance—96. This resulted in 86 articles which were further screened for suitability and 47 were excluded. This resulted in 39 academic articles which were directly related to the topic. In addition, one (01) thesis, 19 reports related to the topic from both local international organizations that were manually searched and included in the study.

Results and Analysis

Gender and Climate Change Education

Climate change is a global phenomenon and its effects on diverse demographic groups are felt differentially (Ebhuoma et al., 2020). As a result, the common urge of environmental threats frequently alters obligations and exposures attributed to both

[1] https://www.theigc.org/blog/gender-equality-through-secondary-education-in-sub-saharan-africa/

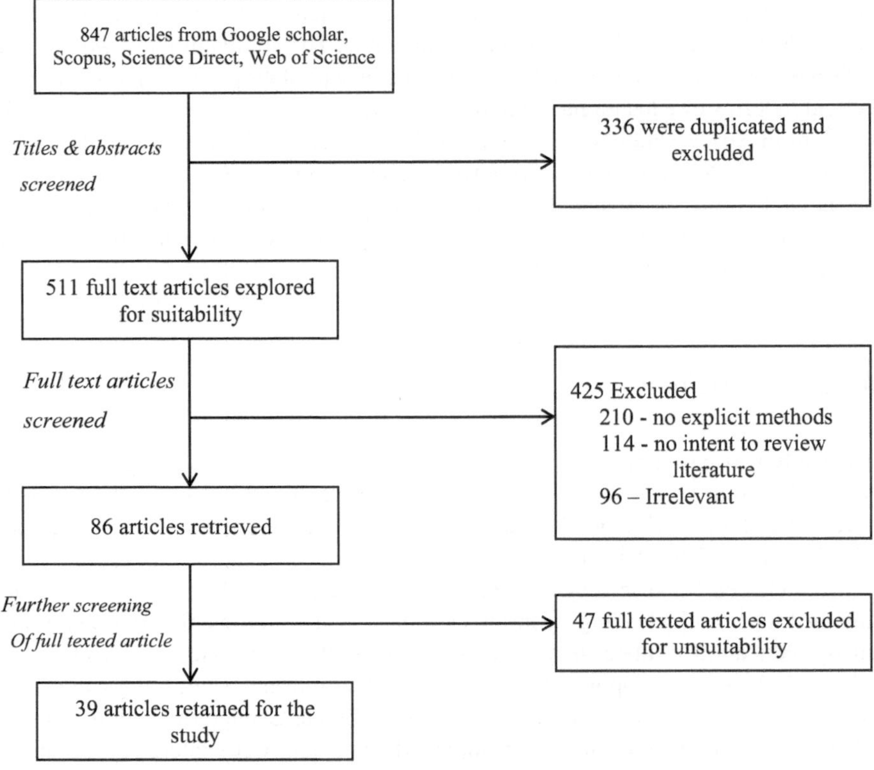

Fig. 8.1 Flow diagram, which illustrates how relevant literature was found and selected

developed and developing countries, urban and rural communities, and between men and women disproportionally. For example, rural communities in sub-Saharan Africa have and share a multifaceted cultural link with their natural environment (Tantoh and Mckay, 2020), which renders women most vulnerable to the effects of climate change. One of the major finds of this review is that women in rural communities of the developing world are mostly affected by the effects of climate change and are excluded from decision-making processes of any important matters. This claim is in consistent with the outcomes of the African Development Bank (AfDB, 2011), which emphasizes that women are marginalized most of the times, even though they produce 80% of food crops in SSA. In contrast with this and despite their contribution to about three-quarters of food production, rising gender inequality has been intensified by increasing poverty, due to different strategies in the use and management of environmental resources (IPCC, 2014). Therefore, it is imperative for existing gender inequalities in relation to climate change to be recognized and resolved urgently, to better mitigate any possible disparity of deliberate intervention (Sims, 2021). Noteworthy is that the effects of climate change on the environment are not gender balanced in general, but rather affect young girls and women more than their male counterparts (Centre for Gender and Disaster, 2020). Thus, gender equality is sorely needed for women to take independent actions and make decisive choices in mitigating the effects of climate change. An implication of this is the possibility that gender equality will facilitate the prevention of gender-based violence, which is a leading item in agendas of ecological resilience programmes (Donkor et al., 2019). These should, therefore, ensure an even participation of men and women in the project cycle as involvements and participation are conceived equitably, to contribute towards achieving greater gender equality outcomes (Sims, 2021).

Climate, Gender Equality and Natural Resources Management

Another outstanding finding from this review is that climate change not only reflects gender imbalances, but also worsens socially-constructed power relations, norms and practices. Rural women for example, are seldom included in decision-making processes regarding the management of resources which they depend on for their livelihood and wellbeing (Tantoh & Mckay, 2020). The nature of their involvement is mostly related to mobilizing the community regarding community development projects and physical work at project sites, such as digging trenches, cleaning and collecting payments (Boateng et al., 2013). This gender prejudice is strengthened by a deeply rooted patriarchy-based socialization, where men are always considered superior to women (Tantoh & McKay, 2020). This condition is disempowering and minimizes women's presence in policymaking positions within organizations, resulting in their exclusion from issues that concern them. This gender inequality has been intensified by rising poverty and diverse governance approaches to the use and manage natural resources (IPCC, 2014). Many argued that women have limited privileges, restricted access to resources and restricted opinion in the community and household decision-making (Lendelvo et al., 2012; Tantoh & Mckay, 2020).

These cultural barriers are translated into the unlikelihood of participating in decision-making processes in matters that affect, particularly rural livelihoods thus, exacerbating living conditions and climate change. Others have argued that when people are fully involved in the management of their resources, it gives them a sense of proprietorship, which is essential for sustainability (Musavengane et al., 2019). It is known that climate change and its effects on natural resources are largely not gender neutral, and that more often will affect women more than males (Centre for Gender and Disaster, 2020). This phenomenon worsens when women are disabled (Pentlow, 2020). In a similar manner, girls with disabilities are marginalized, or totally excluded from social life (Plan International, 2019). Hence, inequalities in the access to natural resources result in increased possibilities for girls and women to live in poverty, and the unlikelihood of accessing education (Djoudi et al., 2016), making them much more vulnerable than men to the effects of climate change (Centre for Gender and Disaster, 2020; Ebhuoma et al., 2020). Thus, threats connected with climate change enhance possibilities of crystalizing gender inequities and even eradicate any progress that might have been made towards gender equality in many developing countries

(Tantoh et al., 2021). This unfair treatment can result in unfortunate consequences for all as women play a unique role in the stewardship of natural resources and support of households and communities. Nevertheless, an exposure to diverse types of environmental transformations has enabled them to develop survival strategies to face these grim circumstances (UNDP, 2011).

It is important to reckon with the fact that a majority of rural women are poor and, mostly subjected to the effects of climate change and environmental degradation (Villamor et al., 2018). This vulnerability is not only due to their lack of empowerment, but also due to the social and economic responsibilities ascribed to them (Tantoh & McKay, 2020). These responsibilities and demands from women leave them with limited time and opportunity for active political participation. Under these circumstances, they are deprived of chances to participate in the decision-making processes that impact their livelihoods and the wellbeing of the environment (Sims, 2021). In contrast, studies on climate change and gender have been critiqued for making sweeping conclusions about the determinants of vulnerability as described by Rao et al. (2019), who conceded that empirical evidence on gender has not been thorough. Therefore, there is a need for an intersectional understanding and a contextualization approach when studying gender-specific changes in mitigative and adaptive contexts. This entails crafting comprehensive principles for each community under study, to integrate gender dimensions directly, into climate change programs.

Gender Education, Climate Change Adaptation and Resilience

One of the most significant findings to emerge from this review is that recent studies devoted to adaptation to climate change highlight the role that adaptation policy could have in terms of poverty eradication, through the reduction of the exposure of vulnerable people (Rao et al., 2019; Tantoh et al., 2022b). The findings revealed existing linkages between climate change and several Sustainable Development Goals (SDGs). Notably, women feel the impacts of climate change mostly due to their use and access to natural resources (World Bank, 2010; Boateng et al., 2013; Fonjong, 2008). At the same time, women acquire valuable local knowledge for managing natural resources in their environment and are capable of responding to climate-related challenges as remarkable improvements agents (Pentlow, 2020; Tantoh & McKay, 2020). In contrast, women are most socially marginalized and economically insecure in several rural communities of the developing world. If mitigation and adaptation are to be effective and efficient, women should be given their share of decision-making positions in natural resource management. Furthermore, they should be involved in the initiation, design, implementation, monitoring and evaluation of rural development projects. This is in line with the 2030 Agenda for Sustainable Development and the 17 SDGs, which espouse the roadmap for progress that is efficient, effective and leaves no one behind (Donkor et al., 2017). Achieving gender equality and women's empowerment are fundamental goals to each one of the 17 DDGs. It is only by safeguarding the rights of women across all these goals that fairness and inclusion, thriving economies and support for the environment now and for the future, can be achieved.

This review found that education can address the current climate crisis, reduce vulnerability and enhance adaptation strategies. Similarly, the findings of Muttarak and Lutz (2014), revealed that education plays a fundamental role in reducing susceptibilities to the negative impacts of environmental degradation and weather-related disasters (Donkor et al., 2017, 2019). Evidence from sub-Saharan Africa, for example, shows that formal education supports the development of cognitive and problem-solving skills, knowledge and risk perception (Sims, 2021). In a similar manner, education is an important factor to raise mutual understanding of the nature of climate crisis, which is vital in prompting local, national and global climate action (Bangay & Blum, 2010; Donkor et al., 2017). Thus, educated people are more successful to respond to climate-related risks. Furthermore, education contributes to reducing vulnerabilities through poverty reduction, enabling social capital and access to information (Sims, 2021). They are susceptible to adequately garnering greater social capital support involving social networks which makes them adaptive and resilient to climate change (Jackson, 2018; Musavengane et al., 2019; Plan International, 2019; UNDP, 2022). Similarly, girls' education has been recognized as one of the most important social and economic elements in reducing vulnerability to the effects of climate change (Blankespoor et al., 2010). This has been emphasized by findings from 125 countries in the developing world between 1980 and 2010, asserting the significant role of female education in reducing climate change (Streissnig et al., 2013). However, although education is widely recognized as a powerful instrument to produce positive change and has been a fundamental component in several developmental efforts, Duncan (2013) argued that there is limited evidence of education being part of the solution to climate emergencies. Nevertheless, girls' and women's education offer extra benefits in supporting climate resilience, which will possibly have more monetary value than a direct investment in climate crisis (Blankespoor et al., 2010). Therefore, it can be argued that the contribution of gender education to environmental sustainability and the role of female education in advancing gender equality has well-known benefits of gender equality in reducing vulnerability. Implementing a gender lens enables the understanding of power imbalances, marginalization and inequality worsened by the climate crisis. This will also facilitate the crafting of measures on how to mitigate and address them (United Nations High Commission for Refugees (UNHCR), 2020). This will further emphasize the essential role of women in spearheading effective and efficient transformation. Furthermore, foundational abilities will be a fundamental resilience element for both the children of today and their future offspring (Smith, 2021). The main weakness of this study was the paucity of adequate primary data on gender and climate education in sub-Saharan African countries. However, secondary data were used to substantiate the claims of the paper.

Conclusion and Recommendations

Most countries in sub-Saharan Africa depend on their natural resource base for a prosperous, economic development. However, climate change continues to affect the environment leading to its irreparable degradation and this trend has modified, particularly rural economies and women. This study sought to investigate the role of gender education in climate change adaptation and effective natural resources management through a systematic review of academic literature. The study found that increasing demands, rising population, and changes in land use patterns coupled with the effect of climate change have exerted enormous pressure on natural resources leading to their overexploitation and degradation. This undermines rural livelihood, particularly, those of women who depend on natural resources for their wellbeing. Hence, climate change affects different demographic groups differently. Rural women, particularly those who are disabled, who are among the most underprivileged and marginalized groups in terms of education. Given the fast pace at which the world is changing, rural women are having to contend with the shortages and difficulties of water, fuel wood, land, while these problems are exacerbated by population growth and their limited level of education. A major challenge that many communities and countries face, however, is how to effectively improve the education of rural women who remain vital in natural resource management. However, studies on adaptive capacity and differentiated-gender vulnerability are limited. Perhaps in an attempt to ensure that the learning crisis is not overshadowed by the climate crisis, many have tried to make the case that global education can be part of the solution to a successful climate crisis. Considerably more work will need to be done to:

- Develop foundational skills in climate education irrespective of other challenges.
- Ensure appropriate systems, services and support for girls and women in sub-Saharan Africa should be a priority for national governments
- Encourage national governments and relevant institutions to put girls and women at the Centre of research that could help discover new approaches to climate change adaptation agricultural practices. particularly among girls and women
- Inspire agricultural risk assessments amongst girls and women, as it is vital to sustainable agricultural production which could help prevent influences on hunger, food security and poverty generally.
- Vote appropriate legal and policy frameworks, which will mean bringing together all necessary legal and policy instruments that can enhance gender adaptation implementation.

Acknowledgements The financial assistance of the National Institute for the Humanities and Social Sciences- Council for the Development of Social Science Research in Africa (NIHSS-CODESRIA) towards this research is hereby acknowledged. Opinions expressed and conclusions arrived at are those of the authors and are not necessarily to be attributed to the NIHSS-CODESRIA.

References

AfDB. (2011). *Climate change, gender and development in Africa.* Retrieved June 25, 2022, from https://www.afdb.org/sites/default/files/documents/publications/climate_change_gender_and_development_in_africa.pdf

Atkinson, H. G., & Bruce, J. (2015). Adolescent girls, human rights and the expanding climate emergency. *Annals of Global Health, 81*(3). Retrieved 20, from https://academicworks.cuny.edu/cgi/viewcontent.cgi?article=1449&context=cc_pubs

Bangay, C., & Blum, N. (2010). Education responses to climate change and quality: Two parts of the same agenda? *International Journal of Educational Development, 30*(4), 335–450. Retrieved June 25, 2022, from https://discovery.ucl.ac.uk/id/eprint/1526915/1/Bangay2010Education359.pdf

Blankespoor, B., Dasgupta, S., Laplante, B., & Wheeler, D. (2010). *Adaptation to climate extremes in developing countries: The role of education.* Policy research working paper no. 5342. World Bank. Retrieved June 25, 2022, from https://documents.worldbank.org/en/publication/documents-reports/documentdetail/328491468340475205/adaptation-to-climate-extremes-in-developing-countries-the-role-of-education

Boateng, J. D., Brown, C. K., & Tenkorang, E. Y. (2013). Socio-economic status of women and its influence on their participation in rural water supply projects in Ghana. *International Journal of Development and Sustainability, 2*(2), 871–890.

Carvajal-Escobar et al., 2008 Carvajal-Escobar, Y., Garcia-Vargas, M., & Quintero-Angel, M. (2008). Women's role in adapting to climate change and variability. *Advances in Geosciences, 14*, 277–280.

Centre for Gender and Disaster. (2020). *Gender and disaster. Bibliography & reference guide* (vol. 1). UCL. https://www.ucl.ac.uk/risk-disaster-reduction/sites/risk-disastereduction/files/gender_disaster_reference_guide_vol_1.pdf

Chigwanda, E. (2016). A framework for building resilience to climate change through girls' education programming. In *The 2016 Echidna global scholars policy brief*. Center for Universal Education at Brookings. https://www.brookings.edu/wp-content/uploads/2016/12/global-20161202-climate-change.pdf

Djoudi, H., Locatelli, B., Vaast, C., Asher, K., Brockhaus, M., & Sijapati, B.B. (2016). Beyond dichotomies: Gender and intersecting inequalities in climate change studies. *Ambio, 45*, 248–262. https://doi.org/10.1007/s13280-016-0825-2

Donkor, F. K., Tantoh, H. B., Ebhuoma, E., & Fullard, S. (2019). Addressing the trilemma of educational trade-offs on Africa's terrain for sustainable development. *CODESRIA Bulletin, 4*, 2017.

Donkor, F. K., Tantoh, H., & Ebhuoma, E. (2017). Social learning as a vehicle for catalysing youth involvement in sustainable environmental management. In *CODESRIA Bulletin 3: Ecologies, economies and societies in Africa.* CODESRIA

Ebhuoma, E. E., Donkor, F. K., Ebhuoma, O. O., Leonard, L., & Tantoh, H. B. (2020). Subsistence farmers' differential vulnerability to drought in Mpumalanga province, South Africa: Under the political ecology spotlight. *Cogent Social Science, 6*, 1792155.

Fonjong, L. N. (2008). Gender roles and practices in natural resource management in the North West Province of Cameroon. *Local Environment, 13*(5), 461–475.

Galvin, M. (2011). Participating in urban myths about women's rural water struggles. *Agenda: Empowering Women for Gender Equity, 25*(2), 87–100

Harding, S. (1998). Women, science, and society. *Science, 281*, 1599–1600.

IPCC. (2014). Climate change 2014: Impacts, adaptation and vulnerability. Part A: Global and sectoral aspects. In C. B. Field, V. R. Barros, D. J. Dokken, K. J. Mach, M. D. Mastrandrea, T. E. Bilir, M. Chatterjee, K. L. Ebi, Y. O. Estrada, R. C. Genova, B. Girma, E. S. Kissel, A. N. Levy, S. MacCracken, P. R. Mastrandrea & L. L. White (Eds.) *Contribution of Working Group II to the fifth assessment report of the Intergovernmental Panel on Climate Change, Intergovernmental Panel on Change (IPCC).* Cambridge University Press.

IPCC. (2021). Summary for policymakers. In V. Masson-Delmotte, P. Zhai, A. Pirani, S. L. Connors, C. Péan, S. Berger, N. Caud, Y. Chen, L. Goldfarb, M. I. Gomis, M. Huang, K. Leitzell, E. Lonnoy, J. B. R. Matthews, T. K. Maycock, T. Waterfield, O. Yelekçi, R. Yu, & B. Zhou (eds.) *Climate change 2021: The physical science basis. Contribution of working group I to the sixth assessment report of the intergovernmental panel on climate change.* In Press.

Jackson, S. (2018). Building trust and establishing legitimacy across scientific, water management and indigenous cultures. *Australasian Journal of Water Resources*, 1–10. https://doi.org/10.1080/13241583.2018.1505994

Jordan, J. C. (2019). *Deconstructing resilience: Why gender and power matter in responding to climate stress in Bangladesh in climate and development*. https://www.academia.edu/42143978/Deconstructing_resilience_why_gender_and_power_matter_in_responding_to_climate_stress_in_Bangladesh_Deconstructing_resilience_why_gender_and_power_matter_in_responding_to_climate_stress_in_Bangladesh

Kwauk, C., Cooke, J., Hara, E., & Pegram, J. (2019). *Girls' education in climate strategies: Opportunities for improved policy and enhanced action in nationally determined contributions*. https://www.brookings.edu/wp-content/uploads/2019/12/Girls-ed-in-climate-strategies-working-paper-FINAL.pdf

Lambrou, Y., & Piana, G. (2006). Gender: the missing component in the response to climate change (p. 46). Food and Agriculture Organization of the United Nations. http://www.fao.org/sd/dimpe1/docs/pe1051001d1b.pdf

Lendelvo, S., Munyebvu, F., & Suich, H. (2012). Linking women's participation and benefits within the Namibian community based natural resource management program. *Journal of Sustainable Development, 5*(12), 2012.

Malala Fund. (2021). *A greener, fairer future: why leaders need to invest in climate and girls' education*. https://assets.ctfassets.net/0oan5gk9rgbh/OFgutQPKIFoi5lfY2iwFC/6b2fffd2c893ebdebee60f93be814299/MalalaFund_GirlsEducation_ClimateReport.pdf

Mbah, M., Ajaps, S., & Molthan-Hill, P. A. (2021). Systematic review of the deployment of indigenous knowledge systems towards climate change adaptation in developing world contexts: Implications for climate change education. *Sustainability, 2021*(13), 4811. https://doi.org/10.3390/su13094811

Musavengane, R., Tantoh, H. B., & Simatele, D. M. (2019). A comparative analysis of collaborative environmental management of natural resources in sub-Saharan Africa: A study of Cameroon and South Africa. *Journal of Asian and African Studies*. https://doi.org/10.1177/0021909618825276

Muttarak, R., & Lutz, W. (2014). Is education a key to reducing vulnerability to natural disasters and hence unavoidable climate change? *Ecology and Society*.

Nalau, J., Becken, S., Schliephack, J., Parsons, M., Brown, C., & Mackey, B. (2018). The role of indigenous and traditional knowledge in ecosystem-based adaptation: A review of the literature and case studies from the Pacific Islands. *Weather, Climate, and Society, 10*(4), 851–865. https://doi.org/10.1175/WCAS-D-18-0032.1

Newman, K., & Lane Smith, S. (2021). *Linking global education and the climate crisis: An alternative approach*. RISE Programme. https://riseprogramme.org/blog/linking-global-education-climate-crisis

Nguyen, T. H., & Ross, A. (2017). Barriers and opportunities for the involvement of indigenous knowledge in water resources management in the Gam river basin in North-East Vietnam. *Water Alternatives, 10*(1), 134–159.

Peek, L., Abramson, D. M., Cox, R. S., Fothergill, A., & Tobin, J. (2018). Children and disasters. In H. Rodríguez, W. Donner, & J. Trainor (Eds.), *Handbook of disaster research. Handbooks of sociology and social research*. Springer. https://doi.org/10.1007/978-3-319-63254-4_13

Pentlow, S. (2020). *Indigenous perspectives on gender, power and climate-related displacement*. Gender Equality Consultant, Cuso International https://cusointernational.org

Plan International. (2019). *Climate Change: Focus on girls & young women.* Plan International Position Paper. https://plan-international.org/publications/climate-change-focus-girls-and-young-women#download-options

Rao, N., Lawson, E. T., Raditloaneng, D. S., & Angula, M. N. (2019). Gendered vulnerabilities to climate change: Insights from the semi-arid regions of Africa and Asia. *Climate and Development, 11*(1), 14–26. https://doi.org/10.1080/17565529.2017.1372266

Ravera, F., Martín-López, B. M., Pascual, U., & Drucker, A. (2016). The diversity of gendered adaptation strategies to climate change of Indian farmers: A feminist intersectional approach. *Ambio, 45*(3), S335–S351.

Sims, K. (2021). *Education, girls' education and climate change: Executive summary.* K4D Emerging Issues Report 29. Institute of Development Studies.

Smith, S. L. (2021). *Linking global education and the climate crisis: An alternative approach.* RISE. Retrieved June 20, 2022, from https://riseprogramme.org/blog/linking-global-education-climate-crisis

Stockdale, N. (2017). *Sub-Saharan Africa world geography: Understanding a changing world, ABC-CLIO, 2017, world geography.* Retrieved July 23, 2022, from worldgeography.abc-clio/Search/Display/1127466

Streissnig, E., Lutz, W., & Patt, A. (2013). Effects of educational attainment on climate risk. *Vulnerability Ecology and Society, 18*(1), 16–40.

Sunderland, T., Achdiawan, R., Angelsen, A., Babigumira, R., Ickowitz, A., Paumgarten, F., Reyes-García, V., & Shively, G. (2014). Challenging perceptions about men, women, and forest product use: A global comparative study. *World Development, 64*(Supplement), 1. https://doi.org/10.1016/j.worlddev.2014.03.003

Tantoh, H. B., & McKay, T. J. M. (2020). Investigating community constructed rural water systems in Northwest Cameroon: Leadership, gender and exclusion. *International Development and Planning Review., 42*(4), 455–478.

Tantoh, H. B., McKay, T. J. M., Donkor, E. F., & Simatele, M. D. (2021). Gender roles, implications for water, land and food security in a changing climate: A systematic review. *Frontiers in Sustainable Food Systems, 5*, 259. https://doi.org/10.3828/idpr.2020.4

Tantoh, H. B., Mokotjomela, T. M., Eromose, E., Ebhuoma, E. E., & Donkor, F. K. (2022a). Factors preventing smallholder farmers from adapting to climate variability in South Africa: Lessons from Capricorn and uMshwati municipalities. *Climate Research, 88*, 1–11. https://doi.org/10.3354/cr01693

Tantoh, H. B., Ebhuoma, E. E., & Leonard, L. (2022b). Indigenous women's vulnerability to climate change and adaptation strategies in Central Africa: A systematic review. In E. E. Ebhuoma, L. Leonard (Eds.), *Indigenous knowledge and climate governance.* Sustainable Development Goals Series. Springer. https://doi.org/10.1007/978-3-030-99411-2_5

Tantoh, H. B. (2021a). Constitutionality and the co-management of water resources in Cameroon. Clean Water and Sanitation. *Encyclopedia of the UN Sustainable Development,* 1–13

Tantoh, H. B. (2021b). Water metering in piped community-based water supply systems: The challenge of balancing social and economic benefits. *Development in Practice.* https://doi.org/10.1080/09614524.2021.1937546

The World Bank. (2010). *Adaptation to climate extremes in developing countries: The role of education.* http://documents1.worldbank.org/curated/en/328491468340475205/pdf/WPS5342.pdf

Tiondi, E. (2000) Women, Environment and Development: sub-Saharan Africa and Latin America. [Tampa Graduate Theses and Dissertations, University of South Florida (USF)]. https://digitalcommons.usf.edu/etd/1549

UNDP. (2008). *UNDP gender equality strategy 2008–2011.* United. Nations Development Programme (UNDP). http://www.iknowpolitics.org/en/knowledge-library/report-white

United Nation High Commission for Refugee (UNHCR). (2020). *Gender, displacement and climate change.* Potsdam Institute for Climate Change Research.

United Nations Development Programme (UNDP). (2011). *Empowered and gendered: Implementing the gender equality strategy*. Background Paper for the Annual Oral Report to the Executive Board. UNDP.

United Nations Development Programme (UNDP). (2022). Retrieved March 5, 2022, from http://www.iknowpolitics.org/en/knowledge-library/report-white-paper/empoweredand-equal-gender-e

Vincent, K. (2007). Uncertainty in adaptive capacity and the importance of scale. *Global Environment Change, 17*, 12–24.

World Economic Forum (WEF). (2022). *The global risks report* (17th ed.). World Economic Forum. SBN: 978-2-940631-09-4.

Chapter 9
Farmers' Climate Change Adaptation Strategies and the Role of Environmental Awareness and Education: A Review on Africa

Mufti Nadimul Quamar Ahmed and Jennifer E. Givens

Abstract Climate change is a global challenge. Even though climate change affects all countries, less-developed countries and poor peasant farmers are especially at risk and have difficulty adapting. Less-developed nations are home to an estimated 500 million small-scale farms, who provide sustenance for approximately two billion people. In addition, it is estimated that these small farms produce nearly 80 per cent of the food consumed in Asia and Sub-Saharan Africa. Existing literature finds Africa is particularly susceptible to the effects of climate fluctuation and change, including rising sea levels, melting glaciers, threatened water supplies, leading to decreased agricultural output, increasing food insecurity, diminished biodiversity, intensifying erosion, drought, and flood. In Africa, smallholder farmers depend on agriculture, which relies on timely rainfall, and thus they increasingly experience the consequences of climate change. This chapter reviews the literature on smallholder farmers in Africa and climate change, focusing on their perceptions of climate change and their adaptation techniques. We also explore how various factors affect smallholder farmers' perceptions of climate change and their adaptation strategies. Finally, we discuss the need to improve awareness and adaptation capacity. Although farmers in various African countries perceive climate change differently, the literature review reveals that they have some views in common. Most farmers observe changes in temperature and rainfall patterns in their area. They also report experiencing increasing floods and droughts and decreasing crop production. Diversification of crops, changing crops, planting drought-resistant crops, incorporating livestock into crop production, shifting the time of agricultural operations, homestead gardening, increasing irrigation, engaging in mixed farming, and migration are common adaptation strategies reported in African countries. We found various household-related factors (gender, age, education, marital status, family size, etc.), farm-related factors (farming experience, size of the farm, etc.), institutional factors

M. N. Q. Ahmed (✉) · J. E. Givens
Department of Sociology, Anthropology and Criminal Justice, Utah State University, Logan, Utah, United States
e-mail: mufti.ahmed@usu.edu

M. F. Mbah et al. (eds.), *Practices, Perceptions and Prospects for Climate Change Education in Africa*, https://doi.org/10.1007/978-3-031-84081-4_9

(access to information, extension contact, etc.), as well as other factors (participation in a social group, training, etc.) are affecting local level adaptation strategies among farmers. We contend that farmers are the "front-liners" in adapting to climate change, and government agencies and other national and international organizations should continue their efforts to support them. Moreover, to contribute to the progress of the United Nations Sustainable Development Goals, farmers' insights and experiences should be integrated in the dominant policies and plans, including coordination of education about climate change and adaptation strategies. Furthermore, we also advocate for emphasizing improving environmental awareness through environmental education particularly in this region to achieve effective adaptation.

Keywords Climate change · Perception · Adaptation · Environmental awareness · Africa

Introduction

Climate change poses the biggest threat to life on Earth and ecological systems (Dhiman et al., 2010; Omann et al., 2009). While both rich and developing nations experience climate change, the poorest nations and their smallholder farmers' livelihoods are hit the hardest because of their immediate dependence on the natural environment and their lack of adaptation resources (Archer et al., 2007; Katharine, 2004; Food and Agriculture Organization (FAO), 2011, 2016). The agriculture industry is responsible for about 3.7 per cent of global Greenhouse gas (GHG) emissions on its own. It is also a major factor in deforestation, responsible for another 7–14 per cent of global GHG emissions (Food and Agriculture Organization (FAO), 2013; Nyang'au et al., 2021). However, this sector is also a major source of sustenance and employment globally. Less-developed nations are estimated to be home to approximately an estimated 500 million small-scale farms, who provide sustenance for approximately two billion people (International Fund for Agricultural Development (IFAD), 2012). In addition, these tiny farms supply over 80 per cent of Asia and Sub-Saharan Africa's food (International Fund for Agricultural Development (IFAD), 2012). Various effects, including rising sea levels, melting glaciers, diminished water supplies, decreased agricultural output and intensifying food insecurity, diminished biodiversity, and intensifying erosion, drought, and flood are becoming common in Africa due to changing climate (De Souza et al., 2015; Mpelasoka et al., 2018; Watts et al., 2015). Marginalized and vulnerable populations in developing countries are concerned about the hazards to agricultural productivity posed by climate change (Koundouri & Nauges, 2005; Isik and Devodas, 2006), and although those who rely on rain-fed agriculture do adopt some coping strategies, they lack adequate resources to spread out the risks associated with agricultural output (Regmi et al., 2020).

Farmers' livelihoods depend on the weather, therefore they use a variety of strategies to combat climate change and preserve their survival. Better climate change understanding can also assist farmers in coping with the effects of climatic variability and change on agricultural output (Kom et al., 2019). As farmers have an in-depth grasp of their local environment and the issues that they experience due to climate change, they may offer useful advice on creating the most effective climate change education and adaptation strategies for fellow farmers. Incorporating the experiences and perspectives of farmers into climate change education helps ensure that the education targeted to farmers is appropriate for their unique contexts. This approach helps to ensure that climate change education is accessible and relevant to farmers, thereby increasing their motivation to participate and implement significant changes to their practices. Therefore, for effective policymaking, it is essential to consider farmers' perspectives on the effects of climate change (Regmi et al., 2020). Furthermore, it is necessary to combine scientific data and indigenous knowledge to guide adaptation and mitigation, as it is widely recognized that indigenous knowledge is an essential component in addressing climate change (Chaudhary & Bawa, 2011; Morton, 2017; Nichols et al., 2004; Speranza et al., 2009). Therefore, to achieve effective and sustainable adaptation policies there is a need to emphasize more on increasing environmental awareness through environmental education.

In this chapter, we explore issues that address the following questions: (a) How do African farmers perceive climate change? (b) How do these farmers cope with the adverse impacts of climate change? (c) What are the major factors that affect their local-level adaptation strategies? (d) What is the role of environmental awareness among farmers, and why is it important to consider farmers' experiences and knowledge in policymaking? This chapter is structured as follows: First, the introduction section (Sect. "Introduction") briefly outlines the background and objectives of this study. In the next section (Sect. "Methodology"), we present the methodology of the study. An overview of climate change's impacts on agriculture in Africa is presented in Sect. "Climate Change Effects on Africa's Smallholder Farmers". Then we present the results and discussion section in different subsections: we review and present how African farmers perceive climate change (subsect. "Understanding Climate Change: African Farmer's Experiences"), we summarize various coping strategies African farmers use to adapt to climate risks (subsect. "Major Adaptation Strategies Employed by African Farmers"), we discuss various factors influencing local-level adaptation strategies among African farmers (subsect. "Major Factors Affecting the Local Level Adaptation Strategies"), and we explain why environmental awareness among farmers is important, how environmental education can play a significant role, and the importance of incorporating farmers' experiences and knowledge into this to develop effective adaptation strategies (subsect. "Role of Environmental Awareness Through Environmental Education"). Finally, in chapter five we present some recommendations for further studies and conclude the chapter.

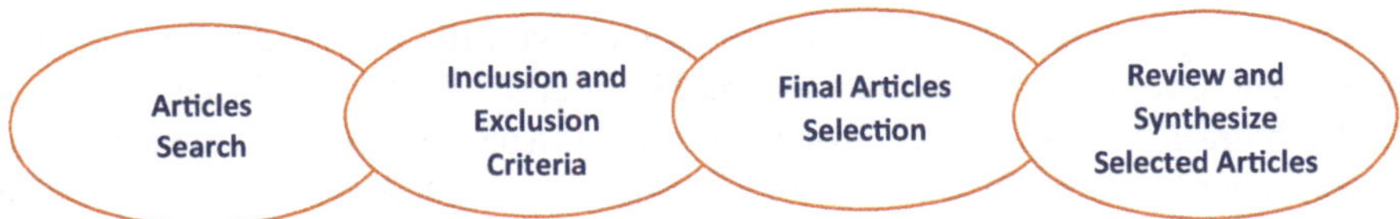

Fig. 9.1 Article review process. *Source* Author's creation (2024)

Methodology

We searched for articles using Web of Science, Scopus, and Google Scholar, to compile the literature review for this book chapter. These databases have also been used in previous research on various environmental and climate change issues (Ahmed et al., 2022, 2024; Haq et al., 2024; Wan et al., 2021). We used a combination of keywords and subject headings to identify articles that address the topic of how African farmers understand and respond to climate change, as well as the most significant challenges they face in doing so. During the literature search, we used certain parameters to narrow our search; for instance, we looked specifically for peer-reviewed English-language studies published between the years 2000 and 2022. Primary and secondary research articles, as well as review papers, were used, while editorials, letters to the editor, and articles written in languages other than English were not. After carefully considering all the articles that met our requirements, we settled on 33 articles to include in this overview. Figure 9.1 summarizes the steps for the final article selection in this chapter.

Climate Change Effects on Africa's Smallholder Farmers

Africa's size and location give it many natural riches and a wide range of climates. However, the continent is also one of the most vulnerable to the effects of climate change, and each country's level of risk varies greatly (Vincent, 2007). According to the Intergovernmental Panel on Climate Change (IPCC, 2007), changes in rainfall patterns and amounts have an impact on soil moisture and soil erosion rates. Crop growth and crop yields are dependent on both of these factors. Unanticipated changes may have a detrimental impact on farmers working on a small scale. Small-scale farmers, who make up most of Africa's farms, would benefit from improvements in resilience in the face of climate change and environmental damage, which threatens food security (Antwi-Agyei et al., 2012; Brown et al., 2007; Mabe et al., 2014; Thomas et al., 2007; UNFCCC, 2007). According to Herrero et al. (2010), climate change had harmful effects on small-scale farmers in Kenya by causing repeated droughts. The Intergovernmental Panel on Climate Change (IPCC) has estimated that a rise of 0.2–0.5 °C every decade will have a significant influence on maize, sorghum, cassava, yam, and cowpea in the West Africa sub-region (Bals et al., 2008; IPCC, 2007). The projected effects on agriculture, including changes in water availability and quality, soil erosion, and crop production, pose immediate and localized economic threats for farmers (Howden et al., 2007; McCarl, 2010). Molua and Lambi (2006) show that climate change will have a negative impact on small-scale farmers in Cameroon by reducing their income from farming. They find a reduction of $0.5 billion in net agricultural profits would occur if temperatures rose by 2.5 °C in Cameroon and there would be a $1.7 billion loss in net revenues if the temperature rises above 5 degrees Celsius. Moreover, there will be a $1.96 billion decline in net revenues if precipitation falls by 7 per cent and a $3.8 billion drop in crop net revenues if precipitation falls by 14 per cent (Molua & Lambi, 2006). In another study conducted in Ghana's Volta area, Tabi et al. (2012) find that climate change has a negative impact on rice growers, which includes deaths of animals, losses of farm capital, heat stress, shortages of water, sluggish development, increasing poverty, and food insecurity. Likewise, multiple climatic forces have negatively affected Nigeria, including rising sea levels in the south, erosion in the southeast, persistent north-central and southwest floods, and growing drought in the north (Anabaraonye et al., 2020). To sum up, these extant studies, as well as many other studies, demonstrate that climate change adversely affects African smallholder farmers (Awazi et al., 2020).

Results and Discussion

Here, we summarize the findings from our literature review in four sub-sections. In the first section, we briefly review how farmers in Africa perceive climate change and variability. Next, we discuss the several ways that farmers have found to deal with the problems they have noticed. Findings on the numerous elements that influence local farmers' adaptation strategies are presented in the third section. Finally, we provide some context and a discussion of why environmental education is particularly pertinent in Africa and why farmers' knowledge and experience are important to consider at the policy level for executing better adaptive strategies among farmers.

Understanding Climate Change: African Farmer's Experiences

There is growing evidence that farmers' understanding of the consequences of extreme weather is an essential first step in their development of adaptation or coping strategies for climate change (Boillat & Berkes, 2013; Deressa et al., 2011; Orlove et al., 2010). Kom et al. (2019) show that 80.5 per cent of smallholder farmers in the Vhembe district, South Africa, were aware of the many facets of climate change. When questioned about the nature of the climate changes they had noticed, most smallholder farmers pointed to the shorter rainy season during the past 35 years and the unequal distribution of rainfall from one location to another. In addition, the majority of respondents opined that average temperatures had been steadily climbing over the course of the previous few years. Another study conducted in South Africa by Gbetibouo and Ringler (2009) revealed similar results.

It is projected that the effects of climate change will have severe negative repercussions for Ghana's environment, economy, and society, particularly on rural farmers whose existence is strongly dependent on rainfall. In the study conducted by Benedicta et al. (2010) on 180 agricultural households, the findings indicated that 92 per cent of the respondents experienced temperature increases, whereas 87 per cent perceived a drop in precipitation throughout the years. Antwi-Agyei and Nyantakyi-Frimpong (2021) also studied smallholder farmers in Ghana. Most of their respondents reported five main climatic events that they observed in their areas. These include irregular precipitation, stronger winds, more frequent flooding, higher temperatures, and the depletion of water sources.

In studying farmers of Ethiopia's Central Rift valley (CRV) region, Adimassu and Kessler (2016) find that the majority believe their crop productivity declined. Moreover, they also observed a significant decline in rainfall in the last few years. In the same vein, Simane et al. (2014) find that farmers in Ethiopia point to shifts in rainfall patterns as a reason for decreased crop production. Similarly, cattle producers in Benin's dry and sub-humid tropical zones also reported changes in local temperature and rainfall patterns (Idrissou et al., 2020).

While the level of awareness varies from country to country, multiple studies conducted across Africa suggest that small-scale farmers are growing more conscious of climate change (Awazi et al., 2020). Studies conducted in Nigeria (Ishaya & Abaje, 2008), South Africa (Gbetibouo, 2009), Rural Sahel (Mertz et al., 2009), Ethiopia (Deressa et al., 2011), all claimed that the effects of climate change were becoming increasingly obvious to small-scale farmers.

Major Adaptation Strategies Employed by African Farmers

African farmers adapt to climate change in a variety of ways. Gbetibouo (2009), Hassan and Nhemachena (2008), and Deressa et al. (2010) noted that African smallholder farmers' primary tactic is diversifying their crops. Other studies reported incorporating livestock into crop production, shifting the time of agricultural operations, homestead gardening and switching crops (Altieri & Koohafkan, 2008; Boko et al., 2007; Easterling et al., 2007; Food and Agriculture Organization (FAO), 2009; Gbetibouo, 2009; Molua & Lambi, 2006). Bojang et al. (2020) studied rice farmers in Gambia, West Africa, and they observed that most farmers are now altering their agricultural schedule to adjust to the weather patterns. Moreover, most farmers stopped old types of farming because they take too long to mature and instead, they use enhanced crop varieties that mature more quickly. In another study, Adimassu et al. (2014) observed that farmers in Ethiopia's CRV have relied on a variety of coping mechanisms, the most prominent of which are the sale of livestock, the receipt of relief aid from Governmental institutions and/or Non-governmental organisations (NGOs) the acquisition of credits, and migration to other places. Similar coping mechanisms used by farmers in CRV or elsewhere in the country have also been noted in other studies (Adimassu & Kessler, 2016; Deressa et al., 2009).

According to Antwi-Agyei and Nyantakyi-Frimpong (2021), smallholder farmers in Ghana employ various tactics to deal with environmental uncertainty and change. Migration, planting drought-resistant crops, more irrigation, engaging mixed farming, and implementing sustainable land and soil management are just a few examples. Another study conducted among northern Benin's cattle farmers reports a wide range of adaptation strategies in response to climate change. These include expanding the types of livestock they raise, combining livestock and crop husbandry, reducing the size of their herds, growing their own forage, practicing transhumance, and engaging in non-farming activities (Idrissou et al., 2020).

Molua and Lambi (2006) found that small-scale farmers in Cameroon use a wide range of traditional adaptation strategies in response to climate change. These include shifting the timing of farming operations, adding more space for planting, taking part in traditional and religious ceremonies, switching crops, growing them in different places, and growing local varieties that only need a short growing season.

According to Oxfam (2008), pastoralists in East Africa have resorted to traditional and adaptive strategies such as migration, broadening herd animal mix, adjusting herd size, supplementing grazing with feed, and collecting rainwater as an alternative to the increasingly unreliable groundwater source.

Major Factors Affecting the Local Level Adaptation Strategies

Although farmers take numerous steps to adapt to the impacts of climate change, they still face significant challenges (Antwi-Agyei & Nyantakyi-Frimpong, 2021). Researchers have found that small-scale farmers' perceptions of multiple factors influence their ability to adjust to climatic shifts and variability (Awazi et al., 2020). Factors such as farmers' education, gender, age, soil fertility, farm size, number of years in the industry, tenure, access to extension services, and availability of finance all have a role in how they adapt to changing environmental conditions (Benedicta et al., 2010). In this section, we classify various factors and discuss them within different categories. The following sections elucidate various factors that significantly affect farmers' adaptation strategies in African countries.

Household-Related Factors

Certain characteristics of the household breadwinner were found to have a statistically significant relationship with both soil and water conservation and seasonal migration. In terms of gender, seasonal migration and water conservation are two adaptation techniques that are more common in male-headed families than in those led by women

(Atinkut & Mebrat, 2016). Women have substantially less access to information, land, and other resources than men do, according to Abaje et al. (2014).

In Nigeria, Enete and Onyekuru (2011) find that farmers' investments in measures to adapt to climate change increased with increasing age. In another study, Enete et al. (2002) noted that as older farmers have more experience compared to younger farmers, they are able to make better production decisions. Obayelu et al. (2014) also reported that older, more experienced farmers adopt soil and water conservation strategies more frequently than opting for better crop varieties, mixed farming, diversification into non-farm sectors, and the timing of planting modifications.

Farmers' access to more options for mitigating the effects of a changing climate and improving their quality of life depends on their level of literacy. If heads of households are literate, they may have many more options to diversify their income as a result of having significantly greater access to information regarding non-farm jobs in comparison to household heads who are illiterate (Adimassu & Kessler, 2016). Moreover, education also increases the likelihood of making creative choices (Getachew et al., 2014). In northern Benin, educated farmers prefer to adopt concentrated livestock feeding and forage fodder growing strategies compared to their counterparts (Idrissou et al., 2020). In addition, the number of years a Nigerian farmer spent in school was found to have a strong and very significant association with the amount of money invested in measures for adapting to climate change (Enete & Onyekuru, 2011).

Marital status is a significant factor in farmers' coping strategies. Marital status influences Ethiopian farmers' coping mechanisms, including relocation to other regions (Adimassu & Kessler, 2016). When crops fail, and there is a lack of food, married people and heads of households are less likely to abandon their homes. This illustrates the strong sense of obligation that married men and women have to provide for their families, regardless of gender (Adimassu & Kessler, 2016).

The size of one's family is also important. Lower-income individuals are less likely to be able to afford hired help, and there may be ethical issues associated with tightly overseeing persons who are not members of the family. Therefore, the value of work done by family members increases (Marenya & Barrett, 2007). Idrissou et al. (2020) did research in Benin and found that farmers who rear cattle sometimes grow crops as well due to the availability of labor. Atinkut and Mebrat (2016) find that larger families have a greater probability of adjusting to climate change and reducing its negative impacts. Other studies also supported this finding (Fatuase & Ajibefun, 2013; Menberu & Yohannes, 2014).

Farm-Related Factors

Farming experience is important for adapting to the impacts of climate change. Farmers who have been in the field for longer periods have a better grasp on how to track and react to environmental shifts, making them more likely to employ adaptation strategies that boost resilience (Alhassan et al., 2019; Arunrat et al., 2017; Mwungu et al., 2018). According to research conducted by Maddison (2006), the views of small-scale farmers in Zimbabwe on climate change differed depending on how long they had been involved in agriculture. More experienced small-scale farmers, particularly those with more than 20 years of experience, were more likely to recognize substantial departures from regular weather patterns (Maddison, 2006). Three to four per cent of small-scale farmers surveyed by Mtambanengwe et al. (2012) in two villages in Zimbabwe indicated they had not seen a change in the climate. These farmers tended to be younger and/or spend more time on activities other than farming. In Ethiopia, Adimassu and Kessler (2016) found that heads of households who had been farming for a longer period were more likely to modify the planting dates of their crops and, as a result, adapt to variations in the amount of rainfall. Shiferaw and Holden (1998) also find a positive correlation between farming experiences and the amount of money invested in new and better agricultural technologies in Ethiopia. Idrissou et al. (2020) investigated the effects of climate change on cattle producers in Benin and found that the more experience a farmer has with the industry, the more effectively they can adjust to the changing climate. Some scholars found similar results in northern Ghana (Mabe et al., 2014) and Nigeria (Obayelu et al., 2014).

In Ethiopia, Atinkut and Mebrat (2016) observed that crop variety and water and soil conservation have been positively and significantly connected with the size of the farm. Families having larger farms are more likely to practice crop diversification. Large farmers can afford to take greater risks with their crops and try out novel types compared to small-scale farmers (Adimassu & Kessler, 2016). A similar result was also reported by Misganaw et al. (2014). Adimassu and Kessler (2016) showed that when compared to small landholding households, households that have a large landholding have a greater chance of gaining access to credit because households with large landholdings have a greater ability to repay their credit the following harvest. Households with larger landholdings per person are also less likely to migrate since they can sustain themselves on the increased production. Furthermore, owning a sizable plot of land increases the possibility of employing other adaptive strategies. Adimassu and Kessler (2016) also noted that in Ethiopia, farmers with larger plots of land were more likely to grow eucalyptus trees, which are more resilient to changes in rainfall than annual crops like wheat and teff.

Institutional Factors

It is vital to have easy access to up-to-date weather reports to prepare effectively and minimize climate-related agricultural losses. Farmers can make better decisions about what crops to plant and when to plant them, according to studies, if they have access to accurate weather forecasts in real-time (Amir et al., 2020; Mutunga et al., 2018). A study by Nyang'au et al. (2021) in Kenya showed that 74 per cent of the households surveyed had access to weather and climate-related information from various sources available in their area. Among them, 34.2 per cent of individuals used this information to make better decisions regarding various farming activities.

Atinkut and Mebrat (2016) observed a significant positive link between extension contact and the possibility of selecting three adaptation measures, which include diversification of crops, conservation of soil and water, and seasonal migration among farmers of Ethiopia. Researchers have shown that farmers who receive information from extension agents know more about the challenges posed by climate change and the steps they may take to address them (Gutu et al., 2012; Nhemachena & Hassan, 2007).

Some Other Factors

Farmers' involvement in social groups has been shown in several studies to have a positive correlation with their adoption of a variety of useful strategies. This is because participation in these groups enables farmers to obtain improved information regarding climate and up-to-date information on crop varieties, both of which will increase their resilience (Stefanovic et al., 2017; Washington-Ottombre & Pijanowski, 2013). According to Idrissou et al. (2020), members of livestock associations benefit from the training provided by development partners like NGOs, agricultural development initiatives, and programs because they learn more about climate variability and its potential effects on their livelihoods through these groups.

Role of Environmental Awareness Through Environmental Education

To address climate change, we must first accept that it is occurring and, second, be cognizant of the harm it can cause. Environmental education is one of the potential strategies that can be utilized in the fight against the negative effects of climate change and the development of new adaptations to those effects. Several studies have demonstrated the urgent need to educate or raise public awareness about the catastrophic implications of climate change (Akerlof et al., 2010; Pandve et al., 2011; Ravera et al., 2016). The 13th Sustainable Development Goal (SDG) of the United Nations is to act quickly to fight climate change and its effects (Ahmed et al., 2022; United Nations (UN), Department of Economic and Social Affairs, 2023). To be more specific, target 3 of SDG-13 encourages increasing education and awareness, as well as building human and institutional capacity, to decrease the impacts of climate change, as well as to mitigate and adapt to it (UN, Sustainable Development Goals, 2023).

Numerous studies demonstrate that raising farmers' understanding of the environmental challenges posed by farming increases the likelihood that they will take steps to alleviate the damages, making environmental awareness a key factor in reducing agriculture's negative effects (Hyland et al., 2015; Story & Forsyth, 2008; Sulewski & Gołaś, 2019;). Maddison (2006) suggests that it is essential for farmers to be aware of changes in the climate attributes, such as temperature and precipitation, that they are experiencing to make appropriate adaptation decisions. Also, other studies show that farmers' understanding, and views of the soil erosion problem caused by climate change have a positive effect on their decisions to take soil conservation measures (Anim, 1999; Araya & Adjaye, 2001).

Some farming activities could also exacerbate climate change. Enete and Onyekuru (2011) find that 52 per cent of Nigerians do not believe that farming is also a cause of climate change in their country. Therefore, they highlight the importance of educating the farmers on the outcomes of some of their practices, which can be a powerful tool for climate change mitigation and adaptation in that region (Enete & Onyekuru, 2011). Moreover, adopting environmentally friendly practices is also aided by exposure to environmental education from credible sources (Adu et al., 2021). Eneji et al. (2020) write about the role of trained environmental educators who can work as extension agents in rural communities to make the farmers aware and knowledgeable regarding climate change. They also suggest that the government should emphasize making practical plans and policies to execute this. Anabaraonye et al. (2020) emphasize the role of NGOs in making rural Nigerian farmers aware of climate change by organizing various seminars, workshops, etc. Moreover, they also emphasize the role of religious institutions, mass media (like radio, television, etc.), distribution of flyers to the farmers, and educational blogs to make people aware of the negative consequences of climate change. According to Anabaraonye et al. (2020), farmers need to understand that removing trees at random causes soil erosion and depletes its nutrients. Therefore, educating farmers about environmental issues would raise their awareness.

The target populations for environmental education efforts include both the present generation and future generations. However, some debate exists on which generation should receive most of the attention (Adu et al., 2021). Regardless, there should be a heightened focus on farmers because they are "front-liners" regarding climate change exposure and adaptation. Governments in developing countries often deploy universal extension services to educate farmers to increase crop yields (Zahra, 2018). In this case, government extension agents provide training and on-site visits to help farmers improve their livelihoods by using the latest research-backed technologies (Anderson & Feder, 2007). However, the education offered frequently does not cater to individual's needs, their level of expertise, their goals, or the level of innovation in their community (van Crowder et al., 1998; Vanclay & Lawrence, 1994; Zahra, 2018). Farmers are the land's custodians; their knowledge of the land, its requirements, and the effects of their agricultural methods can shed light on how to improve sustainability in agriculture.

Farmers' experiences can guide sustainable land management practices, including conserving soil, water, and other natural resources and implementing ecologically friendly farming methods. Sustainable farming methods can be applied, and the land can be cared for in the best way possible if farmers' expertise and experience are included in environmental education programs. Therefore, to create successful strategies to combat climate change and achieve SDG-13, it is essential to understand their viewpoints and experiences. When formulating strategies to combat climate change, considering farmers' perspectives helps make sure that solutions are suited to the realities of their environment and are more likely to be successful. However, it is not solely the responsibility of either the government or non-governmental organizations to educate farmers on the unfavorable effects of climate change and to raise awareness of these issues. Rather, for it to be successful, a holistic strategy is required. There is a need for collaboration between governmental and non-governmental organizations, climate change activists, environmental educators, scholars, academics, political and religious elites, and other influential members of society to raise awareness among farmers and continue efforts to educate them to make them sustainable and more resilient against the threat of climate change. Nyang'au et al. (2021) point out that to better synchronize farmer knowledge with their farming activities, it is important to develop valuable links between farmers and key stakeholders (such as researchers, extension officers, and meteorologists). Similarly, Jellason et al. (2022) recognize that farmers should be involved in policymaking because they have firsthand experience with environmental and climatic shifts over multiple generations and may offer unique insights into these shifts and how best to adapt to them based on their circumstances and priorities. They believe such metrics will help develop individualized, context-specific, and workable approaches to adaptation (Jellason et al., 2022).

Finally, a co-production approach that promotes communication and debate between climate researchers and local or indigenous populations is vital if we are to successfully integrate local or indigenous knowledge with scientific information. This process should foster mutual curiosity and openness on how local knowledge can shape scientific understanding and how scientific frameworks can inform and refine local knowledge (Abu et al., 2019; Filho et al., 2022). Dale and Armitage (2011) suggest direct knowledge co-production, which involves the generating, exchanging, and validating of information through collective research procedures. Additionally, Pooley (2013) justifies this co-production method by saying that when researchers take this approach, they engage with the people who are participating in their research as equal partners with the goal of maintaining respect for all holders of information and their individual methods of knowing.

Conclusions and Recommendations

The repercussions of climate change are visible across all social levels. Farmers, who rely considerably on predictable rainfall and temperatures, are particularly exposed to the consequences of climate change. The agricultural sector not only contributes a significant portion to the GDP but also serves as the primary means of subsistence for millions of people, particularly in the countries of Africa. Therefore, the government and relevant stakeholders should place a high priority on agriculture as well as on smallholder farmers.

The present study is an overview of how African farmers view climate change and what factors affect their adaptation strategies. The impacts of climate change may vary from country to country, therefore further country-specific studies are necessary to enable various stakeholders to work together to create effective policies. In addition, there needs to be more coordination of findings across studies and incorporation of farmers' context-specific experience and knowledge and use of those findings in designing and implementing policy. Integrating environmental education is crucial to achieving the Sustainable Development Goals (SDGs) outlined in the 2030 Agenda for Sustainable Development (Sikhosana, 2022; United Nations, 2015). Prabuddh (2018) argued that sustainable leadership in education is a global priority and that environmental education should be used as a driver for SDGs. Moreover, to achieve SDG-13, especially meeting target 3, farmers' knowledge and experiences should be incorporated into the dominant policies, in addition to the ongoing education to share knowledge about climate change with farmers to co-create effective adaptation strategies. In this chapter, we identified some factors that affect local-level adaptation strategies and several issues. Governments, organizations, and other actors should focus on how to incorporate a better understanding of these findings to make farmers more resilient against climate change. To conclude with a concrete example, one major problem identified by local farmers in African countries identified in the research is a lack of access to timely and accurate weather-related information. Therefore, steps should be taken to improve getting access to this information, especially the early warning system for extreme weather events, and to address other issues identified by the front-line farmers. Continuing to listen to farmers and increasing current research efforts that incorporate farmers' experiences may likely improve both mitigation and adaptation outcomes. We therefore also urge policymakers to increase environmental awareness by promoting environmental education.

References

Abaje, I. B., Sawa, B. A., & Ati, O. F. (2014). Climate variability and change, impacts and adaptation strategies in Dutsin-Ma local government area of Katsina State, Nigeria. *Journal of Geography and Geology, 6*(2), 103–112.

Abu, R., Reed, M. G., & Jardine, T. D. (2019). 'Using two-eyed seeing to bridge Western science and Indigenous knowledge systems and understand long-term change in the Saskatchewan River Delta, Canada. *International Journal of Water Resources Development*. https://doi.org/10.1080/07900627.2018.1558050

Adimassu, Z., Kessler, A., & Stroosnijder, L. (2014). Farmers' strategies to perceived trends of rainfall and crop productivity in the central Rift valley of Ethiopia. *Environment and Behaviour, 11*, 123–140.

Adimassu, Z., & Kessler, A. (2016). 'Factors affecting farmers' coping and adaptation strategies to perceived trends of declining rainfall and crop productivity in the central Rift valley of Ethiopia. *Environmental System Research, 5*(13). https://doi.org/10.1186/s40068-016-0065-2

Adu, I. K., Puthenkalam, J. J., & Antwi, K. E. (2021). Environmental generation framework: A case of environmental awareness among farmers and senior high school students for sustainable development. *Journal of Environmental and Agricultural Studies, 2*(1), 62–78. https://doi.org/10.32996/jeas.2021.2.1.7

Ahmed, M. N. Q., Chowdhury, M. T. A., Ahmed, K. J., & Haq, S. M. A. (2022). Indigenous peoples' views on climate change and their experiences, coping and adaptation strategies in South Asia: A review. In M. F. Mbah et al. (Eds.), *Indigenous methodologies, research and practices for sustainable development, world sustainability series*. https://doi.org/10.1007/978-3-031-12326-9_17

Ahmed, M. N. Q., Givens, J., & Arredondo, A. (2024). The links between climate change and migration: A review of South Asian experiences. *SN Social Science, 4*(64). https://doi.org/10.1007/s43545-024-00864-2

Akerlof, K. D., Berry, P., Leiserowitz, A., Roser-Renouf, C., Clarke, K. L., & Rogaeva, A. (2010). Public perceptions of climate change as a human health risk: Surveys of the United States, Canada and Malta. *International Journal of Environmental Research and Public Health, 7*(6), 2559–2566.

Alhassan, H., Kwaka, P. A., & Adzawla, W. (2019). Farmers choice of adaptation strategies to climate change and variability in arid region of Ghana. *Review of Agricultural and Applied Economics (RAAE), 22*(1), 32–40.

Altieri, M. A., & Koohafkan, P. (2008). Enduring farms: climate change, small-scale and traditional farming communities. In *Environment and development series 6*. Third world network. http://www.fao.org/nr/water/docs/enduring_farms.pdf

Amir, S., Saqib, Z., Khan, M. I., Ali, A., Khan, M. A., Bokhari, S. A., Haq, Z. U. (2020). Determinants of farmers adaptation to climate change in rain-fed agriculture of Pakistan. *Arabian Journal of Geoscience, 13*(19).

Anabaraonye, B., Okafor, J. C., & Hope, J. (2020). Educating farmers in rural areas on climate change adaptation for sustainability in Nigeria. In W. Leal Filho (ed.), *Handbook of climate change resilience*. https://doi.org/10.1007/978-3-319-93336-8_184

Anderson, J. R., & Feder, G. (2007). Agricultural extension. In R. E. Evenson, & P. Pingali (Eds.), *Handbook of agricultural economics, agricultural development: Farmers, farm production and farm markets* (Chap. 44, Vol. 3, p. 234378). Elsevier.

Anim, F. D. K. (1999). A note on the adoption of soil conservation measures in the Northern Province of South Africa. *Journal of Agricultural Economics, 50*, 336–345.

Antwi-Agyei, P., Fraser, E. D. G., Dougill, A. J., Stringer, L. C., & Simelton, E. (2012). Mapping the vulnerability of crop production to drought in Ghana using rainfall, yield, and socioeconomic data. *Applied Geography, 32*, 324–334.

Antwi-Agyei, P., & .Nyantakyi-Frimpong, H. (2021). Evidence of climate change coping and adaptation practices by smallholder farmers in Northern Ghana. *Sustainability, 13*(1308). https://doi.org/10.3390/su13031308

Araya, B., & Adjaye, J. A. (2001). Adoption for farm-level soil conservation practices in Eritrea. *Indian Journal of Agricultural Economics, 56*, 239–252.

Archer, E., Mukhala, E., Walker, S., Dilley, M., & Masamvu, K. (2007). Sustaining agricultural production and food security in Southern Africa: An improved role for climate prediction? *Climatic Change, 83*(3), 287–300.

Arunrat, N., Wang, C., Pumijumnong, N., Sereenonchai, S., & Cai, W. (2017). 'Farmers' intention and decision to adapt to climate change: A case study in the Yom and Nan basins Phichit Province of Thailand. *Journal of Cleaner Production, 143*, 672–685. https://doi.org/10.1016/j.jclepro.2016.12.058

Atinkut, B., & Mebrat, A. (2016). Determinants of farmers choice of adaptation to climate variability in Dera woreda, south Gondar zone, Ethiopia. *Environmental Systems Research, 5*(6). https://doi.org/10.1186/s40068-015-0046-x

Awazi, N. P., Tchamba, M. N., Temgoua, L. F., & Tientcheu-Avana, M. L. (2020). Farmers adaptive capacity to climate change in Africa: Small-scale farmers in Cameroon. In W. Leal Filho, et al. (Eds.), *African handbook of climate change adaptation*. https://doi.org/10.1007/978-3-030-42091-8_9-1

Bals, C., Harmeling, S., & Windfuhr, M. (2008). Climate Change, Food Security and the Right to Adequate Food. Diakone Katastrophenhilfe, Brot fuer die Welt and Germanwatch. https://www.germanwatch.org/sites/default/files/publication/2798.pdf

Benedicta, F., Paul, L., Vlek, A., & Manschadi, M. (2010). Farmers' Perceptions and Adaptation to Climate Change: A Case Study in Sekyedumase District of Ashanti Region, Ghana. In *World food system—a contribution from Europe*. Tropentag, Zurich.

Boillat, S., Berkes, F. (2013) Perception and interpretation of climate change among Quechua farmers of Bolivia: indigenous knowledge as a resource for adaptive capacity. *Ecology and Society, 18*(4). https://doi.org/10.5751/ES-05894-180421

Bojang, F., Traore, S., Togola, A., & Diallo, Y. (2020). Farmers perceptions about climate change, management practice and their on-farm adoption strategies at rice fields in Sapu and Kuntaur of the Gambia, West Africa. *American Journal of Climate Change, 9*, 1–10.

Boko, M., Niang, I., Nyong, A., Vogel, C., Githeko, A., Medany, M., Osman-Elasha, B., Tabo, R., & Yanda, P. (2007) Africa. Climate change 2007: impacts, adaptation and vulnerability. In M. L. Parry, O. F. Canziani, J. P. Palutikof, P. J. van der Linden, C. E. Hanson CE (Eds.), *Contribution of working group II to the fourth assessment report of the Intergovernmental Panel on Climate Change* (pp. 433–467). Cambridge University Press. https://www.ipcc.ch/pdf/assessment-report/ar4/wg2/ar4-wg2-chapter9.pdf

Brown, O., Hammill, A., & Mcleman, R. (2007). Climate change as the 'new' security threat: Implications for Africa. *International Affairs, 83*, 1141–1154.

Chaudhary, P., & Bawa, K. (2011). Local perceptions of climate change validated by scientific evidence in the Himalayas. *Biology Letters, 7*, 767–770. https://doi.org/10.1098/rsbl.2011.0269

Dale, A., & Armitage, D. (2011). Marine mammal co-management in Canada's Arctic: Knowledge co-production for learning and adaptive capacity. *Marine Policy, 35*(4), 440–449. https://doi.org/10.1016/j.marpol.2010.10.019

De Souza, K., Kituyi, E., Harvey, B., Leone, M., Murali, K. S., & Ford, J. D. (2015). Vulnerability to climate change in three hot spots in Africa and Asia: Key issues for policy-relevant adaptation and resilience-building research. *Regional Environmental Change, 15*, 747–753.

Deressa, T. T., Hassan, R. M., Ringler, C., Alemu, T., & Yusuf, M. (2009). Determinants of farmers' choice of adaptation methods to climate change in the Nile basin of Ethiopia. *Global Environmental Change, 19*, 248–255.\

Deressa, T. T., Hassan, R. M., & Ringler, C. (2011). Perception of and adaptation to climate change by farmers in the Nile Basin of Ethiopia. *Journal of Agricultural Science, 149*, 23–31.

Deressa, T. T., Ringler, C., & Hassan, R.M. (2010). *Factors affecting the choices of coping strategies for climate extremes: the case of farmers in the Nile Basin of Ethiopia*. IFPRI discussion paper no.01032. International Food Policy Research Institute, Washington, DC, p. 25. http://ebrary.ifpri.org/cdm/ref/collection/p15738coll2/id/5198.pdf

Dhiman, R. C., Pahwa, S., Dhillon, G. P. S., & Dash, A. P. (2010). Climate change and threat of vector-borne diseases in India: Are we prepared? *Parasitology Research, 1064*, 763–773.

Easterling, W. E., Aggarwal, P. K., Batima, P., Brander, K. M., Erda, L., Howden, S. M. A., Kirilenko, A., Morton. J., Soussana, J. F, Schmidhuber, J., & Tubiello, F. N. (2007). Food, fibre and forest products. Climate Change (2007). Impacts, adaptation and vulnerability. In M. L. Parry, O. F. Canziani, J. P. Palutikof, P. J. van der Linden, C. E. Hanson (Eds.), *Contribution of working group II to the fourth assessment report of the Intergovernmental Panel on Climate Change* (pp. 273–313). Cambridge University Press. https://www.ipcc.ch/pdf/assessment-report/ar4/wg2/ar4-wg2-chapter5.pdf

Eneji, C. V. O., Onnoghen, N. U., Acha, J. O., & Diwa, J. B. (2020). Climate change awareness, environmental education and gender role burdens among rural farmers of Northern Cross River State, Nigeria. *International Journal of Climate Change Strategies and Management, 13*(4/5), 397–415.

Enete, A. A., & Onyekuru, A. N. (2011). Challenges of agricultural adaptation to climate change: Empirical Evidence from Southeast Nigeria. *Tropicultura, 29*(4), 243–249.

Enete, A. A., Nweke, F. I., & Tollens, E. (2002). Determinants of cassava cash income in female headed households of Africa. *Quarterly Journal of International Agriculture, 41*(3), 241–254.

Fatuase, A., & Ajibefun, A. (2013). Adaptation to climate change: a case study of rural farming households in Ekiti State, Nigeria. In *Impacts World, International Conference on Climate Change Effects*. Potsdam.

Filho, W. L., Wolf, F., Totin, E., Zvobgo, L., Simpson, N. P., Musiyiwa, K., et al. (2022). Is indigenous knowledge serving climate adaptation? Evidence from Various African Regions. *Development Policy Review*. https://doi.org/10.1111/dpr.12664

Food and Agriculture Organization (FAO). (2009). Climate change and agriculture policies; How to mainstream climate change adaptation and mitigation into agriculture policies. p. 76. http://www.fao.org/fileadmin/templates/ex_act/pdf/Climate_change_and_agriculture_policies_EN.pdf

Food and Agriculture Organization (FAO). (2011). *Framework programme on climate change adaptation*. Fao-Adapt. http://www.fao.org/docrep/014/i2316e/i2316e00.pdf

Food and Agriculture Organization (FAO). (2013). Climate-smart agriculture sourcebook. Food and agriculture organization of the United Nations. In *Sourcebook on climate-smart agriculture, forestry and fisheries*.

Food and Agriculture Organization (FAO). (2016). *Climate change and food security: risks and responses*. http://www.fao.org/3/a-i5188e.pdf

Gbetibouo, G. A., & Ringler, C. (2009). *Mapping South African farming sector vulnerability to climate change and variability: a subnational assessment*. IFPRI Discussion Paper 00885. Washington, DC.

Gbetibouo. A. G. (2009). *Understanding farmers perceptions and adaptations to climate change and variability: the case of the Limpopo Basin, South Africa* (p. 36). IFPRI discussion paper no. 00849. International Food Policy Research Institute. http://www.ifpri.org/publication/understanding-farmers-perceptions-and-adaptations-climate-change-and-variability.pdf

Getachew, S., Tilahun, T., & Teshager, M. (2014). Determinants of agro-pastoralist climate change adaptation strategies: Case of Rayitu Woredas, Oromiya Region, Ethopia. *Research Journal of Environmental Science, 8*, 300–317.

Gutu, T., Bezabih, E., & Mengistu, K. (2012). Econometric analysis of local level perception, adaptation and coping strategies to climate change induced shocks in North Shewa, Ethiopia. *International Research Journal of Agricultural Science and Soil Science, 2*(8), 347–363.

Haq, S. M. A., Arno, A. T., Lalin, S. A. A., & Ahmed, M. N. Q. (2024). Extreme weather events and expected parental roles in Bangladesh and Beyond: A review of literature. *Weather, Climate, and Society*. https://doi.org/10.1175/WCAS-D-23-0060.1

Hassan, R., & Nhemachena, C. (2008). Determinants of African farmers strategies for adapting to climate change: multinomial choice analysis. *African Journal of Agricultural and Resource Economics, 2*(1), 83–104. http://ageconsearch.umn.edu/bitstream/56969/2/0201per cent20Nhemachenaper cent20per cent26per cent20Hassanper cent20-per cent2026per cent20may.pdf

Herrero, M., Ringler, C., van de Steeg, J., Thornton, P., Zhu, T., Bryan, E., Omolo, A., Koo, J., & Notenbaert, A. (2010). *Climate variability and climate change: impacts on Kenyan agriculture*. International Food Policy Research Institute. https://cgspace.cgiar.org/bitstream/handle/10568/3840/climateVariability.pdf

Howden, S. M., Soussana, J. F., Tubiello, F. N., Chhetri, N., Dunlop, M., & Meinke, H. (2007). Adapting agriculture to climate change. *Proceedings of the National Academy of Sciences, 104*, 19691–19696.

Hyland, J. J., Jones, D. L., Parkhill, K. A., Barnes, A. P., & Williams, A. P. (2015). Farmers' perception of climate change: Identifying types. *Agriculture and Human, 33*(2), 323–339. https://doi.org/10.1007/s10460-015-96089

Idrissou, Y., Assani, A. S., Baco, M. N., Yabi, A. J., Traore, I. A. (2020). Adaptation strategies of cattle farmers in the dry and sub-humid tropical zones of Benin in the context of climate change. Heliyon, *6*. https://doi.org/10.1016/j.heliyon.2020.e04373

International Fund for Agricultural Development (IFAD). (2012). Sustainable small-scale agriculture: feeding the world, protecting the planet. In *Proceedings of the governing council events. In conjunction with the thirty-fifth session of IFAD's Governing Council*. Retrieved February

2012, from https://www.ifad.org/documents/10180/6d13a7a0-8c57-42ec-9b01-856f0e994054.pdf

IPCC. (2007). Climate change: impacts, adaptation and vulnerability. In *Contribution of working group II to the fourth assessment report of the intergovernmental panel on climate change.* Cambridge University Press. https://www.ipcc.ch/pdf/assessment-report/ar4/wg2/ar4-wg2-spm.pdf

Ishaya, S., & Abaje, I. B. (2008). Indigenous people's perception on climate change and adaptation strategies in Jema'a local government area of Kaduna state, Nigeria. *Journal of Geography and Regional Planning, 1*(8), 138–143.

Isik, M., & Devadoss, S. (2006). An analysis of the impact of climate change on crop yields and yield variability. *Applied Economics, 387*, 835–844.

Jellason, N. P., Salite, D., Conway, J. S., & Ogbaga, C. C. (2022). A systematic review of smallholder farmers' climate change adaptation and enabling conditions for knowledge integration in Sub-Saharan African (SSA) drylands. *Environmental Development*. https://doi.org/10.1016/j.envdev.2022.100733

Katharine, V. (2004). Creating an index of social vulnerability to climate change for Africa. *Tyndall Centre for Climate Change Research 56*(41).

Kom, Z., Nethengwe, N. S., Mpandeli, S., Chikoore, H. (2019). Climate change grounded on empirical evidence as compared with the perceptions of smallholder farmers in Vhembe District, South Africa. *Journal of Asian and African Studies,* 1–16. https://doi.org/10.1177/0021909619891757

Koundouri, P., & Nauges, C. (2005). On production function estimation with selectivity and risk considerations. *Journal of Agricultural and Resource Economics*, 597–608.

Mabe, F. N., Sienso, G., & Donkoh, S. (2014). Determinants of choice of climate change adaptation strategies in northern Ghana. *Research in Applied Economics, 6*, 75–94.

Maddison, D. (2006). *The perception of and adaptation to climate change in Africa.* Centre for Applied Environmental Economics and Policy in Africa (CEEPA). Discussion paper no 10. CEEPA, University of Pretoria, Pretoria. http://www.ceepa.co.za/uploads/files/CDP10.pdf

Marenya, P. P., & Barrett, C. B. (2007). Household-level determinants of adoption of improved natural resources management practices among smallholder farmers in western Kenya. *Food Policy, 32*, 515–536.

McCarl, B. A. (2010). Analysis of climate change implications for agriculture and forestry: An interdisciplinary effort. *Climatic Change, 100*, 119–124.

Menberu, T., & Yohannes, A. (2014) Determinants of the adoption of land management strategies against climate change in Northwest Ethiopia. *ERJSSH, 1*(1).

Mertz, O., Mbow, C., Reenberg, A., & Diouf, A. (2009) Farmers' perceptions of climate change and agricultural adaptation strategies in rural Sahel. *Environmental Management, 43*(5), 804–816.

Misganaw, T., Enyew, A., & Temesgen, T. (2014). Investigating the determinants of adaptation measures to climate change: A case of Batii district, Amhara region, Ethiopia. *International Journal of Agricultural Research, 9*(4), 169–186.

Molua, E. L., & Lambi, C. M. (2006). *The economic impact of climate change on agriculture in Cameroon.* In CEEPA discussion paper no. 17.

Morton, J. (2017). Climate change and African agriculture: Unlocking the potential of research and advisory services. In F. Nunan (Ed.), *Making Climate Compatible Development Happen* (pp. 109–135). Routledge.

Mpelasoka, F., Awange, J. L., & Zerihun, A. (2018). Influence of coupled ocean-atmosphere phenomena on the Greater Horn of Africa droughts and their implications. *Science of the Total Environment, 610–611*, 691–702.

Mtambanengwe, F., Mapfumo, P., Chikowo, R., & Chamboko, T. (2012). Climate change and variability: small-scale farming communities in Zimbabwe portray a varied understanding. *African Crop Science Journal, 20*(Suppl 2), 227–241. http://www.bioline.org.br/pdf?cs12041pdf

Mutunga, E., Ndungu, C., & Muendo, P. (2018). Factors influencing smallholder farmers' adaptation to climate variability in kitui county, Kenya. *International Journal of Environmental Sciences & Natural Resources, 8*(5), 155–161.

Mwungu, C. M., Mwongera, C., Shikuku, K. M., Acosta, M., & L€ aderach, P. (2018). *Determinants of adoption of climate-smart agriculture technologies at farm plot level: an assessment from southern Tanzania.* In: Handbook of Climate Change Resilience.

Nhemachena, C., & Hassan, R. (2007). *Micro-level analysis of farmers' adaptation to climate change in southern Africa.* IFPRI Discussion Paper 00714. IFPRI, Washington.

Nichols, T., Berkes, F., Jolly, D., et al. (2004). Climate change and sea ice: Local observations from the Canadian Western Arctic. *Arctic, 57*(1), 68–79.

Nyang'au, J. O., Mohamed, J. H., Mango, N., Makate, C., & Wangeci, A. N. (2021). 'Smallholder farmers' perception of climate change and adoption of climate smart agriculture practices in Masaba South Sub-county, Kisii, Kenya. *Heliyon.* https://doi.org/10.1016/j.heliyon.2021.e06789

Obayelu, O. A., Adepoju, A. O., & Idowu, T. (2014). Factors influencing farmers' choices of adaptation to climate change in Ekiti State, Nigeria. *Journal of Agriculture and Environment for International Development (JAEID), 108*, 3–16.

Omann, I., Stocker, A., & Jäger, J. (2009). Climate change as a threat to biodiversity: An application of the DPSIR approach. *Ecological Economics, 691*, 24–31.

Orlove, B., Roncoli, C., Kabugo, M., & Majugu, A. (2010). Indigenous climate knowledge in southern Uganda: The multiple components of a dynamic regional system. *Climatic Change, 100*, 243–265.

Oxfam. (2008). Survival of the fittest. Pastoralism and climate change in East Africa. Oxfam Briefing Paper 116, 2008. http://www.eldis.org/go/country-profiles&id=39194&type=Document

Pandve, H. T., Chawla, P. S., Fernandez, K., Singru, S. A., Khismatrao, D., & Pawar, S. (2011). Assessment of awareness regarding climate change in an urban community. *Indian Journal of Occupational and Environmental Medicine, 15*(3), 109–112. https://doi.org/10.4103/00195278.93200

Pooley, S. (2013). Historians are from venus, ecologists are from mars. *Conservation Biology, 27*(6), 1481–1483. https://doi.org/10.1111/cobi.12064

Prabuddh, M. (2018). *Environmental education as a driver for sustainable development goals.* https://www.researchgate.net/publication/322835204_Environmental_Education_as_a_Driver_for_Sustainable_Development_Goals_Research_Gate

Ravera, F., Martin-Lopez, B., Pascual, U., & Druker, A. (2016). The diversity of gendered adaptation strategies to climate change of indian farmers: a feminist intersectional approach. *Ambio, 45*(S3), S335–S351. https://doi.org/10.1007/s13280-016-0833-2

Regmi, H. R., Rijal, K., Joshi, G. R., Sapkota, R. P., Thapa, S., & Thapa, G. (2020). Climate change perception among peasants: Role of road infrastructure and cooperatives. *Asian Journal of Science and Technology, 11*(08), 11070–11079.

Shiferaw, B., & Holden, S. (1998). Resource degradation and adoption of land conservation technologies in the Ethiopian highlands: Case study in AnditTid. North Shewa. *Agr Econ, 27*(4), 739–752.

Sikhosana, L. (2022). Clarifying the significance of instructional methodologies for integrating environmental education. *Research in Business & Social Science, 11*(7).

Simane, B., Zaitchik, B. F., Foltz, J. D. (2014). Agroecosystem specific climate vulnerability analysis: application of the livelihood vulnerability index to a tropical highland region. *Mitigation and Adaptation Strategies for Global Change* https://doi.org/10.1007/s11027-014-9568-1

Speranza, C. I., Kiteme, B., Ambenje, P., Wiesmann, U., & Makali, S. (2009). Indigenous knowledge related to climate variability and change: Insights from droughts in semi-arid areas of former Makueni District, Kenya. *Climatic Change, 100*, 295–315. https://doi.org/10.1007/s10584-009-9713-0

Stefanovic, J. O., Yang, H., Zhou, Y., Kamali, B., & Ogalleh, S. A. (2017). Adaption to climate change: a case study of two agricultural systems from Kenya. *Climate and Development* https://doi.org/10.1080/17565529.2017.1411241

Story, P. A., & Forsyth, D. R. (2008). Watershed conservation and preservation: Environmental engagement as helping behavior. *Journal of Environmental Psychology, 28*(4), 305–317.

Sulewski, P., & Gołaś, M. (2019). Environmental awareness of farmers and farms' characteristics. *Problems of Agricultural Economics, 4*(361), 55–81. https://doi.org/10.30858/zer/115186

Tabi, F. O., Adiku, S. G. K., Kwadwo, O., Nhamo, N., Omoko, M., Atika, E., & Mayebi, A. (2012). Perceptions of rain-fed lowland rice farmers on climate change, their vulnerability, and adaptation strategies in the Volta region of Ghana. *Technologies and Innovations for Development*. https://doi.org/10.1007/978-2-8178-0268-8_12

Thomas, D. S. G., Twyman, C., Osbahr, H., & Hewitson, B. (2007). Adaptation to climate change and variability: Farmer responses to intra-seasonal precipitation trends in South Africa. *Climate Change, 83*, 301–322.

UNFCCC. (2007). *Report on the workshop on climate-related risks and extreme events*. Note by the Secretariat. FCCC/SBSTA/2007/7. UNFCCC, Bonn, Germany. http://unfccc.int/resource/docs/2007/sbsta/eng/07.pdf

United Nations (UN), Department of Economic and Social Affairs. (2023). https://sdgs.un.org/goals

United Nations (UN), Sustainable Development Goals. (2023). https://www.un.org/sustainabledevelopment/climate-change/

United Nations. (2015). *United Nations sustainable development summit*. https://www.za.undp.org/content/south_africa/en/home/sustainable-development-goals.html

van Crowder, L., Lindley, W. I., Bruening, T. H., & Doron, N. (1998). Agricultural education for sustainable rural development: Challenges for developing countries in the 21st century. *The Journal of Agricultural Education and Extension, 5*(2), 71–84.

Vanclay, F., & Lawrence, G. (1994) Farmer rationality and the adoption of environmentally sound Practices: a critique of the assumptions of traditional agricultural extension. *European Journal of Agricultural Education and Extension,* 1(1)

Vincent, Ã. K. (2007). Uncertainty in adaptive capacity and the importance of scale. *Global Environmental Change, 17*, 12–24. https://doi.org/10.1016/j.gloenvcha.2006.11.009

Wan, C., Shen, G. Q., & Choi, S. (2021). Underlying relationships between public urban green spaces and social cohesion: A systematic literature review. *City, Culture & Society, 24*, 100383. https://doi.org/10.1016/j.ccs.2021.100383

Washington-Ottombre, C., & Pijanowski, B. C. (2013). Rural organizations and adaptation to climate change and variability in rural Kenya. *Regional Environmental Change, 13*, 537–550. https://doi.org/10.1007/s10113-012-0343-0

Watts, N., Adger, W. N., Agnolucci, P., Blackstock, J., Byass, P., Cai, W., Chaytor, S., Colbourn, T., Collins, M., Cooper, A., et al. (2015). Health and climate change: Policy responses to protect public health. *Lancet, 386*, 1861–1914.

Zahra, F. T. (2018). *Educating farmers to be environmentally sustainable: knowledge, skills and farmer productivity in Rural Bangladesh* [Publicly Accessible Penn Dissertations]. https://repository.upenn.edu/edissertations/3002

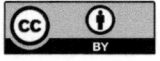

Chapter 10
State Capabilities and Youth Climate Change Education

Chidi Ezegwu and Marcellus Forh Mbah

Abstract Studies on youth education and engagement with climate change issues reveal that while formal education plays a significant role in driving youth engagement, adequate attention has not been given to an effective climate change curriculum and pedagogy that could bring about lasting change. In response, youths have made explicit demands of states and non-state actors at national and international levels to provide them with an effective climate change education that has the potential to lead to appropriate behavioural change or action. This chapter draws on secondary sources to discuss factors affecting the capability of African countries to respond to the demands of youths on climate change. Based on the observable nature of the extractive, regulative and distributive capabilities of many African states, the chapter argues that the continent needs to explore and build efficient and capable institutions to promote context-relevant climate change education by strengthening democracy and good governance. The resulting effect can be seen in an enhanced capacity and empowerment of civil society to act responsibly towards the environment, as effective strategies are implemented to engage young people.

Keywords Climate change education · Youth · State capabilities · Place-based education · Indigenous and local knowledge

Introduction

Scholars agree that climate change is a global and enduring phenomenon with long-term and widespread complexities, requiring a multidimensional approach to address it (Anthony, 2012; Black & Butler, 2014; Harrison, 2023; Kallbekken, 2023; Maron et al., 2016; Schneider & Lane, 2006). The extant literature also indicates that climate

C. Ezegwu (✉) · M. F. Mbah
Manchester Institute of Education, School of Environment, Education and Development, University of Manchester, Manchester, UK
e-mail: ndubuisi.ezegwu@manchester.ac.uk

M. F. Mbah
e-mail: marcellus.mbah@manchester.ac.uk

M. F. Mbah et al. (eds.), *Practices, Perceptions and Prospects for Climate Change Education in Africa*, https://doi.org/10.1007/978-3-031-84081-4_10

change disproportionately impacts different areas of the lives of youths, including education (Karsgaard & Davidson, 2023; Pereznieto et al., 2011). Karsgaard and Davidson (2023) describe how climate change intersects with the well-being of contemporary youths worldwide in diverse ways, impacting their education and life choices. Yet, studies on youth education and participation in climate change actions reveal that while formal education plays a significant role in driving youth engagement, adequate attention has not been placed on an effective climate change curriculum and pedagogy (Eilam, 2022; Kosciulek, 2020; UNDP, 2022b; UNFCCC, 2022). For instance, Eilam (2022) observes that school curricula worldwide have scarcely addressed climate change, and many who pass through educational institutions remain inadequately educated about it. In Pearson's (2021) Global Learner Survey in five countries, 58% of the study participants reported that relevant environmental topics had not been sufficiently taught in schools. The report of a UNESCO (2022) study titled *Youth Demands for Quality Climate Change Education* reveals that the quality of existing climate change education remains in question, and up to 70% of youths who participated in the study had limited knowledge about it. National policies and plans that should communicate states' commitments and capacity to deliver climate change education are also limited. For example, only about 40% of earlier submitted Nationally Determined Contributions (NDC) to the Paris Agreement on climate change made direct references to youth and children, 60% mentioned education in a broad sense, and 23% did not mention youth or children-relevant education at all (UNDP, 2022b).

These observations are particularly critical for Africa, which has an increasing youth population that is projected to constitute over 40% of the low and middle-income school-age population by 2050 and more than a quarter of the global labour force by 2075 (United Nations Economic Commission for Africa, 2017; World Bank, 2023). The observed youth bulge also portends some challenges for Africa. Using data from the International Labour Organisation Statistics, Karkee and O'Higgins (2023) note that over 72 millions of African youths are not in school, training or employment. Regarding climate change, these projections raise questions about how the youths are being prepared for the future and the socioecological environment in which they will live in the coming decades. Barford et al. (2021) caution that young people are more vulnerable to climate and environmental changes but are not adequately educated about them. Consequently, they staged one of the world's largest climate-related protests in September 2019. The youths demand climate change education that will help them comprehend it and take better actions (UNESCO, 2022).

The observed gaps in the provision of climate change education and the demand of youths point to a need to examine how context-relevant climate change education is provided, including the capacity and commitment of the state to do so. This paper discusses the capacity dimension by interrogating African states' capability to provide context-relevant and effective climate change education. It employs Almond's (1965) perspectives on the state's capability to examine Africa's limitations and the ways forward in meeting the demand of youths for the provision of relevant and effective climate change education. It is worth noting at the outset that, while we focus on Africa, only a few selected examples of elements of capability-related issues in some African countries are used to highlight the situation. We do not claim to provide an in-depth discussion of these issues or focus on any particular African country; instead, we draw attention to the element of state capacity that can limit Africa's efforts to meet the demand of youths. In the ensuing section, we briefly discuss the spheres of state capabilities and subsequently examine African states' capacity to promote context-relevant climate change education.

State Capabilities

The demand made by youths for effective climate change education was not made in (and to) a void. It was directed at the international community, which consists of states (including African states) and non-state actors such as development agencies and civil society organisations. Our focus here are the state parties, particularly African states. Article 1 of the Montevideo Convention (1933) defines a state as an entity with a permanent population, a defined territory with a distinct government and the capacity to engage in international relations with other states. States are political systems and entities that act in both international and local environments (Almond, 1956, 1965; Easton, 1957; Pettigrew, 2023). As systems, states (also referred to as nations or countries in this paper) are expected to have capabilities that help them engage effectively in both local and international environments. These capabilities enable them to effectively function, generate and catalyse sustainable development (Almond, 1956, 1965). Almond (1965) identifies five different spheres of capabilities by which the functioning of effective political systems may be analysed. Countries differ in their capabilities, and these tend to influence their respective levels of development and sustainability. We draw on these spheres of capabilities (briefly summarised below; see also Fig. 10.1) to discuss the positioning of African countries in delivering context-relevant climate change education.

According to Almond (1965, p. 196), the concept of capabilities represents "a way of characterising the performance of the political system and of changes in performance, and of comparing political systems according to their performance". The Almond (1965) spheres of capability are summarised below.

a. **Extractive Capability** represents a state's performance in drawing material and human resources from both domestic and international environments to achieve its development goals.
b. **Regulative capability** refers to a state's capacity to control the behaviour of individuals and groups to ensure social order is maintained within its territory and that behaviour conform to the stipulated rules that govern the political system.
c. **Distributive capability** sphere is concerned with the extent to which the state can appropriately allocate goods, services, and opportunities from sources, repositories, and regions of high concentration to locations and populations that need them most. It also considers the state's competence in managing interests and

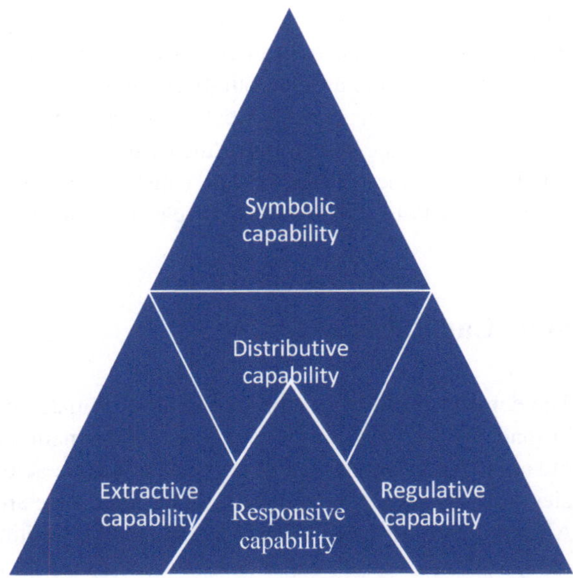

Fig. 10.1 Political system capability triangle

resources to ensure the equitable development of various groups and segments of society.

d. **Symbolic capability** represents the state's ability to reflect its image in the lives and minds of its citizens and effectively demonstrate its policy directions, symbols, and images within the international system.
e. **Responsive capability** refers to the state's capacity to respond effectively to the demands arising from its internal and external environments.

The five spheres of capabilities are interconnected. Strengths or weaknesses in one sphere may be reflected in others due to the interrelated nature of the state's sociopolitical and economic environment. This set of capabilities is linked to system theory (Bang, 2020; Easton, 1957, 1973) and structural–functional perspectives (Harper, 2011; Parsons, 1937) and is used to analyse social institutions and structures. It differs from Amartya Sen's capability approach (CA), which is usually applied to analyse human capability and functioning (Sen, 1993, 1999, 2005).

Like its parent structural–functional theory, one of the key shortcomings of Almond's spheres of capability is their rigidity and conservatism, which limit the framework's ability to account for social change and provide an adequate explanation of the factors that interact in different contexts to produce poverty and economic inequality (Harper, 2011). Sen's CA could be a useful alternative in some contexts, but in relation to our focus, it does not adequately address state-level capabilities. Instead, it centres on human freedom to achieve well-being, emphasising people's capabilities and functionings (Alkire, 2005; Robeyns, 2005, 2021; Sen, 1999, 2005). Nussbaum (2000, 2001, 2004, 2016) developed a list of ten central human capabilities centred on freedom, which state laws ought to promote. However, we find this approach limited for our analysis of state capabilities in delivering climate change education that meets the demands of youths. Given the limitations of Almond's (1965) perspective, we cautiously focus on the first three (extractive, regulative and distributive) in the following section as we discuss various issues that affect the quality of climate change (and broader) education in African countries.

African Capacity to Promote Climate Change Education

There is a shared understanding globally that states ought to provide public goods, such as health services and education to enable their citizens to flourish. Many issues related to sustainable development are connected to the state's capacity to deliver public goods (Ezegwu et al., 2017; Ezegwu, 2020a, b; Khemani, 2019). In this section, we expound on relevant capabilities to discuss African countries' opportunities and challenges in the provision of climate change education in schools which is considered a public good.

Extractive capabilities: The central role of leadership in a political system is to make important decisions that guide how resources are extracted, harnessed, developed and used to achieve national development (Global Infrastructure Hub, 2019). Almond (1965) view a state's performance as evidence of its capability within its environment. Using the extractive capability parameter, we may argue that the capacity of many sub-Saharan African states to harness their human and material resources to advance broader educational development, including climate change education, is relatively limited. They have not been able to mobilise adequate internal and external resources (including foreign aid and international trade) to provide quality education for all. The UNESCO Institute for Statistics (2016) reported that sub-Saharan Africa recorded the highest rates of exclusion in education compared to all other regions of the world, with more than a fifth of sub-Saharan African children aged 6–11 and a third of youths (aged 12–14) being excluded from formal education. Almost 60% of those aged between 15 and 17 were not in school. A similar observation was made in 2019: sub-Saharan Africa had the highest rate of out-of-school children. "Of the 59 million out-of-school children of primary school age, 32 million, or more than one-half, live in sub-Saharan Africa", which also recorded the highest exclusion rate, with 19% of primary school-aged children out of education (UNESCO Institute for Statistics (2019, p. 7). As shown in Table 10.1 the largest proportion of out-of-school children in Africa were those of upper secondary school age (57.5%). The failure of African states to mobilise resources to provide education for children is not just a broken promise to ensure universal access to education; it is a denial of children's basic rights (UNESCO Institute for Statistics, 2016). This failure reveals the state's limited capacity to fulfil the Education for All commitment and advance sustainable development (see Ezegwu, 2020a, b).

The state's failure to provide education for a significant number of primary and secondary school-age children is symptomatic of a broader governance failure. The International Institute for Democracy and Electoral Assistance (IDEA) (2021, p. 1)

Table 10.1 Out-of-school children in Africa

	Out-of-school rate (%)				Out-of-school number (millions)		
	Both sexes	Male	Female	GPIA*	Both sexes	Male	Female
Primary school age	18.8	16.3	21.4	1.24	32.2	14.1	18.1
Secondary school age	36.7	35.3	38.1	1.07	28.3	13.7	14.5
Upper secondary school age	57.5	54.5	60.5	1.10	37.0	17.7	19.3
Combined adolescents and youth of primary, lower secondary and upper secondary age	31.2	28.9	33.6	1.14	97.5	45.5	52.0

* Adjusted gender parity index (GPIA) calculated by dividing the female enrolment ratio by the male enrolment ratio
Source Compiled with data from the UNESCO Institute for Statistics (2019)

observes "a gradual decline in democratic quality in Africa: for the first time in almost a decade, the number of countries designated by the Global State of Democracy (GSoD) Indices as authoritarian now outweighs those deemed to be democratic". This period coincides with a decline in some aspects of development indicators (such as education and human security) across many African countries. For example, Ezegwu, 2020a, b, p. 3) noted that "in 2006, the number of out-of-school children [in Nigeria] stood at around 7.4 million, which increased to about 10.5 million in 2010 and 13.2 million in 2016".

An important aspect of this issue, particularly relevant to climate change, is the capacity of the state to recognise and promote contextually appropriate climate change education frameworks that can enhance deep learning and support a sustainable future. For example, a study by Simpson et al. (2021) across 33 African countries found that the average country-level climate change literacy rate stood at 37%, compared to around 80% across Europe and North America. Zambia is an example of a country with high climate vulnerability and a pressing need for quality climate education interventions. However, the country's National Climate Change Learning Strategy highlights the need to improve staff and institutional capacity to deliver climate change education (Ministry of Land and Natural Resources, Zambia, 2020).

Regulative capability: The relative failure of Africa's governance institutions, noted above has several implications for the continent's development. According to the Global Infrastructure Hub (2019, p. 46), "inclusive institutions with good governance are required to promote and enforce policies that address social inequalities, particularly with regard to under-served and other vulnerable groups". Bad governance and political instability major obstacles to the provision of quality education on a large scale. Between 2020 and 2023, military coups—both successful and attempted—were recorded in Burkina Faso, Central African Republic (CAR), Chad, Djibouti, Gabon, Guinea-Bissau, Madagascar, Mali, Niger and Sudan (Hudson & Towriss, 2023; International Institute for Strategic Studies, 2022; UNDP, 2023a, 2023b). The International Institute for Strategic Studies (2022) Armed Conflict Survey report documented various conflicts in Africa in 2022 and highlighted the involvement of third parties (including state and non-state actors) due to the inability of the affected countries to maintain control, regulate their environment, and ensure human security. It is crucial to note that violent conflicts disrupt education systems (Amnesty International, 2013; Ezegwu et al., 2023; Gersberg et al., 2016). Additionally, coups or political disruptions can lead to the suspension constitutions and the removal of key government officials responsible for maintaining state stability. Ezegwu et al. (2023) highlight the case of Nigeria to demonstrate how high government turnover results in frequent and inconsistent education policy changes. For instance, between 1991 and 1999, Nigeria had eight different federal education ministers across five regimes. Between 1999 and 2015, it had 11 ministers of education, each implementing distinct policies that interrupted their predecessor's education programmes (Obanya, 2011; Gersberg et al., 2016; Ezegwu et al., 2023).

Climate change education is part of the mainstream education system and disruptions due to political conflict and insecurity can ruin its operationalisation and impact. In conflict situations, societal structures often collapse, and governments lose control as the centre cannot often maintain control (Achebe, 1984). According to Hima (2023), the number of people forcibly displaced in Africa in 2022 reached 36 million (mostly women and children), three times higher than similar records over the last decade. Besides displacement, more than 2,000 attacks on education infrastructures were recorded in 14 African countries between 2020 and 2021, with Mali and the Democratic Republic of the Congo being the most affected. Furthermore, threats of attack and armed conflict led to the closure of approximately 7,000 schools in Central Sahel (which includes Mali, Burkina Faso, and Niger), affecting1.3 million children and young people (Hima, 2023). In the Lake Chad Basin region (including Cameroon, Chad, Central African Republic, Libya, Nigeria and Nigeria), Boko Haram (which translates to "Western education is forbidden") and the Islamic State's West Africa Province (ISWP) declared war on formal education. UNDP (2022a, p. 4) notes that Lake Chad Basin countries have struggled with insecurity for decades and, most critically, "Cameroon, Chad, Niger and Nigeria have all been impacted by violent extremism that impedes State functionality" and the ability to provide human security. Where a state cannot ensure security, access to quality education becomes a challenge. If young people cannot access schools, climate change education also remains inaccessible.

Distributive Capabilities: The distributive sphere of analysis evaluates the state's competence in mobilising and equitably distributing public goods equitably and inclusively. Cafaggi and Pistor (2015) also link this to the international context, where state and non-state actors compete for influence and resources in ways that shape domestic development processes. From this perspective, we argue that one of the clearest indicators of African states' incapacity to provide adequate climate change education is the apparent lack of political will to establish African-centred education systems. The extant literature suggests that current curriculums and policies often hinder the provision of context-specific climate change education. Many national education policies have failed to incorporate local and Indigenous perspectives of climate change education and environmental management, including traditional strategies and calendars historically used by Indigenous people (Loya, 2020; Mbah & Ezegwu, 2024; UNESCO, 2018). Loya (2020) explains that educators often claim to maintain neutrality and fairness in teaching, yet many overlook existing power dynamics and how their teaching approaches sustain inequalities in the global education systems. Under this assumed neutrality, educators fail to consider and accommodate learners' lived experiences, sociocultural realities, and local knowledge (Loya, 2020; Mbah & Ezegwu, 2024). Scholars have highlighted how African curriculums remain tied to European models, rooted in flawed colonial education systems that alienated Indigenous cultural values and presented learners with content misaligned with African realities and development needs (Mbah & Ezegwu, 2024; Mbah & Fonchingong, 2019; Scholz et al., 2021). For example, Scholz et al. (2021) found that education policies and plans in Tanzania still reflect 1940s English designs, hindering meaningful integration of new subjects like climate change into the national curriculum. Similarly, Zambia's education plans follow a Western model that does not adequately incorporate the country's cultural values, traditions, and norms (Scholz et al., 2021). Such policies marginalise Indigenous and Local African Knowledge in national and sub-national education processes (Mbah et al., 2021; Opoku & James, 2021). Ajaps and Forh Mbah (2022) argue that rigid curriculum structures create significant barriers to place-based learning, limiting educator's ability to integrate Indigenous and local knowledge into education. Regulatory capability plays a role in enabling education systems to challenge epistemic injustices and reject pedagogies that fail to serve the people.

The pervasive poverty and weak economic development in Africa, widely recognised in the literature as critical barriers to education access (see Gersberg et al., 2016; Ezegwu, 2020a, b), are linked to the distributive capacity of the state—and to some extent, its extractive and regulative capabilities. Corruption undermines the state's ability to equitably extract and distribute its natural and human resources, perpetuating cycles of poverty, underdevelopment, and educational underperformance. The impact of corruption varies across different political systems in Africa. In democratic states, it weakens institutions and oversight bodies responsible for maintaining checks and balances on state powers. In weak democracies and autocratic regimes, corruption fosters ineffective governance as leaders prioritise self-preservation over national development (Duri and Transparency International, 2020). As Khemani (2019) argues, government institutions perform poorly because many internal actors accept mediocrity, believing their misconduct will go unpunished. As a result, African nations struggle to fund transformative education policies, workforce development and technological investments necessary to reduce Africa's vulnerability to climate change. Even Africa's leading economies such as South Africa and Nigeria, face stagnation, while others like Ghana and Ethiopia grapple with mounting debt burdens

(Obe, 2023). Strengthening African states' capabilities is essential to overcoming these limitations and fostering sustainable educational and economic development.

Developing State Capability

Potent ways to deal with weak state capacity include empowerment and reform. Here, we discuss the broader capacity-building needed for sustainable development and (in the following subsection) the steps that need to be taken to improve climate change education for African youths. In doing so, we acknowledge that capacity development takes time and require continuous, progressive engagements. Nonetheless, climate education can be reformed to make it more relevant to youths' needs while simultaneously working to strengthen the state's capacity. While scholars offer different perspectives on how to build state institutional capability (Amin, 2017; Edigheji, 2004; Martinussen, 1998; McGill, 1995), we do not wish to delve into these. Instead, we have identify and briefly discuss practical ways to enhance state capability for delivering climate change education in Africa.

a. **Strengthening democracy and governance**: Political system reforms are essential to developing state capacity. Such reforms must transform governance and restore political power to the people. Ezegwu et al., (2023, p. 127) highlight the "need to deepen democratic practices at all levels to take power from dominant elites that benefited from historical corruption and have used their position to dominate the political space". These reforms are preconditions for fostering objective and broad-based participation, promoting responsible governance, ensuring accountability and need-oriented interventions.
b. **Dealing with corruption and dismantling prebendal structures**: A consistent political will is required to tackle governance and education sector weaknesses—such as corruption and nepotism. Gersberg et al. (2016) detail how corruption as undermined education financing and system effectiveness in Nigeria and recommend transparent frameworks and public sector practices that promote accountability. Ezegwu et al. (2023) recommend dismantling prebendal political structures that lack accountability, noting that policy proclamation alone may not sufficiently address the demand for need-based and context-relevant education.
c. **Addressing root causes of conflict**: Political instability and violent conflicts in many African countries stem from corruption, ethno-religious politics, and struggles over natural resources, among other factors (see Annan, 2004; Aremu, 2010; Williams, 2016). Governments must adopt open, transparent, and inclusive governance practices to address these root causes. Promoting tolerance and ensuring the prudent management of resources will help mitigate grievances, reduce unrest and accelerate development.

d. **Collaboration and partnership**: Capacity development in the education sector requires political will and strong leadership from education managers, policymakers, and curriculum developers. However, it cannot be achieved in isolation, especially given Africa's political, economic and social ties to the global economy. Ajaps and Forh Mbah (2022) emphasise the need for meaningful partnerships and collaborations with international actors to promote context-relevant climate change education. Such collaborations should occur at both vertical and horizontal levels. Vertical collaboration involve international and external partners, while horizontal collaborations focus on intra-African and in-country partnerships. However, care must be taken to ensure that these collaborations align with preserving the autonomy and distinctiveness of the respective knowledge systems in each African state (Mbah & Ezegwu, 2024; Mbah & Fonchingong, 2019; Orlove et al., 2022).
e. **Building staff capacity**: Developing and training state officials and policy leaders is critical to effective governance. As the Global Infrastructure Hub (2019) explains, the attitudes, behaviours, and decisions of those in governance significantly influence system performance. These actors require training, reorientation, and behavioural change to enhance their ability to support state capacity development and inclusively address the needs of the citizenry. Thus, besides dealing with corruption and repressive structures, bespoke capacity building initiatives are necessary to strengthen state administrators and the broader workforce.
f. **Empowerment of the civil society**: An active civil society is crucial in holding public officials accountable. Civil society has a key role in advocating for transparency and utilising information and communication technology to expose unethical practices that weak state capacity (Ezegwu et al., 2023). Apathy and weak or corrupt civil society institutions are among the primary reasons corruption thrives in public spaces. Therefore, mobilisation and empowerment civil society—including academics who produce evidence-based outputs—is essential for enhancing state capability.

A Multi-Dimensional Approach to the Promotion of Effective Climate Change Education for Youths

Besides the broader reforms required to address the various spectrums of capabilities mentioned above, an immediate and more tailored approach to transforming climate change education would require multi-layered engagement. First, African scholars must give critical attention to the production of reliable evidence that can inspire and support political reforms, help civil society organisations engage in policy advocacy and promote relevant programme-based interventions. A literature search suggests that there is a limited amount of published research on the existence, engagement and effectiveness of Indigenous and local knowledge in climate change and environmental education across Africa (Velempini et al., 2018; Zimu-Biyela, 2019; Ajap and Forh Mbah, 2022; Mbah & Ezegwu, 2024). There is also a need for evidence-based policy advocacy to persuade relevant actors and education ministries to reform the general education system and promote Africa-relevant climate change education. While these efforts may not immediately address the identified weak institutional and political capabilities, they could trigger a chain of actions that catalyse capability development and directly inspire positive actions towards Afro-centric climate change education.

People-oriented education sector reforms are crucial for meeting the needs of youths. This makes research and policy collaboration indispensable. While research is necessary to understand local needs and how best to address them, effective policies often rely on reliable evidence. Regarding effective approaches, many African scholars have advocated for place-based education or pedagogy of place. As part of core learning processes, place-based education immerse learners in local and environmental conditions, resources, values, concepts, artefacts, and languages through outdoor learning and interactions with their communities (Velempini et al., 2018; Ferreira, 2020; Ajaps and Forh Mbah, 2022; Mbah & Ezegwu, 2024). According to Ajaps and Forh Mbah (2022), place-based pedagogy involves learners engaging with and understanding their environment, including where they live, the land they farm, and the broader environment associated with Indigenous people. The UK Department for Education (UKDfE, 2023) explains that understanding the world in which learners live involves guiding them to make sense of their immediate and physical environment and community. The depth and frequency of learners' intellectual development tend to increase with their engagement, knowledge and understanding of their environment (or their world), primarily through interactions with their surroundings and people within them. Effective climate change education must provide learners across all levels and disciplines with information about climate change that is relevant to their contexts and intellectually empowering to respond effectively (Greer & Glackin, 2021; Mbah & Ezegwu, 2024; Molthan-Hill et al., 2022).

However, the extant literature suggests that critical place-based approaches, as described above, remain under-explored, along with the existing capacity to deliver them in Africa. Similarly, a UNESCO (2021) report that examined the extent to which climate change issues are integrated into 46 countries' primary and secondary education policies and curriculums notes that there has been little examination of the inclusion of Indigenous and local knowledge in environmental education. This highlights the need to decolonise education policies, curriculums, and teacher training. It also reveals the need to consciously promote learner-centred approaches that provide opportunities to contextualise learners' experiences and adapt learning to their sociocultural backgrounds (Velempini et al., 2018). Such reforms must focus on preparing learners to lead the life, career and change they have reasons to value. This transformation also requires the mobilisation of skilled human resources. Khemani (2019) explains that leaders who propose and pursue reforms must have access to a sufficient pool bureaucrats and state personnel with the expertise to design and implement technically sound policies. At the school level, ensuring the availability of relevant educational resources and materials is essential to facilitate effective teaching and learning in the desired direction.

Conclusion

The mobilisation of youths worldwide to demand appropriate education on climate change implicates states for their failure or limited capacity to help them in their time of need for actionable climate education. It also represents both an indictment and a call on state parties and educators to wake up to their responsibility to educate learners appropriately about climate change. It is an indictment because if states were functioning in all countries (not just in Africa) with relevant sensitivity and capability and providing climate change education appropriately, several youth protests may not have happened. In this chapter, we note that many African states lack adequate capacity to provide the need-based and context-relevant quality education, including climate change education. They are overwhelmed by various factors (e.g. political instability and violent conflicts) that undermine people's peaceful co-existence and freedom to live the lives they have reasons to value. Their situations of governance and sociopolitical and economic progress are marred by corruption, nepotism and prebendal politics.

While youths have the right to be educated appropriately, climate education remains one of the key strategies for promoting sustainable climate action. The potential of African states to effectively provide context-relevant climate change education largely depends on their respective capacities. Each sphere of capacity outlined by Almond is relevant for delivering on this vital right of the youths, but many African states fall short. While we make a strenuous effort to refrain from generalising about the weak institutional capability of most African states, the reality remains that there is little observed in the literature to claim about their efficiency in the provision of formal education and much more the climate change education needed. Among other measures, African countries must muster the political will to decolonise and reform the education system and engage in critical dialogue and collaboration to evolve and deliver African-centred and African-relevant climate change education that recognises local peculiarities and challenges. Scholarly support is essential for generating evidence to guide policy and curriculum alignment.

Although state capability development demands urgent attention, the quest of the youths should not wait until the respective states achieve a reasonable level of capacity. Hence, different states can search for practical strategies for educating the youths and enhancing their climate change action in Africa while also vigorously engaging in capacity development initiatives. Furthermore, investment is needed in research to continuously explore practical approaches and strategies that could work for Africa's climate-resilient via all spectrums of education.

References

Achebe, C. (1984). *Things fall apart*. Anchor Books.

Ajaps, S., & Forh Mbah, M. (2022). Towards a critical pedagogy of place for environmental conservation. *Environmental Education Research, 28*(4), 508–523.

Alkire, S. (2005). Why the capability approach? *Journal of Human Development, 6*(1), 115–135.
Almond, G. A. (1956). Comparative political systems. *The Journal of Politics, 18*(3), 391–409.
Almond, G. A. (1965). A Developmental Approach to Political Systems. *World Politics, 17*(2), 183–214.
Amin, A. (2017). An institutionalist perspective on regional economic development. In *Economy* (pp. 59–72). Routledge.
Amnesty International. (2013). *Keep away from schools or we'll kill you' right to education under attack in Nigeria.* Amnesty International.
Annan, K. (2004). The causes of conflict and the promotion of durable peace and sustainable development in Africa. *African Renaissance, 1*(3), 9–42.
Anthony, R. (2012). Taming the unruly side of ethics: Overcoming challenges of a bottom-up approach to ethics in the areas of food policy and climate change. *Journal of Agricultural and Environmental Ethics, 25*, 813–841.
Aremu, J. O. (2010). Conflicts in Africa: Meaning, causes, impact and solution. *African research review, 4*(4).
Bang, H. P. (2020). David Easton's political systems analysis. In *The sage handbook of political science* (pp. 211–232). SAGE Publications.
Barford, A., Mugeere, A., Proefke, R., & Stocking, B. (2021). *Young people and climate change.* The British Academy.
Black, P. F., & Butler, C. D. (2014). One Health in a world with climate change. *Revue Scientifique Et Technique, 33*(2), 465–473.
Cafaggi, F., & Pistor, K. (2015). Regulatory capabilities: A normative framework for assessing the distributional effects of regulation. *Regulation & Governance, 9*(2), 95–107.
Duri, J., & Transparency International. (2020). *Overview of corruption and anticorruption in sub-Saharan Africa.* Retrieved June 27, 2024, from https://knowledgehub.transparency.org/assets/uploads/kproducts/SubSaharan-Africa_Overview-of-corruption-and-anticorruption-2020.pdf
Easton, D. (1957). An approach to the analysis of political systems. *World Politics, 9*(3), 383–400.
Easton, D. (1973). Systems analysis and its classical critics. *The Political Science Reviewer, 3*, 269.
Edigheji, O. (2004). The African state and socio-economic development: An institutional perspective. *African Journal of Political Science, 9*(1), 84–103.
Eilam, E. (2022). Climate change education: The problem with walking away from disciplines. *Studies in Science Education, 58*(2), 231–264.
Ezegwu, C., Adedokun, A. O., & Ezegwu, C. (2017). Street children, integrated education and violence in northern Nigeria. In *Education and extremisms* (pp. 60–73). Routledge.
Ezegwu, C., Okoye, D., & Wantchekon, L. (2023). Impacts of political breaks on education policies, access and quality in Nigeria (1970–2003). *Research on improving systems of Education.*
Ezegwu, C. (2020a). *Almajri and the Eluding Hope of Accessing Basic Education in Nigeria during COVID-19.* University College London Blog. Retrieved June 12, 2020, from https://blogs.ucl.ac.uk/ceid/2020/06/12/ezegwu/
Ezegwu, C. (2020b). *Masculinity and access to basic education in Nigeria.* Lancaster University. https://doi.org/10.17635/lancaster/thesis/1072
Ferreira, J. G. (2020). Student perceptions of a place-based outdoor environmental education initiative: A case study of the "Kids in Parks" program. *Applied Environmental Education & Communication, 19*(1), 19–28.
Gersberg, A., Rai, S., Ezegwu, C., Nnodu, I., Ojo, A., Panguru, A. Nugroho, C., Elacqua, G., & Alves, F. (2016). *Comparative Review of Basic Education Reforms: Nigerian Country Case Study*. EDOREN and UBEC.
Global Infrastructure Hub. (2019). *Inclusive infrastructure and social equity practical guidance for increasing the positive social outcomes of large infrastructure projects.* Global Infrastructure Hub
Greer, K., & Glackin, M. (2021). 'What counts' as climate change education? Perspectives from policy influencers. *School Science Review, 103*(383), 16–22.
Harper, D. (2011). *Structural-functionalism: Grand theory or methodology.* University of Leicester.

Harrison, R. T. (2023). W(h)ither entrepreneurship? Discipline, legitimacy and super-wicked problems on the road to nowhere. *Journal of Business Venturing Insights, 19*, e00363.
Hima, H. (2023). Learning on the move: Resetting the agenda for education and learning in conflict-affected settings. United States of America. Retrieved January 9, 2024, from https://policycommons.net/artifacts/3857471/learning-on-the-move/4663421/
Hudson, A., & Towriss, D. (2023). *Two more coups in Africa: similarities, differences, and what comes next*. Retrieved July 2, 2024, from https://www.idea.int/blog/two-more-coups-africa-similarities-differences-and-what-comes-next
International Institute for Democracy and Electoral Assistance. (2021). *The state of democracy in Africa and the middle east 2021: Resilient democratic aspirations and opportunities for consolidation*. International IDEA.
International Institute for Strategic Studies. (2022). *The armed conflict survey 2022: Sub-Saharan Africa regional analysis*. Retrieved January 14, 2024, https://www.iiss.org/online-analysis/online-analysis//2022/11/acs-2022-sub-saharan-africa
Kallbekken, S. (2023). Research on public support for climate policy instruments must broaden its scope. *Nature Climate Change, 13*(3), 206–208.
Karkee, V., & O'Higgins, N. (2023). *African youth faces pressing challenges in the transition from school to work*. ILO. Retrieved January 14, 2024, from https://ilostat.ilo.org/african-youth-face-pressing-challenges-in-the-transition-from-school-to-work
Karsgaard, C., & Davidson, D. (2023). Must we wait for youth to speak out before we listen? International youth perspectives and climate change education. *Educational Review, 75*(1), 74–92.
Khemani, S. (2019). *What is state capacity?* World bank policy research working paper, p. 8734.
Kosciulek, D. (2020). *Strengthening youth participation in climate-related policymaking*. Policy briefing, 225.
Loya, K. I. (2020). Creating inclusive college classroom: Granting epistemic credibility to learners. In *Teaching and learning for social justice and equity in higher education: foundations* (pp. 117–135).
Maron, M., Ives, C. D., Kujala, H., Bull, J. W., Maseyk, F. J., Bekessy, S., & Evans, M. C. (2016). Taming a wicked problem: Resolving controversies in biodiversity offsetting. *BioScience, 66*(6), 489–498.
Martinussen, J. (1998). The limitations of the World Bank's conception of the state and the implications for institutional development strategies. *IDS Bulletin, 29*(2), 67–74.
Mbah, M. F., & Ezegwu, C. (2024). The decolonisation of climate change and environmental education in Africa. *Sustainability, 16*(9), 3744.
Mbah, M., & Fonchingong, C. (2019). Curating Indigenous knowledge and practices for sustainable development: Possibilities for a socio-ecologically-minded university. *Sustainability, 11*(15), 4244.
Mbah, M., Johnson, A. T., & Chipindi, F. M. (2021). Institutionalising the intangible through research and engagement: Indigenous knowledge and higher education for sustainable development in Zambia. *International Journal of Educational Development, 82*, 102355.
McGill, R. (1995). Institutional development: A review of the concept. *International Journal of Public Sector Management, 8*(2), 63–79.
Molthan-Hill, P., Blaj-Ward, L., Mbah, M. F., & Ledley, T. S. (2022). Climate change education at universities: Relevance and strategies for every discipline. *Handbook of climate change mitigation and adaptation* (pp. 3395–3457). Springer International Publishing.
Montevideo Convention on the Rights and Duties of States. (1933, 26 December).
Nussbaum, M. C. (2000). 'Aristotle, politics, and human capabilities: A response to Antony', Arneson, Charlesworth, and Mulgan. *Ethics, 111*(1), 102–140.
Nussbaum, M. C. (2001). *Women and human development*. Cambridge University Press.
Nussbaum, M. C. (2004). On hearing women's voices: A reply to Susan Okin. *Philosophy & Public Affairs, 32*, 193.

Nussbaum, M. C. (2016). Introduction: Aspiration and the capabilities list. *Journal of Human Development and Capabilities, 17*(3), 301–308.
Obanya, P. (2011). *Politics and the dilemma of meaningful access to education: The Nigerian story. CREATE pathways to access. Research monograph No. 56*.
Obe, A. V. (2023). *Africa in 2023: Continuing political and economic volatility*. Chatham House. Retrieved January 14, 2024, from https://www.chathamhouse.org/2023/01/africa-2023-continuing-political-and-economic-volatility
Opoku, M. J., & James, A. (2021). Pedagogical model for decolonising, indigenising and transforming science education curricula: A case of South Africa. *Journal of Baltic Science Education, 20*(1), 93–107.
Orlove, B., Dawson, N., Sherpa, P., Adelekan, I., Alangui, W., Carmona, R., ... & Wilson, A. (2022). *ICSM CHC white paper I: Intangible cultural heritage, diverse knowledge systems and climate change. Contribution of knowledge systems group I to the international co-sponsored meeting on culture, heritage and climate change*. Discussion Paper, p. 103 ICOMOS & ISCM CHC.
Parsons, T. (1937). *The structure of social action*. Free Press.
Pearson Global Learner Survey. (2021). *Making the grade for climate education: i learned about the weather, but not climate change*. Retrieved January 14, 2024, from https://plc.pearson.com/sites/pearson-corp/files/pearson/future-of-learning/global-learner-survey/2021/climate/gls-2021-one-pager-climate.pdf
Pereznieto, P., Harper, C., & Jones, N. (2011). *Youth vulnerabilities and adaptation exploring the impact of macro-level shocks on youth: 3F crisis and climate change in Ghana, Mozambique and Vietnam*. Overseas Development Institute.
Pettigrew, J. (2023). *Robber noblemen: A study of the political system of the Sikh Jats*. Taylor & Francis.
Robeyns, I. (2005). The capability approach: A theoretical survey. *Journal of Human Development, 6*(1), 93–117.
Robeyns, I. (2021). The capability approach. In *The Routledge handbook of feminist economics* (pp. 72–80). Routledge
Schneider, S. H., & Lane, J. (2006). An overview of 'dangerous' climate change. *Avoiding Dangerous Climate Change, 7*(11).
Scholz, W., Stober, T., & Sassen, H. (2021). Are urban planning schools in the global south prepared for current challenges of climate change and disaster risks? *Sustainability, 13*(3), 1064.
Sen, A. (1993). Capability and well-being. *The Quality of Life, 30*, 270–293.
Sen, A. (1999). *Development as freedom*. Knopf.
Sen, A. (2005). Human rights and capabilities. *Journal of Human Development, 6*(2), 151–166.
Simpson, N. P., Andrews, T. M., Krönke, M., Lennard, C., Odoulami, R. C., Ouweneel, B., & Trisos, C. H. (2021). Climate change literacy in Africa. *Nature Climate Change, 11*(11), 937–944.
The Ministry of Land and Natural Resources, Zambia. (2020). *National climate change learning strategy*. Republic of Zambia Ministry of Land and Natural Resources.
UK Department for Education. (2023). Development Matters Non-statutory curriculum guidance for the early years foundation stage. Retrieved January 14, 2024, from https://assets.publishing.service.gov.uk/media/64e6002a20ae890014f26cbc/DfE_Development_Matters_Report_Sep2023.pdf
UNDP. (2022b). *Elevating meaningful youth engagement for climate action*. UNDP.
UNDP. (2023a). *Gender-responsive climate change actions in Africa*. UNDP.
UNDP. (2022a). *Conflict analysis in the Lake Chad Basin 2020–2021: Trends, developments and implications for peace and stability*. UNDP.
UNDP. (2023b). *Soldiers and citizens: Military coups and the need for democratic renewal in Africa*. UNDP.
UNESCO. (2018). *The report of the UNESCO expert meeting on indigenous knowledge and climate change, Nairobi, Kenya, 27–28 June 2018*. United Nations Educational, Scientific and Cultural Organization.
UNESCO. (2022). *Youth demands for quality climate change education*. UNESCO.

UNESCO Institute for Statistics. (2016). *Leaving no one behind: How far on the way to universal primary and secondary education?* UNESCO.
UNESCO Institute for Statistics. (2019). *New methodology shows that 258 million children, adolescents and youth are out of school.* UIS Fact Sheet No. 56, September 2019.
UNESCO. (2021). Learn for our planet: A global review of how environmental issues are integrated in education. UNESCO.
UNFCCC. (2022). Africa climate week 2022: Youth affiliated event. UNFCC. Retrieved January 14, 2024, from https://unfccc.int/sites/default/files/resource/Youth%20Affiliated%20Event%20ACW22%20Summary%20Report.pdf
United Nations Economic Commission for Africa. (2017). *Social development policy division. Population and youth section. Africa's youth and prospects for inclusive development: regional situation analysis report.* UN. ECA. https://hdl.handle.net/10855/24011
Velempini, K., Martin, B., Smucker, T., Ward Randolph, A., & Henning, J. E. (2018). Environmental education in southern Africa: A case study of a secondary school in the Okavango Delta of Botswana. *Environmental Education Research, 24*(7), 1000–1016.
Williams, P. D. (2016). *War and conflict in Africa.* John Wiley & Sons.
World Bank. (2023). Investing in youth, transforming Africa. Retrieved January 14, 2024, from https://www.worldbank.org/en/news/feature/2023/06/27/investing-in-youth-transforming-afe-africa
Zimu-Biyela, N. (2019). Using the school environmental education programme (SEEP) to decolonise the curriculum: Lessons from Ufasimba Primary School in South Africa. *International Journal of African Renaissance Studies-Multi-, Inter-and Transdisciplinarity, 14*(1), 42–66.

Part III
Regional and Country Case Studies

Chapter 11
Antecedents of Climate Change Literacy in sub-Saharan Africa

Tabani Ndlovu and Sihle Ndlovu

Abstract While climate change has brutally impacted many African communities, particularly rural communities often inhabited by less-educated subsistence farmers, there seems to be a missing link that ought to map the calamities of climate change with human activities on one hand and remedies to reduce the adverse impacts of climate change on the other hand. This can be traced back to climate change literacy and conceptualisation of this phenomenon, leading to lacklustre climate change mitigation practices in the sub-Saharan African sub content, especially among the vulnerable communities that bear the bulk of the brunt of this phenomenon. Despite many education institutions across sub-Saharan Africa now offering climate change literacy education, the concept remains somewhat abstract, lacking practicability as many graduates face the harsh realities of securing elusive jobs and earning a living as opposed to engaging in broad and futuristic climate change mitigation endeavours. This chapter focuses on sub–Saharan Africa's rural populations who live off the land and are predominantly characterised by low education and resource levels, rendering them unable to actively participate in, and influence climate change initiatives in their localities. The chapter argues that the framing of climate change literacy alienates the very communities it seeks to engage among sub-Saharan Africa's rural and often poor communities. The chapter identifies the socio-economic realities of sub Saharan Africa's rural people which limit their ability to participate effectively in shaping their localities, rendering the current climate change and sustainability discourse elitist. To that effect, the chapter contributes to literature in the area of climate change literacy among vulnerable communities in sub-Saharan Africa and proposes the use of a sustainability discourse ladder as an inclusive framework for translating climate change issues into a format relatable to the targeted groups.

Keywords Climate change literacy · Climate change · sub-Saharan Africa · Sustainable leadership · Climate change discourse ladder

---

T. Ndlovu (✉)
Faculty of Business, Higher Colleges of Technology, Abu Dhabi, UAE
e-mail: tndlovu@hct.ac.ae

S. Ndlovu
GARD Division, Higher Colleges of Technology, Abu Dhabi, UAE
e-mail: sndlovu@hct.ac.ae

M. F. Mbah et al. (eds.), *Practices, Perceptions and Prospects for Climate Change Education in Africa*, https://doi.org/10.1007/978-3-031-84081-4_11

Introduction

As at 6 August 2024, the total population of Africa was estimated to be about 1,498, billion (Worldometer, 2021), rendering Africa the second most populous continent behind Asia. In 2021, the African population was estimated to be growing at a rate of about 2.45% per year, which positioned Africa as the fastest growing continent across the entire globe (Adesete et al., 2023; Fotso-Nguemo et al., 2023). This rapid growth, while representing a potentially sizeable future market, is often credited to poor knowledge of and limited access to family planning resources and knowledge (Adesete et al., 2023). Notably, the high birth rates tend to be prevalent in poor communities which are often associated with lower education levels (Caldwell & Caldwell, 1990; Seidu et al., 2023). Therefore, as may be expected, the growth is closely linked to challenges emanating from vicious cycles fuelled by low education levels, scarce resources and often high unemployment levels (Bratton, 2008; Yaya et al., 2021). The high population growth among Africa's poor and disadvantaged fuels the cycle of poverty and its antecedents such as unemployment, hence the link between high birth rates and poor African communities (Caldwell & Caldwell, 1990; Yaya et al., 2021). Coincidentally, the scenario of poverty, unbridled population growth and resource constraints mean climate change literacy in these communities tends to be lacklustre, impractical or insufficient. This renders these communities ill-prepared for climate change calamities.

This chapter seeks to contribute to the climate change discourse, largely focusing on the African continent, specifically, sub-Saharan Africa. Specifically, the chapter draws insights from a study conducted in 2022 by the authors on climate change readiness in Africa (Ndlovu & Ndlovu, 2022). The study sought views from a sample of influential climate change conference participants drawn from a number of countries across sub-Saharan Africa on a variety of issues around sustainability, climate change, climate change literacy, cultural barriers inhibiting effective climate change preparedness among others. The main findings of the study indicated predominantly low levels of preparedness to mitigate climate change among sub-Saharan Africa's poor rural communities. It could be inferred from the results that rural areas were largely inhabited by less resource-endowed communities whose main focus was eking a living from agricultural activities. They utilised predominantly traditional agricultural methods and had little resources to translate any climate change knowledge into meaningful practical activities and therefore suffered a disproportionately high brunt of climate change calamities repeatedly.

The Context: Poor Communities in Rural Africa

The sub-Saharan Africa population is largely concentrated in rural areas. Most of the people live well below the poverty datum line (Lipton, 2023).sub-Saharan Africa is home to about 50.7% of the world's poorest people (Thurlow et al., 2019. Life in these communes is characterised by high unemployment levels and a struggle for survival (Neumayer, 2011; Khine & Langkulsen, 2023). In contrast to the above, Africa recorded some of the globe's highest primary school enrolment rates standing at well over 80% in 2020 (UNESCO, 2024). The high primary school enrolment rates however falter as one begins to look at higher and tertiary education levels in Sub Saharan Africa (Asongu & Odhiambo, 2019; Obasuyi & Rasiah, 2019). Experts posit that the disparities emanate from the fact that less-educated people in rural communities often lack the resources and or knowledge to engage fully with mainstream economic endeavours (Sharma & Jaiswal, 2018). The failure to gainfully engage with mainstream economic activities also extends to engagement with climate change issues as people's focus is mainly on eking out a living from harsh economic environments (Moshtari & Safarpour, 2024). The poor rural landscapes in sub-Saharan Africa are often characterised by poor infrastructure making them difficult to access and further limiting the inhabitants from actively participating in broader economic and climate change-related issues which tend to be driven from major cities (da Silva, 2023). Most of sub-Saharan Africa's rural populations are therefore not actively participating in topical economic and climate change discourse issues. Instead, their daily survival is premised on more immediate subsistence issues (Magesa et al., 2023). Most of the inhabitants rely on subsistence agriculture despite years of persistent droughts emanating from climate change (Asher & Novosad, 2020).

According to Neumayer (2011) there is a symbiotic link between inequality and unsustainability with one fuelling the other in a deadly cycle. In this discussion, the link alluded to above suggests that the high poverty levels alienate poor people from engaging in sustainability and climate change issues, further driving them into unsustainable activities and fuelling a continuation of the same. The discourse on futuristic and somewhat elitist climate change issues is pushed back as they may not resonate with the immediate struggle for survival (Perry et al., 2023). The lack of engagement by poor communities is regrettable as these are the very people worst affected by climate change. Climate change-related disasters have a higher and adversely disproportionate effect on poor communities (Adesete et al., 2023; Ayanlade et al., 2023). The rather bleak situation necessitates coordinated climate change awareness strategies to address antecedents of climate change literacy which if not addressed, threaten to derail attempts to emancipate the rural poor of sub-Saharan African communities (Nwokolo et al., 2023). Climate change literacy and leadership are crucial to mobilise and engage alienated rural communities to elicit their active participation (Acikalin et al., 2024). While climate change issues are often conflated to synonymously include sustainability, social justice, governance issues among others, this often confuses strategies to be taken to mitigate adverse impacts of the phenomenon. The subsections below examine climate change, focusing on various strands and implications for climate change literacy in Africa, south of the Sahara.

Environmental Sustainability and Climate Change

Environmental sustainability, which closely relates to climate change, focuses on preserving natural resources such as water, air, food, energy and land to meet the needs of current and future generations (Goodland, 1995). To achieve environmental sustainability, resources must be used sparingly to ensure they are not wantonly depleted. Most people living in rural sub-Saharan Africa survive from subsistence agriculture (Oluwatayo & Ojo, 2016). The farming methods used unfortunately do very little to preserve the land (Abdulai et al., 2023). With each year of use, the land gradually degrades, leading to lower yields in subsequent years (Barbier & Hochard, 2018). Some farmers then use chemicals such as fertilizers to improve yields. This unfortunately pollutes the soil and water draining from such soils (Dimkpa et al., 2023).

For many farmers, fertilizers are unaffordable, so they suffer from little to no yields in successive farming years (Oluwatayo & Ojo, 2016). As education levels and land use knowledge tends to be low, soils are not managed effectively leading to soil erosion (Barbier & Hochard, 2018). The degradation of the land causes deforestation which in turn reduces the tree populations that are crucial in absorbing carbon dioxide and mitigating climate change. With little knowledge and scarce resources, the rural communities have little at their disposal to mitigate climate change and nurture the soils that are their source of livelihood (Dimnwobi et al., 2023). Sadly, the act of eking out a living from the soil has become the source of land degradation in rural sub-Saharan communities (Magesa et al., 2023). Arguably then, climate change literacy needs underlying issues to be addressed if it is to take root.

Some agricultural training is being offered both through higher education institutions as well as through the training of farmers (Amede et al., 2023). The lack of resources means that education and training efforts do not go beyond the classroom and thus remain tokenistic at best. It can be argued therefore that the climate change discourse remains accessible and relevant to those with access to resources (Bedeke, 2023). Rural farmers also keep livestock which are said to produce methane gas, another catalyst fuelling greenhouse gas emissions and consequently, climate change (Barbier & Hochard, 2018). A key question therefore is how less-resourced rural communities in sub-Saharan Africa can participate actively in climate change mitigation efforts as well as how the climate change debate can be framed in a way which is accessible to the seemingly alienated rural communities. This chapter contends that there are key antecedents of climate change literacy that need to be addressed if disparate communities of poor farmers in Africa are to be roped in to be part of the climate change mitigation strategies through climate change literacy. Notably though, there are other significant contributors towards deteriorating weather conditions in sub-Saharan Africa as noted in the discussion below.

Commercial Land Use and Exploitation of Resources in Africa

Mining operations have sprouted up in many rural sub-Saharan communities, further compounding land use challenges. Companies from various parts of the world have set up shop and exploited the land, often leaving it unusable for agricultural and other subsistence purposes (Lee, 2017). Taking a cue from international mining operations, local artisanal miners have also traversed the land, employing unsafe practices and further degrading the land (Jegede, 2016; Zvarivadza, 2018). Evidence shows that where mining operations have taken place, the land remains unusable, often characterised by craters, soils polluted by mining chemicals and rocky terrain that poses challenges for the locals and their livestock (Mniga, 2024).

Added to the environmental challenges, in some areas, mining companies are alleged to be abusing workers, often paying bribes to officials so that they look the other way (Chanakira et al., 2019). Such practices have left the poor workers without any form of protection or representation. Questions arise here as to whether climate change literacy has found its way into the players concerned and if so, what could be done to further involve the different stakeholders for more meaningful differences to be made.

The exploitation of land by foreign outfits in sub-Saharan Africa dates back to the colonial era where the sub-continent was parcelled out to different colonial powers and resources extracted and shipped out to Europe (Asafa, 2015; Marschall, 2016). Despite gaining independence, many African countries still see their mineral resources exploited by foreign companies, notably from China (Pigato & Tang, 2015). The concern both pre and post-independence in many countries is that once mining operations have ceased, local communities are left with unusable land (Muhirwa et al., 2023). The poor rural communities are not consulted on the use of their land even though they bear the brunt of the resulting environmental challenges (Zanini et al., 2023). Leadership is often missing on how best to balance different stakeholder interests, with many community leaders feeling powerless to challenge the multinational behemoths which often come with blessings from top political leaders (Dechasa & Desta, 2023). At best, local communities are forced to accept unskilled jobs from companies mining their land, rendering them unable to influence land use practices (Brown, 2019). The discourse on climate change continues to focus on seemingly long term issues, using a language that does not often resonate with poor local communities and thus alienates the very people meant to be engaged (Cheda, 2024). This suggests that climate change literacy could be either missing or ineffective as people remain powerless to make any meaningful changes in dealing with remnants of mining operations.

It is important to point out that the challenges of sustainable land use in Sun Sahara Africa have multi-pronged and often complex causes. Blame cannot be attributed to mining companies alone. Lack of clear local government policies and poor framing of the discourse are also contributing factors. Seeing no other ways to sustain themselves, many people in sub-Saharan countries have resorted to artisanal mining

(Obodai et al., 2023). Without up-to-date land use knowledge and tools, the artisanal miners employ basic methods that scour the land and leave it prone to erosion (Djibril et al., 2017; Hagos et al., 2016; Moyo et al., 2018). When the rains come, the soil is washed away leaving gullies and contributing to deforestation (Jønsson & Fold, 2011). From the scenario painted above, it is clear that local poor people face tough choices—leave the land alone and starve or engage in rudimentary mining operations to survive. If these people were equipped with appropriate education and resources, perhaps their impact on the land would be different (Amoako et al., 2023). It can be argued that proactive climate change literacy and leadership can significantly change land use practices in sub-Saharan Africa.

The absence of proactive and appropriately framed climate change literacy initiatives leaves communities in rural sub-Saharan Africa reeling from continued droughts which lead to poor agricultural yields. This is increasingly forcing many people to turn to artisanal mining, an area where they are even less equipped to operate sustainably. The result is that whatever action these communities take to survive, they are exacerbating the effects of climate change in their communes (Muhirwa et al., 2023). Poor rural communities are rendered powerless, often focusing merely on getting menial jobs to survive (Carstens & Hilson, 2009). The poor land use practices perpetuate a cycle of poverty and renders the rural communities as victims of resultant climate change calamities requiring intervention from higher authorities in their countries (Way, 2016). The discourse on climate change literacy in rural sub-Saharan African communities needs underlying leadership and education issues to be put in place.

Arguably then, the discussion of climate change literacy remains elitist and remote to many poor communities whose focus is eking out a living. Climate change literacy needs to relate to people's day-to-day issues to be meaningful (Rakshit et al., 2023). With limited resources, many of Africa's poor communities unfortunately endure harsh realities caused by climate change (Ahmadalipour et al., 2019; Weber et al., 2020). Subsistence reliance on the land leaves little space available for the affected communities to engage with seemingly nebulous and theoretical climate change concepts that may not be directly relatable to people's day-to-day lives (Sharma & Jaiswal, 2018). This chapter posits that with a different approach and framing of the climate change discourse, rural communities could be enabled to engage better.

Economic Influences on Climate Change Literacy in Africa

From a sustainability point of view, rural communities ought to be engaged on how they can sustainably fend for themselves while preserving resources for future generations (Kuhlman & Farrington, 2010; Matekenya et al., 2021). This suggests that issues of survival should not be divorced from long term use of natural resources as this is intricately linked to climate change. It is advisable to initiate programs that financially emancipate previously disenfranchised communities so that they can actively participate in broader climate change discourse issues. Opportunities that enrich some while causing harm to the environment or sidelining other stakeholders should be avoided in an inclusive, stakeholder-oriented approach. The sad reality on the ground is that narrow self-interests of a few are prioritised and pursued even at the expense of the environment or the rights of others (Rikani et al., 2023). This may be caused by lack of resources, low education levels which are recipes for poor governance systems and weak or corruptible leadership (Sharma & Jaiswal, 2018). Prevailing policy guidelines in many sub-Saharan countries may not be fit for purpose to curtail emerging climate change and sustainability vices. Low education levels often lead to limited employment opportunities which in turn means disenfranchisement from actively participating in wealth creation endeavours and payment of taxes (Rikani et al., 2023). Limited employment opportunities compromise people's propensity to voice their concerns and actively engage in the current climate change discourse (Madhou & Sewak, 2019). To correct this, people need to be equipped with both resources as well as climate change literacy that is anchored in their day-to-day environments.

There is light at the end of the tunnel though as sub-Saharan's population is increasingly getting younger (Baah-Boateng, 2016). The younger generations may engage more actively with increased emphasis of sustainability curricula in institutions of higher learning across the sub-continent (Chigunta, 2017; Moshtari & Safarpour, 2024). According to ILOSTAT (2021), the sub-Saharan Africa unemployment rate stood at 6.629% in 2020, a figure which aggregates rather big disparities between high unemployment in some Southern African countries such as Eswatini, Namibia, South Africa, Sudan and Zimbabwe pitted against low unemployment rates in African states such as Rwanda, Tanzania and Uganda (Fields, 2023).

Climate Change and Social Sustainability in sub-Saharan Africa

Social sustainability discussions centre on delivering social justice and emancipation of people. Despite social sustainability increasingly featuring in sustainability discourse in Africa, understanding of this by both communities and organisational players is still at infancy (Denu et al., 2023). According to ADEC Innovations (2021), the concept of social sustainability comprises initiatives to engender justice, fairness and equity to deliver fair working environments, access to health facilities, affordable hygienic living conditions, equality and diversity, inclusive empowerment of communities and promotion of a volunteering culture. All these ideals are desirables that should be a common standard for any human being. The reality especially in the sub-Saharan region is that many people are invariably denied the above rights (Dada et al., 2023). With many people being financially excluded from partaking in wealth creating endeavours, this reduces their social status and renders them incapable of engaging in participating in social justice issues. Further, lack of climate change education or irrelevant climate change education renders these people incapable of actively participating in climate change discourse to emancipate their communities from the harsh grip of climate change (Chitimira & Warikandwa, 2023).

Economic emancipation therefore acts as a key building block, without which people are unable to participate in social justice issues (Ibid, 2023). With sub-Saharan Africa hosting over 70% of the world's poor (Ajide & Dada, 2023), the region is a focal point for poverty and strife, and by extension, the majority of people here lack the capacity and wherewithal to engage in meaningful climate change response (Beegle et al., 2016; Bolarinwa et al., 2021).

Poor economic conditions coupled with fledgling democratic and governance systems means that a lot of people may not be equipped to stand up for themselves and hold their leaders to account (Milanovic, 2023). The climate change narrative should be framed in a way that engages people from the grassroots to address their core day-to-day survival issues. Presenting a standard "catch all" message will likely not resonate with poor, disenfranchised sub-Saharan communities. This chapter argues that the climate change discourse in sub-Saharan Africa needs to address underlying issues of poverty, access to resources in order to bring sustainability high enough in poor people's orders of hierarchy.

Education in Africa and Implications for Climate Change Literacy

Through education, there is hope that the climate change discourse could be adapted to mobilise and emancipate rural sub-Saharan populations (Ajaps, 2023). Notwithstanding the recognised role of education, the education penetration rate still remains low in sub-Saharan Africa (Dembélé & Oviawe, 2007). The sparse distribution of populations across the sub-continent's rural and often inaccessible landscapes makes it difficult for the governments to provide education for all (Moshtari & Safarpour, 2024). Consequently, generations of less educated and remote communities perpetuate with no hope to break the cycle unless people are resettled (De Schutter et al., 2023). Coupled with low education, the same communities lack access to well-developed healthcare systems and viable economic opportunities.

Where education is available, sometimes it fails to be meaningful to address prevailing sustainability issues (Lipton, 2023). Research has shown that in many instances, curricula is poorly designed, focusing on just reading and writing as opposed to the acquisition of the twenty-first century skills (Moshtari & Safarpour, 2024; Neumayer, 2011; Thurlow et al., 2019). Therefore, despite possessing education certificates, graduates may still be unable to seek entrepreneurial opportunities and instead, continue to focus on the search for elusive jobs. Where employment opportunities are available, they offer menial wages. Limited earnings mean that parents are not able to send their children to good schools thus sentencing them to the same fate and the cycle continues.

As a way out of the perpetual cycles of low education, poverty, limited employment opportunities, many people move to urban areas (Tacoli et al., 2015). Unfortunately, the receiving urban towns and cities are themselves not well equipped to house the influx of migrants. This results in the mushrooming of slums and a life of squalor for many migrants.

Rural to Urban Migration in sub-Saharan Africa

The move from rural areas to sub-Saharan Africa's cities has not changed the economic and social prospects of many migrants (Østby, 2016). Many of the migrants fail to find work and end up living in slums with poor living conditions and lives which are far worse than those they escaped from in the rural areas (Gould & Prothero, 2023). Slums of Kibera in Kenya, Khayelitsha and Gugulethu in South Africa are examples of the foiled hopes and dreams of many migrants. It would seem that efforts to escape poverty instead land the migrants to further poverty but this time in an urban setting (Chen et al., 2023). Yet again, the framing of the climate change discourse eludes the migrants, despite their move to the urban areas. To change the lives of these people, this chapter argues that a focus on entrepreneurship education can engender a new culture of creating work as opposed to looking for work (Fomba et al., 2023). The curricula needs refocusing to equip graduates with twenty-first century skills while reframing the climate change discourse to ensure it is centred on people's day to day challenges.

Mobile Telephony and Digitisation in Africa

Mobile phone penetration rates have soared in sub-Saharan Africa (Asongu et al., 2018). This presents huge opportunities for disseminating information in general and climate change education in particular. In some communities, mobile telephony has ushered in a new era of money transfers and banking services, facilitating easy trade between people and communities (Asamoah et al., 2020). Notwithstanding this, there is still a big challenge of availability of electricity to charge mobile telephony gadgets and ancillary systems supporting such (Amankwah-Amoah et al., 2018; Asamoah et al., 2020). Again sub-Saharan Africa is missing out on available solar resources which should power up all devices if requisite infrastructure was availed. The mobile telephony penetration rates present a new entrepreneurial eco system that could create employment and trade opportunities around selling airtime, charging devices, repairs and servicing among others. Overall, it is very encouraging to see the high penetration rates of mobile telephony in sub-Saharan Africa although the jury is still out on the effectiveness of this in advancing education in general and climate change literacy in particular (Nwokolo et al., 2023).

Climate Change and Gender Differences

There are still concerns around the equitable treatment of women versus men in in some parts of sub-Saharan. Gender injustices and the role of women renders females in poor rural communities more susceptible to lower access to education and opportunities (Van den Broeck et al., 2023). Lack of access to education is closely linked to gender inequalities and poverty (Bassey & Bubu, 2019). To fully emancipate people in rural settings, gender imbalances have to be reversed, with education in general and climate change emphasizing on the roles of both men and women. General education and climate change literacy in particular have the potential to create opportunities for both male and females but particularly bring up women to take up their places among their male counterparts. To achieve this, education must tackle some of the cultural roots that permeate today's gender imbalances. This must empower women so that they are not dependent on their male folks and can stand on their own (Devonald et al., 2024). The journey towards true sustainability therefore requires emancipation of all people. Education sits at the core of any emancipation strategies and needs to be reframed to equip people with twenty-first century skills with a focus on addressing real life issues and gender imbalances.

The Climate Change Discourse Ladder: A Prpoposal

The foregoing discussions argued that education in general and climate change literacy in particular exclude many of sub-Saharan Africa's less educated rural poor, and therefore needs to be reframed for relevance and currency. The climate change literacy discourse is arguably disconnected from the realities of the people targeted and therefore does not resonate with the aspirations of these people. The issues and examples highlighted in the current climate change discourse tend to be long term, thereby not addressing poor people's immediate survival concerns and thus failing to engage them on adoption of mitigation strategies. In response, this chapter proposes a climate change discourse ladder to help relate climate change literacy to poor and less educated people's peculiar life circumstances (see Ndlovu & Ndlovu, 2022). The proposal is for:

(i) A reframed climate change discourse structured in the form of a Climate Change Discourse Ladder to localise climate change issues by relating them to people's personal day-to-day lives, conjuring initiatives of what can be done to mitigate climate change effects. The approach starts at an individual level, gradually broadening out to seek more collective solutions at community, regional, national and international levels for a truly collaborative approach. Figure 11.1 proposes the climate change discourse ladder, a framework for translating climate change issues into personal contexts and gradually broadening these out to local community issues and eventually to broader multi stakeholder and multinational or even global issues.

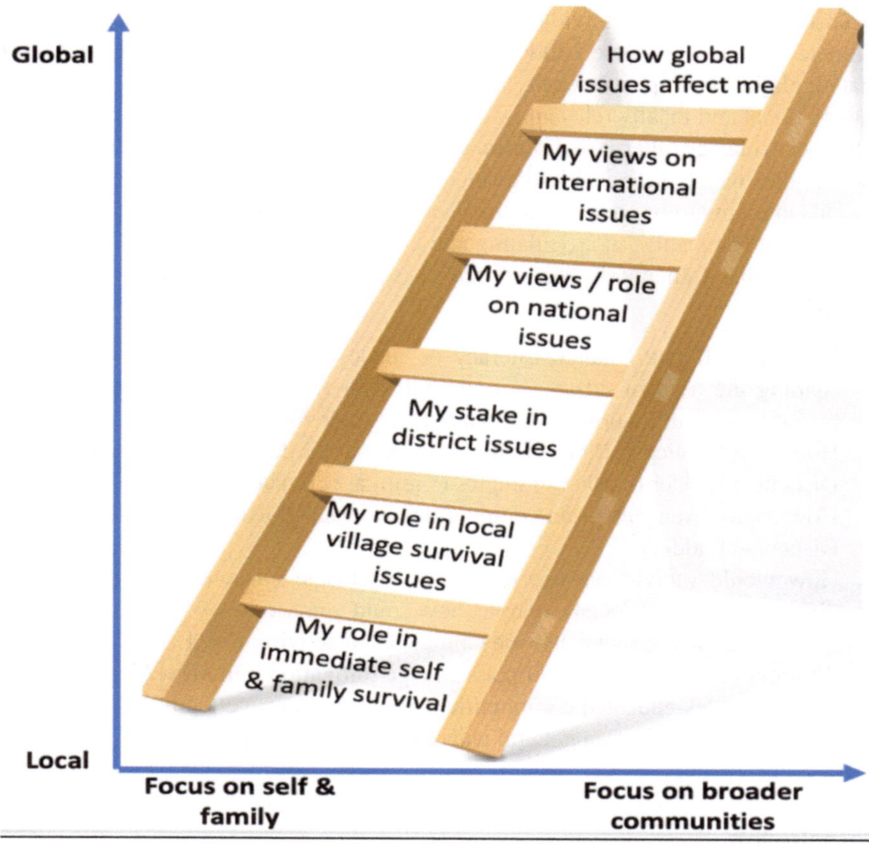

Fig. 11.1 The climate change discourse ladder. *Source* Ndlovu and Ndlovu (2022)

(ii) Proactive and engaging climate change leadership to rope in disenfranchised communities,
(iii) Meaningful and practical education fostering the acquisition of 21st century skills that go beyond just learning to read and write.
(iv) Climate change literacy that fosters community connectedness, harnessing the collaborative power of otherwise disparate community groups into a cohesive force operating in unison.

The Climate Change Literacy Ladder posits that to begin with, sustainability efforts need to be embraced locally at a personal level, with a particular focus on the individual and their family. From this local level base, the framework expands to include the wider community efforts and interests. The individual and community efforts then need to be integrated, thereby culminating in a jointly shared, mutual and collaborative response to the sustainability agenda. Such an approach acknowledges sustainability as a shared responsibility, which requires efforts from all community stakeholders but fosters ownership at an individual level because of relatability of sustainability to the individual and their family survival issues. Therefore, the Climate Change Discourse Ladder attempts to provide a sustainability framework that highlights the importance of relating otherwise abstract climate change concepts into personal and locally relevant issues. The framework is by no means perfect but attempts to engage otherwise disenfranchised people and communities by reframing the climate change discourse into a language that ordinary people can relate to.

This attempts to re-envision approaches to engage communities that have traditionally not been active in articulating and promoting climate change issues. While proposing this framework, readers are encouraged to ponder on the following questions:

1. How can Climate Change Literacy be made mandatory for leaders who are shaping the lives of different communities in Africa, parcelling out land and resources to multinational firms but leaving their communities at risk?
2. How could policy makers in sub-Saharan Africa deploy the Climate Change Discourse Ladder to drive strategies to mitigate climate change effects?
3. How could disenfranchised communities be reached using the Climate Change Discourse Ladder?
4. How would individuals use the Climate Change Discourse Ladder in their different contexts? As an example, how could you use it?
5. How could Governments use the Climate Change Discourse Ladder to mobilise different groups to develop climate change mitigation initiatives?
6. How could les-educated communities be engaged more effectively on issues to do with sustainability in general and climate change in particular?

Conclusion

This chapter highlighted the intricate links between sub-Saharan Africa's poor subsistence economies and the impact this has on climate change literacy. The chapter highlighted the potentially vicious cycles of poverty and repetitive calamitous climate disasters that can perpetuate in the poor sub-Saharan communities unless intervention measures are initiated. It argued that the current framing of the climate change discourse renders it elitist and inaccessible to less educated and often poor communities who ought to benefit from a better understanding of how to combat the adverse effects of climate change. Because of limited understanding of climate change issues and lack of resources, poor communities are less prepared for climate change-related calamities and suffer adversely higher proportions from such phenomenon. The chapter proposed a Climate change discourse ladder, aimed at scaffolding climate change literacy and unpacking this in bite-size chunks accessible to the vulnerable communities in sub-Saharan Africa.

References

Abdulai, A. R., Tetteh Quarshie, P., Duncan, E., & Fraser, E. (2023). Is agricultural digitization a reality among smallholder farmers in Africa? Unpacking farmers' lived realities of engagement with digital tools and services in rural Northern Ghana. *Agriculture & Food Security, 12*(1), 11.

Acikalin, ŞN., Esra, S. A. R. I., & Ercetin, ŞŞ. (2024). Role of education in awareness on climate change. *Current Perspectives in Social Sciences, 28*(1), 56–63.

ADEC Innovations. (2021). Retrieved August 11, 2024, from https://www.adecesg.com/resources/faq/what-is-social-climatechange/

Adesete, A. A., Olanubi, O. E., & Dauda, R. O. (2023). Climate change and food security in selected sub-Saharan African countries. *Environment, Development and Sustainability, 25*(12), 14623–14641.

Ahmadalipour, A., Moradkhani, H., Castelletti, A., & Magliocca, N. (2019). Future drought risk in Africa: Integrating vulnerability, climate change, and population growth. *Science of the Total Environment, 662*(1), 672–686.

Ajaps, S. (2023). Deconstructing the constraints of justice-based environmental sustainability in higher education. *Teaching in Higher Education, 28*(5), 1024–1038.

Ajide, F. M., & Dada, J. T. (2023). Poverty, entrepreneurship, and economic growth in Africa. *Poverty & Public Policy, 15*(2), 199–226.

Amankwah-Amoah, J., Osabutey, E. L., & Egbetokun, A. (2018). Contemporary challenges and opportunities of doing business in Africa: The emerging roles and effects of technologies. *Technological Forecasting and Social Change, 131*(1), 171–174.

Amede, T., Konde, A. A., Muhinda, J. J., & Bigirwa, G. (2023). Sustainable farming in practice: Building resilient and profitable smallholder agricultural systems in sub-Saharan Africa. *Sustainability, 15*(7), 5731.

Amoako, C., Adarkwa, K. K., & Koranteng, K. A. (2023). Survival now, sustainability later: The emerging artisanal mining and the dying agricultural livelihoods in the Akyem Abuakwa traditional area of Ghana. *Environment, Development and Sustainability, 25*(2), 1645–1666.

Asafa, J. (2015). The triple causes of African underdevelopment: Colonial capitalism, state terrorism and racism. *International Journal of Sociology and Anthropology, 7*(3), 75–91.

Asamoah, D., Takieddine, S., & Amedofu, M. (2020). Examining the effect of mobile money transfer (MMT) capabilities on business growth and development impact. *Information Technology for Development, 26*(1), 146–161.

Asher, S., & Novosad, P. (2020). Rural roads and local economic development. *American Economic Review, 110*(3), 797–823.

Asongu, S. A., & Odhiambo, N. M. (2019). Basic formal education quality, information technology, and inclusive human development in sub-Saharan Africa. *Sustainable Development, 27*(3), 419–428.

Asongu, S. A., Nwachukwu, J. C., & Aziz, A. (2018). Determinants of mobile phone penetration: Panel threshold evidence from sub-Saharan Africa. *Journal of Global Information Technology Management, 21*(2), 81–110.

Ayanlade, A., Smucker, T. A., Nyasimi, M., Sterly, H., Weldemariam, L. F., & Simpson, N. P. (2023). Complex climate change risk and emerging directions for vulnerability research in Africa. *Climate Risk Management, 40*, 100497.

Baah-Boateng, W. (2016). The youth unemployment challenge in Africa: What are the drivers? *The Economic and Labour Relations Review, 27*(4), 413–431.

Barbier, E. B., & Hochard, J. P. (2018). Land degradation and poverty. *Nature Climate Change, 1*(11), 623–631.

Bassey, S. A., & Bubu, N. G. (2019). Gender inequality in Africa: A re-examination of cultural values. *Cogito, 11*(3), 21–36.

Bedeke, S. B. (2023). Climate change vulnerability and adaptation of crop producers in sub-Saharan Africa: A review on concepts, approaches and methods. *Environment, Development and Sustainability, 25*(2), 1017–1051.

Beegle, K., Christiaensen, L., Dabalen, A., & Gaddis, I. (2016). *Poverty in a rising Africa.* World Bank Publications.

Bolarinwa, S. T., Adegboye, A. A., & Vo, X. V. (2021). Is there a nonlinear relationship between financial development and poverty in Africa? *Journal of Economic Studies.*

Bratton, M. (2008). Poor people and democratic citizenship in Africa. In *Poverty, participation, and democracy: A global perspective* (pp. 28–64).

Brown, M. B. (2019). *Africa's choices: After thirty years of the World Bank.* Routledge.

Caldwell, J. C., & Caldwell, P. (1990). High fertility in sub-Saharan Africa. *Scientific American, 262*(5), 118–125.

Carstens, J., & Hilson, G. (2009). Mining, grievance and conflict in rural Tanzania. *International Development Planning Review, 31*(3), 301.

Chanakira, D. K., Mujere, J., & Spiegel, S. (2019). Traditional leaders and the politics of labour recruitment in Zimbabwe's platinum mining industry. *The Extractive Industries and Society, 6*(4), 1274–1281.

Cheda, T. (2024). The implication of extractive industries' operation on human and environmental rights of vulnerable groups in pursuit of environmental sustainability in Africa. In *Human rights and the environment in Africa* (pp. 353–371). Routledge.

Chen, M., Huang, X., Cheng, J., Tang, Z., & Huang, G. (2023). Urbanization and vulnerable employment: Empirical evidence from 163 countries in 1991–2019. *Cities, 135*, 104208.

Chigunta, F. (2017). Entrepreneurship as a possible solution to youth unemployment in Africa. *Laboring and Learning, 10*(2), 433–451.

Chitimira, H., & Warikandwa, T. V. (2023). Financial inclusion as an enabler of United Nations sustainable development goals in the twenty-first century: An introduction. *Financial inclusion and digital transformation regulatory practices in selected SADC countries: South Africa, Namibia, Botswana and Zimbabwe* (pp. 1–22). Springer International Publishing.

da Silva, A. (2023). *The road to success* (Doctoral dissertation, Northern Arizona University).

Dada, J. T., Ajide, F. M., & Arnaut, M. (2023). Income inequality, shadow economy and environmental degradation in Africa: Quantile regression via moment's approach. *International Journal of Development Issues, 22*(2), 214–240.

De Schutter, O., Frazer, H., Guio, A. C., & Marlier, E. (2023). *The escape from poverty: Breaking the vicious cycles perpetuating disadvantage.* Policy Press.

Dechasa, R. D., & Desta, H. M. F. (2023). Mining operations and marginalized communities: Environmental racism in Ishmael Beah's radiance of tomorrow. *Journal of Equity in Sciences and Sustainable Development, 6*(1), 43–59.

Dembélé, M., & Oviawe, J. (2007). Introduction: Quality education in Africa: International commitments, local challenges and responses. *International Review of Education/Internationale Zeitschrift für Erziehungswissenschaft/Revue Internationale de l'Education*, 473–483.

Denu, M. K., Bentley, Y., & Duan, Y. (2023). Social sustainability performance: Developing and validating measures in the context of emerging African economies. *Journal of Cleaner Production, 412*, 137391.

Devonald, M., Jones, N., Iyasu Gebru, A., & Yadete, W. (2024). Rethinking climate change through a gender and adolescent lens in Ethiopia. *Climate and Development, 16*(3), 176–186.

Dimkpa, C., Adzawla, W., Pandey, R., Atakora, W. K., Kouame, A. K., Jemo, M., & Bindraban, P. S. (2023). Fertilizers for food and nutrition security in sub-Saharan Africa: An overview of soil health implications. *Frontiers in Soil Science, 3.*

Dimnwobi, S. K., Okere, K. I., Onuoha, F. C., & Ekesiobi, C. (2023). Energy poverty, environmental degradation and agricultural productivity in sub-Saharan Africa. *International Journal of Sustainable Development & World Ecology, 30*(4), 428–444.

Djibril, K. N. G., Cliford, T. B., Pierre, W., Alice, M., Kuma, C. J., & Flore, T. D. J. (2017). Artisanal gold mining in Batouri area, East Cameroon: Impacts on the mining population and their environment. *Journal of Geology and Mining Research, 9*(1), 1–8.

Fields, G. S. (2023). The growth–employment–poverty nexus in Africa. *Journal of African Economies, 32*(Supplement_2), ii147–ii163.

Fomba, B. K., Talla, D. N. D. F., & Ningaye, P. (2023). Institutional quality and education quality in developing countries: Effects and transmission channels. *Journal of the Knowledge Economy, 14*(1), 86–115.

Fotso-Nguemo, T. C., Weber, T., Diedhiou, A., Chouto, S., Vondou, D. A., Rechid, D., & Jacob, D. (2023). Projected impact of increased global warming on heat stress and exposed population over Africa. *Earth's Future, 11*(1), e2022EF003268.

Goodland, R. (1995). The concept of environmental climate change. *Annual Review of Ecology and Systematics, 26*(1), 1–24.

Gould, W. T., & Prothero, R. M. (2023). Space and time in African population mobility. In *People on the move* (pp. 39–50). Routledge.

Hagos, G., Sisay, W., Alem, Z., Niguse, G., & Mekonen, A. (2016). Participation on traditional gold mining and its impact on natural resources, the case of Asgede Tsimbla, Tigray, Northern Ethiopia. *Journal of Earth Sciences and Geotechnical Engineering, 6*(1), 89–97.

ILOSTAT. (2021). Retrieved August 11, 2024, from https://data.worldbank.org/indicator/SL.UEM.TOTL.ZS?locations=ZG

Jegede, A. O. (2016). The environmental and economic implications of the climate change and extractive industry nexus in Africa. *Environmental Economics, 7*(4), 95–103.

Jønsson, J. B., & Fold, N. (2011). Mining 'from below': Taking Africa's artisanal miners seriously. *Geography Compass, 5*(7), 479–493.

Khine, M. M., & Langkulsen, U. (2023). The implications of climate change on health among vulnerable populations in South Africa: A systematic review. *International Journal of Environmental Research and Public Health, 20*(4), 3425.

Kuhlman, T., & Farrington, J. (2010). What is sustainability? *Sustainability, 2*(11), 3436–3448.

Lee, C. K. (2017). *The specter of global China: Politics, labor, and foreign investment in Africa.* University of Chicago Press.

Lipton, M. (2023). Why poor people stay poor. In *Rural development* (pp. 66–81). Routledge.

Madhou, A., & Sewak, T. (2019). Examining structural unemployment in sub-Saharan Africa: Empirical evidence from unobserved components. *Open Economies Review, 30*(5), 895–904.

Magesa, B. A., Mohan, G., Matsuda, H., Melts, I., Kefi, M., & Fukushi, K. (2023). Understanding the farmers' choices and adoption of adaptation strategies and plans to climate change impact in Africa: A systematic review. *Climate Services, 30*, 100362.

Marschall, S. (2016). *The heritage of post-colonial societies* (pp. 347–363). Routledge.

Matekenya, W., Moyo, C., & Jeke, L. (2021). Financial inclusion and human development: Evidence from sub-Saharan Africa. *Development Southern Africa, 38*(5), 683–700.

Milanovic, B. (2023). The great convergence: Global equality and its discontents. *Foreign Affairs, 102*, 78.

Mniga, M. (2024). *Recolonizing Africa: An ethnography of land acquisition, mining, and resource control.* Taylor & Francis.

Moshtari, M., & Safarpour, A. (2024). Challenges and strategies for the internationalization of higher education in low-income East African countries. *Higher Education, 87*(1), 89–109.

Moyo, F., Ndlovu, T., Francis, B., & Ncube, T. M. (2018). The effects of artisanal mining on irrigation farming-the case of Umzinyathini Irrigation Scheme in Umzingwane District, Southern Matabeleland, Zimbabwe. *African Journal of Public Affairs, 10*(2), 139–162.

Muhirwa, F., Shen, L., Elshkaki, A., Hirwa, H., Umuziranenge, G., & Velempini, K. (2023). Linking large extractive industries to sustainable development of rural communities at mining sites in Africa: Challenges and pathways. *Resources Policy, 81*, 103322.

Ndlovu, T., & Ndlovu, S. (2022). The Primacy of People's Socio-Economic Issues as Antecedents of Sustainability Framing in Africa. *Management and leadership for a sustainable Africa, volume 1: Dimensions, practices and footprints* (pp. 37–57). Springer International Publishing.

Neumayer, E. (April, 2011). Climate change and inequality in human development. *UNDP-HDRO occasional papers.*

Nwokolo, S. C., Eyime, E. E., Obiwulu, A. U., & Ogbulezie, J. C. (2023). Africa's path to sustainability: Harnessing technology, policy, and collaboration. *Trends in Renewable Energy, 10*(1), 98–131.

Obasuyi, F. O. T., & Rasiah, R. (2019). Addressing education inequality in sub-Saharan Africa. *African Journal of Science, Technology, Innovation and Development, 11*(5), 629–641.

Obodai, J., Mohan, G., & Bhagwat, S. (2023). Beyond legislation: Unpacking land access capability in small-scale mining and its intersections with the agriculture sector in sub-Saharan Africa. *The Extractive Industries and Society, 16*, 101357.

Oluwatayo, I. B., & Ojo, A. O. (2016). Is Africa's dependence on agriculture the cause of poverty in the continent? An empirical review. *The Journal of Developing Areas, 50*(1), 93–102.

Østby, G. (2016). Rural–urban migration, inequality and urban social disorder: Evidence from African and Asian cities. *Conflict Management and Peace Science, 33*(5), 491–515.

Perry, M., Okot, A., Mfitumukiza, D., Pullanikkatil, D., Kayendeke, E., Mwesigwa, G., Thakadu, O., & Muwanika, V. (October, 2023). Survival vs. sustaining: A multidisciplinary inquiry of the environmental dilemma in rural Uganda. In *Natural resources forum*. Wiley.

Pigato, M., & Tang, W. (2015). *China and Africa: Expanding economic ties in an evolving global context*.

Rakshit, B., Jain, P., Sharma, R., & Bardhan, S. (2023). An empirical investigation of the effects of poverty and urbanization on environmental degradation: The case of sub-Saharan Africa. *Environmental Science and Pollution Research, 30*(18), 51887–51905.

Rikani, A., Otto, C., Levermann, A., & Schewe, J. (2023). More people too poor to move: Divergent effects of climate change on global migration patterns. *Environmental Research Letters, 18*(2), 024006.

Seidu, A. A., Ahinkorah, B. O., Anjorin, S. S., Tetteh, J. K., Hagan, J. E., Jr., Zegeye, B., Adu-Gyamfi, A. B., & Yaya, S. (2023). High-risk fertility behaviours among women in sub-Saharan Africa. *Journal of Public Health, 45*(1), 21–31.

Sharma, G., & Jaiswal, A. K. (2018). Unclimate change of climate change: Cognitive frames and tensions in bottom of the pyramid projects. *Journal of Business Ethics, 148*(2), 291–307.

Tacoli, C., McGranahan, G., & Satterthwaite, D. (2015). *Urbanisation, rural-urban migration and urban poverty*. Human Settlements Group, International Institute for Environment and Development.

Thurlow, J., Dorosh, P., & Davis, B. (2019). Demographic change, agriculture, and rural poverty. *Sustainable Food and Agriculture*, 31–53.

UNESCO. (2024). Primary school enrolment rates in sub-Saharan Africa. Retrieved August 11, 2024, from https://uis.unesco.org/en/topic/education-africa

Van den Broeck, G., Kilic, T., & Pieters, J. (2023). Structural transformation and the gender pay gap in sub-Saharan Africa. *PLoS ONE, 18*(4), e0278188.

Way, S. A. (2016). *Examining the links between poverty and land degradation: From blaming the poor toward recognising the rights of the poor* (pp. 47–62). Routledge.

Weber, T., Bowyer, P., Rechid, D., Pfeifer, S., Raffaele, F., Remedio, A. R., Teichmann, C., & Jacob, D. (2020). Analysis of compound climate extremes and exposed population in Africa under two different emission scenarios. *Earth's Future, 8*(9), 1473.

Worldometer. (2021). Retrieved August 11, 2024, from https://www.worldometers.info/world-population/africa-population/#google_vignette

Yaya, S., Yeboah, H., & Udenigwe, O. (2021). Demography, development and demagogues. Is population growth good or bad for economic development? In *Beyond free market* (pp. 109–124). Routledge.

Zanini, M. T. F., Migueles, C. P., Gambirage, C., & Silva, J. (2023). Barriers to local community participation in mining projects: The eroding role of power imbalance and information asymmetry. *Resources Policy, 86*, 104283.

Zvarivadza, T. (2018). Climate change in the mining industry: An evaluation of the National Planning Commission's diagnostic overview. *Resources Policy, 56*(1), 70–77.

BY

Chapter 12
Critical Thinking in Sustainable Business: Examining Pragmatic and Ethical Issues for Supporting Biodiversity (Eco-centrism) in Business Education

Helen Kopnina, Mariusz Baranowski, Liz Nantunda, Wilson Muyinda Mande, Tatjana Radovanovic, and Zaina Gadema

Abstract This research aims to examine the case for mainstreaming the positioning of ecocentric perspectives in university business school settings. We argue for an urgent need to reorient anthropocentric normative framings of sustainability in business and management education to ecocentric ontologies and epistemologies within pedagogical praxes of design and delivery. We draw upon two examples from university business schools' sustainability programmes in Africa and The United Kingdom as comparative cases to illuminate the commonalities and differences in ontological and epistemological characteristics. Uganda and England are contextually different, whether it be geography, economy(ies), or how these contexts, in turn, shape/influence pedagogical approaches and practices in each of those business schools. Findings show that Education for Sustainable Development Goals

---

H. Kopnina (✉) · L. Nantunda · Z. Gadema
Newcastle Business School, Northumbria University, Newcastle, UK
e-mail: helen.kopnina@northumbria.ac.uk

L. Nantunda
e-mail: liz.nantunda@northumbria.ac.uk

Z. Gadema
e-mail: zaina.gadema@northumbria.ac.uk

M. Baranowski
Faculty of Sociology, Department of Sociology of Social Stratification, Adam Mickiewicz University, Poznań, Poland
e-mail: mariusz.baranowski@amu.edu.pl

W. M. Mande
Nkumba University, Entebbe Nkumba, Entebbe, Uganda
e-mail: mandewm@nkumbauniversity.ac.ug

T. Radovanovic
Engineering Faculty, Seoul National University, Seoul, South Korea

M. F. Mbah et al. (eds.), *Practices, Perceptions and Prospects for Climate Change Education in Africa*, https://doi.org/10.1007/978-3-031-84081-4_12

(ESDGs) increasingly feature within courses across 'sustainable' management-centric pedagogies such as 'sustainable marketing' and 'sustainable supply chain management'. However, these normative framings of sustainability pedagogies were found to potentially negate the inclusion of 'deep green' rooted concepts, ontologies and epistemologies in business and management education. In contrast to these normative approaches, we present and analyse opportunities for critically evaluating sustainability education that centres on biodiversity through ecopedagogy and eco-literacy.

Keywords Biodiversity · Business education · Eco-literacy · Education for Sustainable Development Goals (ESDGs) · Environmental education

Introduction: Sustainable Development Versus Environmental Sustainability

Ecosystem decline has experienced a notable surge in acceleration over the past few decades, as documented by reputable sources such as the International Union for Conservation of Nature (IUCN, 2022) and the Intergovernmental Science-Policy Platform on Biodiversity and Ecosystem Services (IPBES, 2019). IUCN demonstrates that the number of critically endangered or extinct species is increasing exponentially, including, in 2022 alone, giant tortoise, Bramble Cay Melomys, Baiji, Spix's macaw, Western black rhinoceros, the Ivory-billed woodpecker and the Australian mountain mist frog (Morton, 2022). The root causes of ecosystem decline include growth in human population and associated increase in production, exacerbated by global structural inequalities (IPBES, 2019; IUCN, 2022).

Does the current business university curriculum sufficiently address the significant issue of ecosystem decline within the field? This inquiry examines the extent to which educational programs within business schools are equipped to tackle and comprehensively engage with the complexities surrounding terminal disappearances as central tenets within the broader domains of ethical and social dimensions of contemporary business practices.

Half a century ago, the Limits to Growth report (Meadows et al., 1972) sounded a cautionary note, followed by the Tbilisi Declaration, endorsed by the United Nations Environmental Program (UNEP) and the United Nations Educational, Scientific and Cultural Organization (UNESCO), which strongly promoted environmental education (UNESCO, 1977).

This educational paradigm aimed to inform and inspire students, fostering ecological literacy, often termed eco-literacy (Van Matre, 1978). The Tbilisi Declaration outlined specific goals and principles for environmental education, emphasizing ecological knowledge, cultivating awareness of environmental challenges, nurturing sensitivity to the need for conservation, and actively equipping individuals with the skills to address environmental issues (UNESCO, 1977). A decade later, the World

Conference for Environment and Development (WCED) convened by the United Nations (UN) in 1983, introduced the conceptual definition of sustainable development in the seminal publication of the Brundtland Report, 'Our Common Future' (WCED, 1987). According to critics, this report downplays the fundamental drivers of environmental degradation—the growth economy (IPCC, 2022; Moranta et al., 2022; Washington, 2020), and "inadequate attention is paid to the question of how growth can be decoupled from biodiversity loss" (Otero et al., 2020, 1).

Nevertheless, the conventional curriculum targets Education for Sustainable Development (ESD) and Education for Sustainable Development Goals or ESDGs (Chiang & Chen, 2022; Weybrecht, 2022). The 'goals' paradoxically tend to marginalise ecological restoration or nonhuman species' rights to flourish (Kopnina, 2020; Tallberg et al., 2022; Visseren-Hamakers, 2020). Ontological framings of the Anthropocene are usually driven from a humanist perspective, as "most sustainability education promotes humans as the primary change agents and environmental stewards" (Taylor, 2017, 1461).

The Chartered Association of Business Schools (CABS, 2021) identifies schools along the gradient of what they see as their contribution to the 'public good'. Most of it is ESDG with no mention of biodiversity, aside from a single mention in the Foreword by Paul Polman at the University of Oxford: "We urgently need to start living within our **planetary boundaries**, before we reach irreversible tipping points that do permanent damage to our **biodiversity** and critical ecosystems—nature's guarantors of happy, healthy and prosperous societies" (CABS, 2021, 2).

Perhaps one of the reasons planetary boundaries and biodiversity are not reflected in Business and Management School curriculums is the normative paradox of responsible management education. Educators adopting this perspective typically frame social and ethical values through a 'responsible business' curriculum. Economic theory is the 'go to' ontological domain, emphasising freedom of choice, and the capitalist profit-motive (Moosmayer et al., 2019). Some researchers believe that SDGs are of central importance in business schools precisely because they do not challenge existing corporate practices (Chiang & Chen, 2022; Seto-Pamies & Papaoikonomou, 2020).

Considering the primary foci of the SDGs is placed upon economic welfare and social justice, they have, to date, as broad framings, proved woefully inadequate for addressing biodiversity loss and halting habitat decline (IPBES, 2019; IUCN, 2022). Standardised pedagogies "do not provide the paradigm shift that is needed to respond to the implications of the Anthropocene" (Taylor, 2017, 1461). Yet, SDGs are readily embraced by business schools (Dean et al., 2018; Rashid, 2019; Fang & O'Toole, 2023), perhaps precisely because they do not challenge the growth and profit paradigm. Since the business case for biodiversity conservation, highlighted in the Leaders' Pledge for Nature, calls for considering the economics of biodiversity loss, the question of market mechanisms emerges. Panwar et al. (2023) reflect that businesses still must realise that few of their present strategies are adequate in addressing biodiversity loss. Acknowledging this shortcoming, it was noted that without degrowth and a more nature-inclusive, ecocentric or Earth-centred approach, ecosystem decline is likely to continue (Hankammer et al., 2021; Piccolo et al., 2022;

Taylor et al., 2020; UN, 2022; Washington et al., 2018). Yet, as Bobulescu (2021) asserts, business schools continue to fail to teach students to think critically beyond normative conceptualisations and framings of sustainable development.

In addition to discussing the normative framings of sustainable development and issues of biodiversity, ecopedagogy and eco-literacy, this chapter discusses the approaches to teaching (about) biodiversity at Northumbria University in the United Kingdom (UK) and Nkumba University, in Uganda, East Africa. These universities were chosen because they offer a contrast between how biodiversity-related issues are taught in developed and developing countries with various histories of involvement with biodiversity protection and education. An important distinction is the identification of two geographically, political and education-policy different points of analysis, one being in the Global North and another in the Global South. Both global areas are separated by starkly differing rates of economic development.

As a more developed country (MDC), the UK is long known to grapple with the numerous side-effects of development, encompassing high rates of urbanisation, consumption, production, rapid and continued biodiversity loss. The UK, in 2024 is ranked 18th in the World for its HDI ranking (Human Development Index). Uganda, in contrast, formerly categorised as a least developed country (LDC) in 2022, by 2024, moves to the categorisation of a 'lower-middle income' country, having achieved substantive improvements in its HDI ranking, having climbed 7 places from 166 in 2022 to 159 in 2024 for the 193 countries assessed by the United Nations Development Programme (UNDP) (UNDP, 2024). Some pockets of wilderness remain in Uganda, but these spaces stay at risk of significant decline rates (Iritié, 2015; Rosales, 2008). Most of the United Kingdom's biodiversity finds itself within protected areas (The Environmental Audit Committee, 2022), where 75% of zoos under the Animal Ambassador Encounters (AAEs) umbrella encourage human interaction through education that supports close contact with animals, compared to over 50% of Uganda's wildlife resources are outside designated protected areas, mostly in privately owned land transformed for grazing and agriculture purposes, an urgent concern for protection and change (NEMA, 2016; Spooner et al., 2021).

In the United Kingdom, over a hundred biodiversity-related programs are offered at the Master's level (https://www.mastersportal.com/study-options/268927102/biodiversity-conservation-united-kingdom.html). All universities in the UK are accountable to 'independent' institutions that are nevertheless guided by government policy for higher education. Examples would be Universities UK (UUK) or the Office for Students (OFS). UK is a 'knowledge' economy, an institutionally structured MDC. This explains the emphases in education programmes, whether it be a focus on STEM, for instance, or the prevalence of reductive and normative biases in business models for sustainability, typically espousing techno-centric forms of intervention/models of sustainability.

In Uganda, some universities such as Makerere University Kampala (Mak) focusing on PhD programmes such as 'Plant Breeding and Biotechnology in 2008 and Agricultural and Rural Innovations in 2012' as well as Uganda Martyrs University (UMU) focusing on 'Agroecology and Food Systems in 2015'. Likewise, programmes at the Master level in MUK focusing on plant Breeding and Seed systems in 2008, and UMU's Monitoring and Evaluation In 2016 were initiated. The eco-pedagogy in Universities tries to integrate the Government's priority development agenda as set out in the National Vision 2025, linking to the Centre of Excellence for Climate Smart Agriculture and Biodiversity Conservation (SABC).

This chapter responds to the deeply problematic nature of a sustainable development framework for addressing the global environmental crisis through (business) education. A few alternative visions for environmental sustainability have been developed, some by independent scientific bodies, governments, and non-governmental organizations. The 'Leaders' Pledge for Nature' promised to step up global ambition for biodiversity with UNESCO (n.d.). The UN Harmony with Nature report (UN, 2022) states: "Loss of biodiversity, desertification, climate change and the disruption of natural cycles are among the costs of our disregard for Nature and the integrity of its ecosystems and life-supporting processes".

This chapter argues that SDG-oriented teaching in business schools needs rethinking. Below, we review several promising initiatives for safeguarding the environment, each connected, directly or not, to businesses, through the concept of eco-literacy. As opposed to the anthropocentric framing of the environment in business and management education, eco-literacy stimulates the development of interdisciplinary ecological knowledge. At the same time, ecopedagogy supports didactic strategies for well-being within the limits of planetary boundaries (Whiteman et al., 2013).

Following this review, we discuss the relevance of these initiatives for business education, highlighting differences between the dominant ESDGs and more species-inclusive, ecosystem- and biodiversity-oriented transformative education. In the analysis sections, we argue that both an ethical re-orientation towards ecosystem-centred approaches to nature and a pragmatic re-orientation to education is needed, i.e., to address some of the root causes of ecosystem decline, such as human population growth and increase in production.

From the Leaders' Pledge for Nature Positive Approaches

The Leaders' Pledge for Nature highlights the fact that environmental degradation and social and economic well-being are interrelated. It states:

> Science clearly shows that biodiversity loss, land and ocean degradation, pollution, resource depletion, and climate change are accelerating at an unprecedented rate. This acceleration is causing irreversible harm to our life support systems and aggravating poverty and inequalities as well as hunger and malnutrition. (Leaders' Pledge for Nature, 2021, np)

In response to this Pledge 'Creating a Nature-Positive Future' (UNDP et al., 2021) outlines the need for diverse measures to halt biodiversity loss. But what does 'nature positive' biodiversity conservation look like, and can it help slow or reverse biodiversity decline?

Critical Pedagogy and Ecopedagogy

Critical pedagogy (Freire, 2000), applied in business education (Bobulescu, 2021) encourages lecturers and students to challenge existing power structures, becoming aware of and questioning societal hegemonies. These types of education include promoting critical thinking (particularly on economic growth), the tools needed to understand non-linear relationships in complex systems, embracing an environmental ethic, making/taking active responsibility, and generating lifelong/career commitments to environmental activism (OECD, n.d.). If we were to consider the ambition of the Nature Needs Half (NNH) movement, much more activism in education and beyond would be needed (Kahn, 2008).

Related to critical pedagogy, ecopedagogy enables critical thinking about paradoxes and trade-offs, such as the push for economic development and simultaneously the desire to protect the environment (Gaard, 2008; Misiaszek, 2017). This involves critical eco-literacy focused on understanding environmental issues (Kahn, 2008; Orr, 2024). The method used real-life business processes, critical thinking, and role-playing games to develop students' sustainability knowledge and skills vital to promoting sustainable business futures. A more extensive and integrated example of this type of ecopedagogy initiative is described below using the example of Northumbria University's Newcastle Business School and Nkumba University.

Case Studies: Ecopedagogy in Business Schools

Nkumba University, Kampala, Uganda

At Nkumba University, within the School of Sciences (SCOS) and School of Business Administrations, various contemporary issues related to environmental sciences, ethics, business, and life are linked to the development of sustainability for biodiversity. Since 2015, the two schools have adopted topics on biodiversity merged from environment, ethics, and business. First, at level 3, a diploma in Environment Impact Assessment is studied, enabling learners to integrate knowledge into Level 6, Bachelor's degree programmes underpinning areas such as Environment Management, Fisheries Management and Technology, Tourism Operations Management, Wildlife and Forestry Management, Agribusiness and Public Health. Thereafter, at Level 7, an advancement to study Master of Science in Environmental Health and Master of Science in Natural Resources Management is prescribed. Therefore, the mentioned courses focus on biodiversity and sustainability, discussing challenges like land use conflicts, pollution, contamination of water resources, laws on environmental protection, and other ethical issues related to biodiversity. Describing biodiversity structure, in this context, there is classification, function, evolution and distribution of all living things. Focusing on waste management, environmental impact assessment, environmental health, wetland ecology, and environmental disaster assessment to enrich the curriculum.

Furthermore, business administrative mitigation programmes in place link to develop ethical concepts, styles, African culture context, obstacles, and ethical consideration on foreign aid, as well as an amalgamation of 'environmental', 'ethics' and 'business' associated with nature and origins of environmental crisis, degradation, and risks impacting biodiversity and business in Uganda. Although biodiversity issues are indeed discussed in various programmes, there is a need to make biodiversity a cross-cutting course unit, not only for business ethics but also across all disciplines.

Given the highlighted courses incorporated in programmes at all levels, various Eco-pedagogy teaching methods are considered for the effective impartation of knowledge to learners to construct required outcomes. The methods include lectures, discussions, demonstrations, and Socratic. For instance, field trips are an interesting approach to bringing awareness and understanding to learners due to their learning styles (Ketlhoilwe & Velempini, 2021). For example, places visited include national parks, game reserves, forests, and hills where they observe and learn about biodiversity, including biodiversity loss. These approaches used at Nkumba help to develop critical thinking, awareness, or eco-centrism and to criticise the growth approach to development credence to Ketlhoilwe and Velempini's innovative disruptive pedagogical of learning.

At Nkumba, students take field trips that enable them to use an observation approach, where year 1 and 2 students visit the Uganda Wildlife Conservation Education Centre. The students are asked to write journals on studied lecture topics about animals (including lions, chimpanzees, lions, leopards, and crocodiles, to mention but a few), and plants (16 Ugandan medicine plants). This observation method is in line with Kolb (2014), who identified that we learn through experience when we reflect on how we do things, thus enhancing our knowledge and skills as we process information and apply it well. Students are led to reflect consciously on how they think about biodiversity by writing journals after observing and exploring distinct ways of protecting the environment, species, and associated risks. Each observation field trip takes a week, and then, on return, students are to compile reports from written journals, which guide the commencement of lectures. Other pedagogy methods incorporated include visiting national and game reserves, including Kidepo, Murchison Falls, Queen Elizabeth, and Bwindi Impenetrable Forest. On day one, orientation about the place is given to students about how to conduct themselves in this place. On day 2, tracking a selected species is done depending on its visibility. As a result, it gives students an insight into the degree of impact biodiversity loss could have on the environment. Nkumba upholds eco-pedagogy as it seeks to produce graduates who are convinced that humans and nature survive together. The combination of lecturers and educational visits to the biodiversity sites is intended to enable learners to see the biodiversity issues and consider their short and long-term implications. From a business perspective, there is usually a focus on Forest conservation, water (fisheries), material (timber), and risk management (waste management). Here, it was essential to connect indicators of eco-literacy, exemplified by Griffiths' (2018) win–win strategy for conserving biodiversity with an initiative of people balancing nature use.

Biodiversity Structure, Function and Interaction

At Nkumba University, Biodiversity is taught as part of the Bachelor of Science degree in Environment Sciences and Bachelor of Science in Wildlife Resources Management. Each of these is a three-year degree programme. The course on biodiversity structure and interaction is offered to enable participants to (a) discuss biodiversity in totality; (b) appreciate the ethical approach to protecting, managing and preserving biodiversity; (c) learn the use of indigenous knowledge along sound biological lines (that is, the two complementing each other). The learning outcome of this course includes: (i) students to acquire techniques of sustainable biodiversity use and management; (b) students to integrate local knowledge and expertise in biodiversity conservation; (c) students to identify various forms of biodiversity and its utilisation by human beings; (d) students to critically evaluate biodiversity structure, function and interaction in wildlife management; (e) students to undertake rights-based education and research to influence development policy.

The topics covered in this course include (1) the tropical environment and its biodiversity; (2) hot deserts and their biodiversity; grassland and their primary production; (3) savannah and population dynamics; (4) biodiversity of the savannah; (5) biodiversity of lakes, energy flow and biogeochemical cycling; (6) wetland biodiversity and succession process; (7) tropical rain forests and biodiversity; (8) mountain, zonation and community gradients; (9) global ecology; (10) biodiversity and climate change; (11) causes, consequences and solutions to biodiversity threats; (12) attributes and values of biodiversity.

Northumbria University, Newcastle upon Tyne, United Kingdom

At Northumbria University's Newcastle Business School, several corporate sustainability and business ethics courses are given. Since 2020, there has been a move to include a practical awareness of the link between business and biodiversity through inclusion in the blended learning modules. In undergraduate education in the UK, the sixth form spans educational Levels 4 to 6. Business, Economy and Society at level 5, Fostering Sustainable Futures at level 6, Leadership for Responsible Change at level 7 (Masters level), and a Bachelor's Degree Apprenticeship module Leadership for Responsible Change. These modules emphasised the importance of assessing the potential sustainability impact in times of strategic or emergent change; however, both modules were primarily focused on SDGs and climate. They also focused on change brought about through an identified need by companies to re-organise working conditions, implement growth strategies (sic!), become more competitive, be more technologically advanced, or be forced to act through new legislation. However, standard change models taught at Northumbria lacked emphasis on the need to change concerning biodiversity loss.

Due to the recognition of biodiversity value and extinction risks, an intervention module employing ecopedagogy, highlighting threats to biodiversity and ways to address it actively was set up as the undergraduate level 6 module titled Strategic Leadership for Responsible Change. This module is designed to help the students develop their strategic and leadership skills, linking the subjects of business, leadership, environmental ethics, sustainability, and economic development. In this course, building on level 1 to 5 courses in sustainability and business ethics, different theoretical frameworks, ethical dilemmas, as well as the practice of environmental and corporate governance are discussed. This course focuses on considerations of justice, development, and sustainability that influence resource use, social equality, and biological conservation in the context of sustainable business. These questions are raised: If an alternative path to economic development cannot be found, how can businesses help overcome sustainability challenges? The course considers alternatives to conventional sustainability approaches and ecologically benign models of production, particularly degrowth economy, steady-state economy, Cradle-to-Cradle, and circular economy.

The module incorporates interactive strategies, such as the debate centred around the proposition: <u>Poverty reduction can be decoupled from an increase in the consumption of natural resources</u>. The students are asked to define the terms poverty and natural resources and think about what comprises these resources and where they are extracted from (e.g., land, forest, or sea that house biodiversity). Also, a discussion of Nature Needs Half is initiated with students, including addressing major criticisms:

(1) Feasibility and Land Use Conflict: This concept might clash with existing land uses for agriculture, urban development, and infrastructure.
(2) Economic Impacts: Critics argue that dedicating half of the Earth's land and water to conservation could have adverse economic consequences, potentially limiting opportunities for resource extraction and economic growth in some regions.
(3) Local Communities: Conservation efforts should consider the needs of local communities.
(4) Trade-offs and Alternatives: Critics argue that it may be more practical to focus on targeted conservation efforts, prioritize critical areas, and sustainable land management practices.

The students also engaged in the corporate role-play exercise titled the Shell game, assuming roles of stakeholders such as managers, consumers, shareholders, and civil society representatives, but also of non-human species (for example, Polar bears). The students are then presented with the Leaders' Pledge for Nature, UN's Harmony with Nature, or NNH, and asked to think how these could relate to the oil company's operations. At the mock corporate meeting, various stakeholders discussed the topic of Shell's deliberations as to whether to drill oil in the Arctic. As ecopedagogy also seeks to activate students as citizens and empower them, the roles in the Shell game can also be reversed, and the "Polar bear" needs to be prepared to represent the Dutch or English government, or a CEO of Shell, in a lively and thought-provoking discussion. Part of the discussion loops back to the Pledge, reflecting on how to go about 'Creating a Nature-Positive Future'.

Another key area this module focuses on is the relationship between business and biodiversity. The more "classic" business teaching approach to eco-literacy includes indirect effects of production on biodiversity. The cases of closed-loop production and the life cycle assessments of products about their 'biodiversity footprint' are presented through lectures. The students are taught about the limitations of circularity, as, for example, food, clothes, and other consumables cannot be infinitely reused (Kopnina & Poldner, 2022; Kopnina et al., 2022a, 2022b) (see Table 12.1).

Table 12.1 Comparison of topics within the given educational designs

	Addressed through				
	Curriculum themes (in alphabetical order)	Northumbria University	Nkumba University	Ecopedagogy	Eco-literacy
1	Addressing climate change and carbon reduction	X	X	X	X
2	Addressing root causes of biodiversity loss	X	X		X
3	Attention to alternative economic models	X		X	
4	Critical thinking	X	X	X	
5	Enhancing action competencies	X		X	
6	Explicit attention to biodiversity		X		X
7	Explicit inclusion of non-human stakeholders	X	X		X
8	Sustainable supply chains, logistics	X			X
9	Waste management, circularity	X	X		X

Discussion: Critical Thinking, Ecocentrism and Degrowth in Business Education

Even though the term degrowth is increasingly used in social debates, it is still largely invisible in (business) education. Bobulescu (2021) states that critical pedagogy is needed to develop the necessary skills for questioning economic growth business practices. Such critical pedagogies can include eco-literacy or the Anthropocene-attuned 'common worlds' pedagogies expanding the limits of humanist framings (Taylor, 2017). Ecocentrism has not been reflected in educational practice or research, both in the ethical sense of recognising intrinsic values of nature or biodiversity and in the practical sense of devising protection strategies. Considering the above reservations about the contradictory SDGs, what would 'nature positive' (https://www.naturepositive.org/) initiatives look like in a broader societal context and education?

One proposed solution to counter the over-use of natural resources is the so-called circular production model, originating from *Cradle to Cradle: Remaking the Way We Make Things*, an influential book by McDonough and Braungart (2002). This book builds upon earlier ideas of industrial ecology (Graedel, 1996) and promotes closed-loop or circular system strategies that support production that does not require a new input of natural resources, ideally decoupling economic activity from resource use, which deserves attention. The 10-R circular economy hierarchy refers to priorities in production cycles (Potting et al., 2017). The first R of Refuse, or avoid making or buying new products, seems most effective in achieving degrowth that could benefit nature. The second R is Rethink, which implies a radical revision of current business models, followed by Reduce, Re-use, Repair, Refurbish, Remanufacture, Repurpose, Recycle, and Recover.

In education, acknowledgement of certain limitations of the buzz terms popular in corporate and business education, such as circular economy, needs to be checked and tempered, and other initiatives, namely, degrowth, need to be highlighted (Kopnina & Padfield, 2021). At the same time, the literature indicates that degrowth questions whether the growing economy, dependent on increased energy and material throughput, can sustain the welfare of future generations (Hankammer et al., 2021), degrowth rarely enters business school curricula. However, within business school curricula, the topic of absolute decoupling circularity has only recently started to emerge in business management, operations, logistics, and supply chain management education (Kopnina, 2020).

Kopnina and Padfield (2021), for example, discuss the circular economy's 10-R hierarchy within the context of business education. The students were able to identify areas that are unlikely to be 'nature-positive', such as packaging and transport or take-back and repair options. These case studies employed in critical education demonstrated clear shortcomings, demonstrating that circularity in practice fails to provide absolute decoupling of resource use from corporate profit (Kopnina & Padfield, 2021). Learning to engage with product-to-service shift initiatives, dematerialisation, and degrowth strategies promises benefits not just in terms of product innovations for young corporate leaders, and a larger benefit to society, but also for the vision of the shared earth (Kopnina & Poldner, 2022). The so-called well-being economy (UN, 2022), the steady-state economy and degrowth (Smith et al., 2021), and dematerialization (Dietz & O'Neill, 2013), the production-to-service shift (Kopnina & Poldner, 2022) all purport to counter the built-in obsolescence (Webster, 2021) and address larger issues such as biodiversity loss and climate change.

Linking Business and Conservation Through Education

Concerning educational implications, initiatives like Nature Needs Half (NNH) mean that both ethical and pragmatic interventions to the presently SDG-dominated curricula, especially at business schools, are needed (Bobulescu, 2021). The Principle for Responsible Management Education (PRME) was meant to overhaul business school teaching systems (Bartlett et al., 2020). However, there is little evidence that PRME addressed ecocentric values and moral obligations to nature (Piccolo et al., 2022), a great challenge in critical thinking for business school graduates.

The broader ecocentric conservation movement considers three basic principles: (1) wildlife refers to living beings with their intrinsic value; (2) other species have a right to continued existence, free from anthropogenic pressures; and (3) habitat destruction is the leading cause of biodiversity loss (Locke, 2014; Noss & Cooperrider, 1994; Wilson, 2016). This implies that interspecies justice, also called ecojustice (Kopnina, 2016; Cafaro et al., 2017; Kopnina et al., 2018; Washington et al., 2018; Crist et al., 2021; Kopnina, Mahammad, & Olareru, 2022a, 2022b) needs to be considered.

Sustainability (and Business) Education: From Eco-literacy to ESDG

ESD and curriculum interventions during the Decade of Education for Sustainable Development (DESD) helped to firmly integrate sustainable development in policy and educational practice (UNESCO, 2005). Consequently, the ESDG (UNESCO, 2017), including the fourth goal of "Quality Education" supports the knowledge and skills needed to promote sustainable development.

Going against the grain, interdisciplinary scholars argue the focus should be on environmental sustainability and not on further economic development, both in policy and education (e.g., Orr, 2024; Otero et al., 2020; Washington, 2015). As Wackernagel et al. (2017, 1) point out, "Ranking high on the SDG index strongly correlates with high per person demand on nature (or 'Footprints'), and low ranking with low Footprints, making evident that the SDGs as expressed today vastly underperform on sustainability."

Population growth, a major factor in the increasing consumption of resources, is ignored in the SDGs (Washington, 2015). Ironically, while population concern is often criticised as neo-Malthusian, arguing that it is the wealthy capitalist elites that are to blame for unsustainability, it is perhaps precisely "oligarchs, capitalists and free-market economists" who "gain most from the denial of population growth" (Maynard, 2021, 26). Indeed, these groups may "have a vested interest in a growing population, seeing expanding markets for their goods and services, boosting consumerism globally and seeding exaggerated fears… that without fresh cohorts of young people as labour, social services and pension funds will collapse" (Maynard, 2021, 26).

While people in poor (and populous) countries have a minimal environmental impact compared to the 'privileged elites', blaming the supposedly homogenous group of Western/Northern consumers ignores evidence of growing middle classes in developing countries or migration from low-consumption to high-consumption countries (https://databank.worldbank.org/source/wealth-accounts). Saliently, the very core of social justice principles should enable social and economic mobility. Equal division of resources with the same net consumption is still going to result in over-consumption unless degrowth is experienced by all consumers (Hickel, 2020).

While the critique of capitalism is well-placed, capitalism is arguably no better than socialism concerning former communist or socialist countries, such as the former Russian Federation, China, or Vietnam, on environmental degradation, production, etc. When it comes to the environment, the results might not be as different. It is a critique of industrialism and economic growth that needs to be highlighted. Whether or not the economic pie is divided equally, it is still consumed and limited. Moreover, data on the consumption of meat, for example, in developing countries such as (traditionally vegetarian) India, do not show that its consumption is significantly lower than in Western countries, where vegetarianism is on the rise (Nezlek & Forestell, 2020; Sexton et al., 2022; Sundet et al., 2023). There is also a rise in voluntary childlessness in Western countries (Cain, 2013; Lockwood et al., 2020). Thus, blaming one part of the world or humanity for unsustainable practices appears too simplistic, and ethical issues such as poverty and hunger are unlikely to be solved by mere calls for global equality and justice.

A report by Herrington (2020) comparing the Limits to Growth (Meadows et al., 1972) World3 model simulations with real-world data has concluded that there is close alignment between the world's current path and two specific scenarios, BAU2 (Business as Usual revised) and CT (Comprehensive Technology). Although we cannot use the World3 model as a predictive mechanism, it certainly supports understanding the system dynamics and patterns of our current trajectory in different categories such as Population, Industrial Output, Pollution, and Ecological Footprint, among others (Herrington, 2020; Perlman, 1983). These simulations not only help portray the interrelation between environmental degradation and social and economic well-being, as highlighted in the Leader's Pledge to Nature, but could provide a useful lens for faculty and students in business programs to examine the intricate systems in which business and ecology reside (Tallberg et al., 2022).

As for the different countries' context presented in the studies, the larger discussion for follow-up research could be how sustainable development definitions, praxes and epistemological frameworks tend to contain a series of normative architectures of design and delivery dominated by a Global North narrative of sustainability and sustainable development concepts. This gap or injustice(s) occur in space–time-place localities that in terms of a pragmatic 'adaptation response' may hinder/help dependent on a host of conditionalities, including, not least, power-structure-agency leverage differentials that play out as can be seen, in climate change talks (e.g. Annex I countries = mitigation techno-centric/socio-technical climate responses, such as adaptation and reliance on Global North support, presently less prominent in comparative education context.

Conclusion

Findings show that Education for Sustainable Development Goals (ESDGs) is increasingly featured within Business School degree programme courses across 'sustainable' management-centric pedagogies such as 'sustainable marketing' and 'sustainable supply chain management'. The case studies above presented a small sample opening a wider discussion through a critical political-ecology lens, pluralism and de-centering narrative that builds on local/indigenous and participatory knowledge-nexus contextualising Global South contexts and considerations of human population growth. Ergo, the historical timeline and context of the 'Green Growth' and/or the 'De-Growth' movements have failed to shift the needle so far from the 'status quo' to 'radical system change'.

Normative framings of sustainability pedagogies were found to potentially negate the inclusion of 'deep green' rooted concepts, ontologies and epistemologies in business and management education. While ESDG is often equated with sustainability education in both Uganda and England's business schools, substantive gaps prevail in terms of ecocentrism thinking and pedagogical praxes. We call for university business schools to harness ecocentric conceptualisations of sustainability to improve 'eco-literacy' within and across contemporary didactic strategies for more nuanced, ethically driven, environment-inclusive education provision.

References

Bartlett, P. W., Popov, M., & Ruppert, J. (2020). Integrating core sustainability meta-competencies and SDGs across the silos in curriculum and professional development. In G. Nhamo & V. Mjimba (Eds.), *Sustainable development goals and institutions of higher education. sustainable development goals series* (pp. 71–85). Springer.

Bobulescu, R. (2021). Wake up, managers, times have changed! A plea for degrowth pedagogy in business schools. *Policy Futures in Education, 20*(2), 188–200. https://doi.org/10.1177/14782103211031499

CABS. (2021). Chartered Association of Business Schools—Business schools and the public good. https://cabs-199e2.kxcdn.com/wp-content/uploads/2021/06/Chartered-ABS-Business-Schools-and-the-Public-Good-Final-1.pdf

Cafaro, P., Butler, T., Crist, E., Cryer, P., Dinerstein, E., Kopnina, H., Noss, R., Piccolo, J., Taylor, B., Vynne, C., & Washington, H. (2017). If we want a whole earth, nature needs half. A reply to 'Half-Earth or Whole Earth? Radical ideas for conservation, and their implications. *Oryx—The International Journal of Conservation, 53*(1), 400.

Cain, M. (2013). *The childless revolution.* Diversion Books.

Chiang, M., & Chen, P. (2022). Education for sustainable development in the business programme to develop international Chinese college students' sustainability in Thailand. *Journal of Cleaner Production, 374*, 134045.

Crist, E., Kopnina, H., Cafaro, P., Gray, J., Ripple, W. J., Safina, C., Piccolo, J. J., et al. (2021). Protecting half the planet and transforming human systems are complementary goals. *Frontiers in Conservation Science, 91*, 1–9. https://doi.org/10.3389/fcosc.2021.761292

Dean, B. A., Gibbons, B., & Perkiss, S. (2018). An experiential learning activity for integrating the United Nations Sustainable Development Goals into business education. *Social Business, 8*(4), 387–409.

Dietz, R., & O'Neill, D. (2013). *Enough is enough: Building a sustainable economy is a world of finite resources.* Berrett-Koehler Publishers.

Fang, J., & O'Toole, J. (2023). Embedding sustainable development goals (SDGs) in an undergraduate business capstone subject using an experiential learning approach: A qualitative analysis. *The International Journal of Management Education, 21*(1), 100749.

Griffiths, V. F. (2018). *Win-win? Balancing people's uses of nature with biodiversity No Net Loss* (Doctoral dissertation, University of Oxford).

Freire, P. (2000). *Pedagogia da Terra.* Peiropolis.

Gaard, G. (2008). Toward an ecopedagogy of children's environmental literature. *Green Theory & Praxis: THe Journal of Ecopedagogy, 4*(2), 11–24.

Graedel, T. E. (1996). On the concept of industrial ecology. *Annual Review of Energy and the Environment, 21*(1), 69–98.

Hankammer, S., Kleer, R., Mühl, L., & Euler, J. (2021). Principles for organizations striving for sustainable degrowth: Framework development and application to four B Corps. *Journal of Cleaner Production, 300*, 126818.

Herrington, G. (2020). Update to limits to growth: Comparing the World3 model with empirical data. *Journal of Industrial Ecology, 25*(3), 614–626. https://doi.org/10.1111/jiec.13084

Hickel, J. (2020). *Less is more: How degrowth will save the world.* Random House.

IPBES. (2019). Media Release: Nature's Dangerous Decline 'Unprecedented' Species Extinction Rates 'Accelerating'. *Intergovernmental Science-Policy Platform on Biodiversity and Ecosystem Services (IPBES).* https://www.ipbes.net/news/Media-Release-Global-Assessment

IPCC. (2022). *Climate Change 2022: Mitigation of climate change: Full report.* IPCC. https://report.ipcc.ch/ar6wg3/pdf/IPCC_AR6_WGIII_FinalDraft_FullReport.pdf

Iritié, B. G. J. J. (2015). Economic growth and biodiversity: An overview. Conservation policies in Africa. *Journal of Sustainable Development, 8*(2), 196–208.

IUCN. (2022). *International union for conservation of nature.* https://www.iucn.org/

Kahn, R. (2008). From education for sustainable development to ecopedagogy: Sustaining capitalism or sustaining life. *Green theory & praxis: The Journal of Ecopedagogy, 4*(1).

Ketlhoilwe, M. J., & Velempini, K. (2021). Wilding educational policy: The case of Botswana. *Policy Futures in Education, 19*(3), 358–371.

Kolb, D. A. (2014). *Experiential learning: Experience as the source of learning and development.* FT Press.

Kopnina, H. (2016). Half the earth for people (or more)? Addressing ethical questions in conservation. *Biological Conservation, 203*, 176–185.

Kopnina, H. (2020). Education for the future? Critical evaluation of education for sustainable development goals. *Journal of Environmental Education, 51*(4), 280–291.

Kopnina, H., & Padfield, R. (2021). (Im)possibilities of circular production: Learning from corporate case studies of (un)sustainability. *Environmental and Sustainability Indicators, 12*, 100161.

Kopnina, H., & Poldner, K. (2022). *Circular economy: Challenges and opportunities for ethical and sustainable business*. Routledge.

Kopnina, H., Washington, H., Gray, J., & Taylor, B. (2018). The 'future of conservation' debate: Defending ecocentrism and the Nature Needs Half movement. *Biological Conservation, 217*, 140–148.

Kopnina, H., Boatta, F., Baranowski, M., & de Graad, F., et al. (2022a). Does waste equal food?: Examining the feasibility of circular economy in the food industry. In H. Lehmann (Ed.), *The impossibilities of the circular economy* (pp. 11–22). Routledge.

Kopnina, H., Mahammad, N., & Olareru, F. (2022b). Exploring attitudes to biodiversity conservation and Half-Earth vision in Nigeria. *Biological Conservation, 272*, 109645.

Leaders' Pledge for Nature. (2021). United to reverse biodiversity loss by 2030 for sustainable development. https://www.leaderspledgefornature.org/wp-content/uploads/2021/06/Leaders_Pledge_for_Nature_27.09.20-ENGLISH.pdf

Locke, H. (2014). Nature needs half: A necessary and hopeful new agenda for protected areas in North America and around the world. *The George Wright Forum, 31*, 59–371.

Lockwood, B., Powdthavee, N., & Oswald, A. J. (2020). Are environmental concerns deterring people from having children? IZA Discussion Paper No. 15620, SSRN: https://ssrn.com/abstract=4241599 or https://doi.org/10.2139/ssrn.4241599

Maynard, R. (2021). Overpopulation denial syndrome. *The Ecological Citizen, 5*(1), 23–28.

McDonough, W., & Braungart, M. (2002). *Cradle to cradle: Remaking the way we make things*. North Point Press.

Meadows, D., Meadows, D., Randers, J., & Behrens, W. (1972). *The limits to growth*. Universe Books.

Nezlek, J. B., & Forestell, C. A. (2020). Vegetarianism as a social identity. *Current Opinion in Food Science, 33*, 45–51.

Misiaszek, G. W. (2017). *Educating the global environmental citizen: Understanding ecopedagogy in local and global contexts*. Routledge.

Moranta, J., Torres, C., Murray, I., Hidalgo, M., Hinz, H., & Gouraguine, A. (2022). Transcending capitalism growth strategies for biodiversity conservation. *Conservation Biology, 36*(2), e13821.

Moosmayer, D. C., Waddock, S., Wang, L., Hühn, M. P., Dierksmeier, C., & Gohl, C. (2019). Leaving the road to Abilene: A pragmatic approach to addressing the normative paradox of responsible management education. *Journal of Business Ethics, 157*, 913–932.

Morton, A. (2022). Australia's mountain mist frog declared extinct as red list reveals scale of biodiversity crisis. *The Guardian*. https://www.theguardian.com/australia-news/2022/dec/10/australias-mountain-mist-frog-declared-extinct-as-red-list-reveals-biodiversity-crisis

NEMA. (2016). National biodiversity strategy and action plan II (2015–2025): *Theme: Supporting transition to a middle income status and delivery of sustainable development goals*. NEMA, nema.go.ug.

Noss, R., & Cooperrider, A. (1994). *Saving nature's legacy: Protecting and restoring biodiversity*. Island Press.

OECD. (n.d.). Global competency. www.oecd.org/education/Global-competency-for-an-inclusive-world.pdf

Orr, D. (2024). Education and the great transition? *The Ecological Citizen, 7*(1), epub-095.

Otero, I., Farrell, K. N., Pueyo, S., Kallis, G., Kehoe, L., Haberl, H., Plutzar, C., Hobson, P., García-Márquez, J., Rodríguez-Labajos, B., & Martin, J. L. (2020). Biodiversity policy beyond economic growth. *Conservation Letters, 13*(4), e12713.

Panwar, R., Ober, H., & Pinkse, J. (2023). The uncomfortable relationship between business and biodiversity: Advancing research on business strategies for biodiversity protection. *Business Strategy and the Environment, 32*(5), 2554–2566.

Perlman, F. (1983). Against his-story, against leviathan. Retrieved August 14, 2023, from theanarchistlibrary.org

Piccolo, J., Taylor, B., Washington, H., Kopnina, H., Gray, J., Alberro, H., & Orlikowska, E. (2022). "Nature's contributions to people" and peoples' moral obligations to nature. *Biological Conservation, 270*, 109572.

Potting, J., Hekkert, M. P., Worrell, E., & Hanemaaijer, A. (2017). Circular economy: Measuring innovation in the product chain. *Planbureau voor de Leefomgeving* (PVL, Netherlands Environmental Assessment Agency), 2544. https://www.pbl.nl/en/publications/circular-economy-measuring-innovation-in-product-chains

Rashid, L. (2019). Entrepreneurship education and sustainable development goals: A literature review and a closer look at fragile states and technology-enabled approaches. *Sustainability, 11*(19), 5343.

Rosales, J. (2008). Economic growth, climate change, biodiversity loss: Distributive justice for the global north and south. *Conservation Biology, 22*(6), 1409–1417.

Seto-Pamies, D., & Papaoikonomou, E. (2020). Sustainable development goals: A powerful framework for embedding ethics, CSR, and sustainability in management education. *Sustainability, 12*(5), 1762. https://doi.org/10.3390/su12051762

Sexton, A. E., Garnett, T., & Lorimer, J. (2022). Vegan food geographies and the rise of Big Veganism. *Progress in Human Geography, 46*(2), 605–628.

Smith, S. J. T., Baranowski, M., & Schmid, B. (2021). Intentional degrowth and its unintended consequences: Uneven journeys towards post-growth transformations. *Ecological Economics, 190*, 107215.

Spooner, S. L., Farnworth, M. J., Ward, S. J., & Whitehouse-Tedd, K. M. (2021). Conservation education: Are zoo animals effective ambassadors and is there any cost to their welfare? *Journal of Zoological and Botanical Gardens, 2*(1), 41–65.

Sundet, Ø., Hansen, A., & Wethal, U. (2023). Performing meat reduction across scripted social sites: exploring the experiences and challenges of meat reducers in Norway. *Consumption and Society*, 1–20.

Taylor, A. (2017). Beyond stewardship: Common world pedagogies for the Anthropocene. *Environmental Education Research, 23*(10), 1448–1461.

Taylor, B., Chapron, G., Kopnina, H., Orlikowska, E., Gray, J., & Piccolo, J. (2020). The need for ecocentrism in biodiversity conservation. *Conservation Biology, 34*(5), 1089–1096.

Tallberg, L., Välikangas, L., & Hamilton, L. (2022). Animal activism in the business school: Using fierce compassion for teaching critical and positive perspectives. *Management Learning, 53*(1), 55–75.

The Environmental Audit Committee. (2022). https://publications.parliament.uk/pa/cm5802/cmselect/cmenvaud/136/136-report.html

UN. (2022). Harmony with Nature: Report of the Secretary-General. See https://documents-dds-ny.un.org/doc/UNDOC/GEN/N22/444/24/PDF/N2244424.pdf?OpenElement

UNDP. (2024). Human development summary capturing achievements in the HDI and complementary metrics that take into account gender gaps, inequality, planetary pressures and multideminsional poverty. https://hdr.undp.org/data-center/specific-country-data#/countries/GBR

UNDP, SCBD & UNEP-WCMC. (2021). *Creating a nature-positive future: The contribution of protected areas and other effective area-based conservation measures.* UNDP.

UNESCO. (1977). Intergovernmental conference on environmental education, Tbilisi, USSR, 14–26 October 1977: final report. https://unesdoc.unesco.org/ark:/48223/pf0000032763

UNESCO. (2005). United Nations decade of education for sustainable development (2005–2014) framework for the international implementation scheme. 32 C/INF.9. http://unesdoc.unesco.org/images/0013/001311/131163e.pdf

UNESCO. (2017). Education for sustainable development goals. https://www.sdg4education2030.org/education-sustainable-development-goals-learning-objectives-unesco-2017

UNESCO. (n.d.). Biodiversity. https://www.unesco.org/en/biodiversity

Van Matre, S. (1978). *Sunship earth.* American Camping Association.

Visseren-Hamakers, I. J. (2020). The 18th sustainable development goal. *Earth System Governance, 3*, 100047.

Washington, H. (2015). *Demystifying sustainability: Towards real solutions*. Routledge.
Washington, H. (2020). *What can I do to help heal the environmental crisis?* Routledge.
Washington, H., Piccolo, J., Chapron, G., Gray, J., Kopnina, H., & Curry, P. (2018). Foregrounding ecojustice in conservation. *Biological Conservation, 228*, 367–374.
WCED. (1987). Our common future: report of the world commission on environment and development. Geneva, UN-Dokument A/42/427
Webster, K. (2021). A circular economy is about the economy. *Circular Economy and Sustainability, 1*(1), 115–126.
Weybrecht, G. (2022). Business schools are embracing the SDGs–but is it enough?–How business schools are reporting on their engagement in the SDGs. *The International Journal of Management Education, 20*(1), 100589. https://doi.org/10.1016/j.ijme.2021.100589
Whiteman, G., Walker, B., & Perego, P. (2013). Planetary boundaries: Ecological foundations for corporate sustainability. *Journal of Management Studies, 50*(2), 307–336.
Wilson, E. O. (2016). *Half-earth. Our planet's fight for life*. Liveright Publishing.

Chapter 13
Climate Change Education Through Agroecology Curricula in Sierra Leone

Bruno Borsari

Abstract Global climate change is the keystone anthropogenic factor that for decades has been debated and analyzed from various levels of governance in an effort of achieving a sustainable future. Yet, increasing greenhouse gas emissions (GHGs) continue impacting the livelihoods of millions of people, forcing human displacement and mass migrations. Often these events trigger crop failures leading to food shortages. Their causes are multifaceted and prompted by drastic changes to the environment like floods, uncontrollable fires and/or soil erosion, which expand desertification, and biodiversity loss. Industrial agriculture is the leading human activity for GHGs, that is exacerbating climate change on a planetary scale, with effects on vast regions of the global south. Nonetheless, farming remains a primary economic activity. In Africa it is carried out primarily through peasant agriculture, which is practiced with a priority of providing food for local communities. Thus, most African farming is intertwined with traditional approaches to agriculture that can be sensitive to a conservation of natural resources, from which agriculture depends. Curricula in agroecology can be transformative and have potential for realizing an effective implementation of climate change education. This assessment study triangulated qualitative data that were collected through a document analysis review, observations, and case study about rice intensification (SRI). Recommendations from this evaluation yielded an original model that could serve stakeholders in Sierra Leone and more countries to employ agroecology as a vehicle to spur programs in climate change education.

Keywords Africa · Agroecology · Climate change · Education · Food security · Sierra leone · Sustainability

---

B. Borsari (✉)
Department of Biology, Winona State University, Winona, MN, USA
e-mail: bborsari@winona.edu

M. F. Mbah et al. (eds.), *Practices, Perceptions and Prospects for Climate Change Education in Africa*, https://doi.org/10.1007/978-3-031-84081-4_13

Introduction

Climate change is a multifaceted and complex phenomenon that has always had significant effects on human livelihood and economies (Bunce et al., 2010), as most of these are maintained only within acceptable, environmental boundaries (Adger et al., 2003). Therefore, climate change impacts have been studied in public health (Coates et al., 2020), agriculture and resource management (Borsari et al., 2014), and have emerged in recent years as relevant instruction themes for education in tropical regions (Randell & Gray, 2019), demonstrating how intertwined climate is to every aspect of human life. For these reasons, global climate change is deserving of much attention at all levels of governance, internationally (IPCC, 2022). Its implications for human and ecosystems' health have been debated and analyzed for decades, in an effort of reducing greenhouse gas (GHGs) emissions, with limited successes (Borsari et al., 2014). The scope of achieving GHGs reductions is to restore the planetary balance that was lost through industrialization and population growth, but also with uncontrolled expansions of industrial agriculture worldwide (Foley et al., 2011). The challenges of lessening GHG emissions to more sustainable levels continue impacting the livelihoods of billions of people (especially in developing countries), causing human misery and desertification, while amplifying mass migrations (IPCC, 2022). The nexus between climate change, agriculture, and biogeochemical cycles (including the water cycle), have been shaping the physical attributes of ecosystems for thousands of years yet, the scale of climate change-induced disturbances has created distressing uncertainties for the future of the biosphere (Rockström & Gaffney, 2021). The rising frequencies of livestock and crop losses that every year affect swelling numbers of farmers around the world, especially in Africa stand as an undeniable reminder that climate change is real (Rosenzweig & Parry, 1994). These climate-change-induced crops failures have amplified food insecurity, human migrations, and a depopulation of rural settings, expanding further the size of African cities and stressing urban environments with many new challenges (Bayar & Aral, 2019; Burrows & Kinney, 2016). People of west African nations have not been spared from this unfortunate trend, due to an increasing expansion of deserts, dwindling crop yields, biodiversity loss and water shortages, that every year threaten food security a step further (IPCC, 2019).

The impacts of a growing human population associated with themes like global climate change, education, or sustainable development require new paradigms for agricultural practices and instruction (Borsari, 2023; Onwueme et al., 2008). There is a need for reforming education in Africa and making it more accessible to all, relevant, and action-driven to abate the challenges posed by climate change. Also, accentuating the urgency of adapting curricula (especially those in the agricultural sciences), is another legitimate priority for ensuring conservation of natural resources and human communities within ecosystems. Agriculture remains a pivotal, anthropic activity that must reckon with the ecological complexity of farms and adjust to specific climatic and environmental conditions (Gliessman, 2023; Altieri et al., 2015). For example, it is no longer conceivable to assume that the same agronomic techniques may be universally applicable, together with homogeneous pedagogies of instruction in agriculture, with the expectation of optimizing success equally, when managing complex agroecosystems. Additionally, it may not be possible to achieve effective outcomes from instructional efforts in Africa if these remain oblivious of considering the cultural knowledge of local communities about their crops and farming methods (Borsari, 1999). Agriculture affects everybody's life and access to food is a sacrosanct human right! Undoubtedly, a conversion of ecosystems into farming systems spurred population size, enabled settlement of humankind and the rise of civilizations. Yet, for the last sixty years the success of the 'green revolution' has been heralded worldwide, as the model of farming capable of eradicating malnutrition and starvation. However, despite its promises the industrial agriculture model that emerged from the green revolution, has fulfilled its ambitious goals only partially, promising an eradication of famines without admitting its predictable limitations (Borsari, 2023). The 'agribusiness' approach to an industrialization of food production remains emphatic in modern farming and among philanthropic organizations like the Alliance for the Green Revolution in Africa (AGRA), continue to persuade African countries that an intensification of agriculture through the agribusiness model is the only way to avert food insecurity (Thompson, 2012). However, a recent assessment published by a large group of African and German non-governmental organizations, pointed out that AGRA has failed in its intent to resolve the problem of hunger and food insecurity, in this world region (Aijuka et al., 2021).

Nonetheless, this *status quo* in food production remains dominant and financially supported; although it is oblivious to the fact that industrial agriculture continues to remain highly dependent on energy and water uses, while being the driving force of the planetary, climate change crisis (Kerr, 2012). Paradoxically, the 'new green revolution' of present times, is not immune from economic losses caused by climate change events yet, the brunt of these catastrophes continues to distress multitudes of peasant growers in the developing world, victims of this capitalist, neocolonial approach to farming (Patel, 2012). Therefore, there is an urgent need for disentangling agriculture from the greedy plans of corporative, foreign investors while revisiting education and curricula to prepare future leaders in Africa to resolve present and foreseeable climate change challenges through a sustainable agriculture (Borsari, 1999), which driven by agroecological approaches, enables farmers also reclaiming their food sovereignty (Altieri et al., 2015).

Agricultural education and curricula could play a pivotal role in fostering awareness about climate change education on a broadest spectrum in Africa, as this continues threatening its food systems. This is evident also, as the long-term productivity of small, family farms are increasingly threatened by market policies uncontrollable by single producers (Patel, 2012), and by resource uses that often accelerate environmental degradation, with higher levels of (GHG) emissions, that amplify the consequences of climate change on lands and people.

According to recent studies by some agricultural scientists, there are increasing pressures to double agricultural outputs before the end of the twenty-first century (Dube et al., 2016; Foley et al., 2011; Ickowitz et al., 2019). This need for intensifying food production is justified by an expansion of the human population size, that is expected to reach the 10 billion marks by 2050. Although maximizing food production may be in its appearance necessary to prepare coping with increasing food insecurity trends, agribusiness has been criticized as linear and even ineffective for resolving the chronic nutrition crises that continue affecting large segments of humanity (Borsari, 2023; Rasmussen et al., 2018). These problems concern most directly Africa and its people because this is a continent with a high proportion of land degradation, highly weathered soils and with higher-than-average rates of population growth. African agriculture is primarily a subsistence human activity, which is done by peasant farmers, whose priority is providing food for their own communities. Therefore, most farming in Africa is intertwined with traditional knowledge and approaches to agriculture, that are aware about conserving resources because these are necessary to support local food production (Borsari, 1999). For these reasons, agroecology, which is a science, a set of practices that rely also on the wisdom of indigenous knowledges and a social movement (Wezel et al., 2009), is proposed in this work as the focus of curriculum design and implementation, aimed at fostering an effective action-driven education about climate change in Africa. Curricula in agroecology are multidisciplinary and can transform the life of all learners thus, becoming adaptable for various didactic approaches in climate change education (Ebel et al., 2020).

The purpose of this work consisted in looking at the past to document the cause/s of the current adoption, or lack of adoption of agroecology in climate change education in Sierra Leone. Inevitable constraints affected this work, however. These consisted in the difficulty of locating literature about the topic under study, that referred to this country. Also, the ten-year civil war of the 1990s, the Ebola virus outbreak of 2014–15 and the Covid-19 pandemic of 2020, have been tragic events that caused incalculable losses. However, through this assessment study the author attempted to shed light on the present status of climate change education in Sierra Leone, which may assist leaders and stakeholders in this west African nation to move on, toward a sustainable development.

Sierra Leone as Research Context

Sierra Leone is a west African country of 73.326 km^2 (Latitude: 7°–10° N. Longitude: 10°–13° W.), adjacent to the Atlantic Ocean and bordering Guinea and Liberia (Fig. 13.1). Its tropical climate and diverse ecosystems range from savanna to rainforests and the country is inhabited by and a varied population of about 7,092,113 according to the census of 2015 (SSL, 2017). The capital city is Freetown, and the country is administratively organized in five regions, that are further subdivided into 16 districts (Nat. Elect. Commission, 2018). Sierra Leone is a constitutional republic with a unicameral parliament and a directly elected president who serves a five-year term for a maximum of two terms.

Agriculture in Sierra Leone is devoted primarily to fulfill the food needs of local communities. Most Sierra Leonians are engaged directly in agriculture, whose profits

Fig. 13.1 Location of Sierra Leone in Africa with major cities around the country, including its capital, Freetown

accounted for 58% of the gross domestic product (GDP) in 2007 (Wren Media, 2007). Although the country is rich in minerals and more natural resources, agriculture remains the largest employer with 80% of the population working in this sector (König, 2008). Rice (*Oryza sativa L.*) is the most important food crop of this west African nation, and it is disconcerting to learn from a recent simulation study conducted by van Oort and Zwart (2018), that irrigated rice yields in west Africa could soon decrease between 45 and 15%, due to a reduced photosynthetic activity occurring at higher temperatures, induced by climate change. Hence, the need of strengthening education to mitigate climate change in Africa is most urgent and its nexus to food production and security is obvious to anyone who lives, or has lived in Sierra Leone.

Education in Sierra Leone

Education in Sierra Leone is mandatory and free for all children in government-sponsored public schools for six years at primary level (students are usually 6–12 years old), for three additional years in junior secondary education and in senior secondary schools, for students between 13 and 18 years old (Wang, 2007). A chronic dearth of schools and adequately prepared teachers made any implementation of curricula almost impossible especially, during the civil war of 1991, leaving two thirds of the adult population of the country illiterate (Human Development Report, 2009). This ten-year conflict resulted in the destruction of more than a thousand primary schools, and despite their reconstruction that led to doubling students' enrollment between 2001 and 2005 since the end of the war (McDermott & Allen, 2015), education services in Sierra Leone continue to be affected by a plethora of challenges (Mai, 2019).

Regarding higher education, Sierra Leone has three universities of which Fourah Bay College has the longest tradition in teachers' preparation; it was founded in 1827 in Freetown and is the oldest college in west Africa (Jones-Parry, 2006). Instead, Njala University, located in Bo District was established as the Njala Agricultural Experimental Station in 1910 and became a university in 2005 (Njala University College, 2007), the agricultural school of Sierra Leone. The third and newest one is the University of Makeni, in the northern province of the country, which was established in 2005 and was granted university status in August 2009 (University of Makeni, 2018).

During this 35-year timeframe Sierra Leone has been affected by significant socio-political catastrophes as the already mentioned civil war and major public health crises, that have further impoverished the small west-African country and reduced extremely, education opportunities for its citizens.

Methodology

Sierra Leone is the African country considered in this work because this is where the author lived and taught agriculture at the St. Joseph Vocational Institute of Lunsar, in the late 1980s, while assisting this school and nearby communities to achieve better food security, through some agronomic methods that were inclusive of ecological principles and practices (Borsari, 1996). Diverse modalities of instruction for the proposed curricula are presented and these span from elementary to higher education, including programs in rural extension for farmers.

The nature of this work was archival and thus, nonexperimental. As suggested by Patten (2004), through this casual-comparative research, a comprehensive document analysis review examined data to understand the dynamics of 35 years (from 1987 to 2022) of education history in Sierra Leone, complemented by field notes from observations conducted by the author, between 1987 and 1989. According to Glesne and Peshkin (1992), field notes can be valid qualitative data on reflective thoughts and interpretations about what is being experienced by researchers. The documents were identified through the ERIC database, which is the largest search engine dedicated to literature in education. Keywords like climate change education, curriculum, Sierra Leone, Africa were employed to retrieve the documents. The AGRICOLA database instead was employed to gather documents about curricula in the agricultural sciences, and rice cultivation practices in Sierra Leone, which is the dominant, cultivated crop. Also, an analysis of the case-study about the system of rice intensification (SRI) was considered. Creswell and Creswell (2018) pointed out that case studies are descriptive endeavors focusing on searching solutions to specific problems that occur in explicit research contexts. Additional advantages of using case studies relate to the fact that these provide a better context for certain topics to be studied (Miles, 2015). The SRI case study added more qualitative data that allowed for a triangulation of the results. This effort attempted to explain farming patterns in Sierra Leone, while developing new hypotheses and paradigms, supportive of an employment of agroecology education to assist in mitigating the unpredictable effects caused by climate change on food production and security.

Observations data were collected between 1987 and 1989 at the St. Joseph Vocational Institute (SJVI) in Lunsar and surrounding areas in the Northern Province of Sierra Leone to document farming practices. Primary purpose of these notes (including photos) was to leave an information record about what had been accomplished at the SJVI farm, that would have served an adjustment of future teachers joining the agriculture program at this school.

The seventeen goals for sustainable development (SDGs) launched by the United Nations in 2015 in pursuit of Agenda 2030, constituted the theoretical underpinnings of this work.

Significance of the Study

The successes obtained by peasant organizations like 'La Via Campesina' in Latin America have been of great inspiration to design study programs in agroecology, that rely heavily on the sharing of knowledges (Martínez-Torres & Rosset, 2010), where this is passed from learner to learner, horizontally, rather than vertically (teacher to learner only). The experience gained by the author in this small, yet culturally and ecologically diverse country of west Africa is a microcosm where an initial exposure to agroecology has begun to expand, making this knowledge employable beyond the borders of Sierra Leone. According to Kellaghan and Stufflebean (2003), the education emphasis of this work could serve to classify this report also as a needs assessment study that culminated with an articulation of specific recommendations. Thus, applications of the results emerging from this chapter could become feasible (with specific adaptations), for an implementation of curricula in climate change education, in many more African countries.

Results

Document Analysis Review

The literature search conducted with the ERIC database yielded thirty-four documents, and these served to shed light on the history and the present status of education in Sierra Leone. Many challenges have been affecting the education system of this west African nation. Among these are most significant the long-armed conflict of the 1990s that besides the tragic loss of human life destroyed most schools and their already modest infrastructure (Rossini, 2007). Berton (1999) pointed out that the very limited resources for education in Sierra Leone remain inadequate and that despite conjunct efforts of missionaries and their organizations these did not fulfill the schooling needs of the country, especially in remote, rural areas. Poverty, geographic isolation, and gender inequalities in Africa (Adekunle Adeyeye & Ighorojeh, 2019; Muhanguzi, 2019) constitute additional barriers to education, and Sierra Leone is not exempt from these constraints (World Bank, 2020).

Thus, reorganizing the whole education system remains a top priority for Sierra Leone because for too long learning relied on a rigid, top-down approach that followed since colonial times, the structure of British education (Berton, 1999). The latter did not consider an education in agriculture worthy enough to ensure up-scale mobility in its society even though agriculture in Sierra Leone employs about two thirds of its labor force (International Trade Administration, 2021). According to Ebun-Cole (1992), this tendency of dislike for agricultural education muffled an enthusiasm for inquiry and creativity among many adult learners. On the contrary, it is well known that in extension education where farmers are the recipients of instruction, Knowles' theory of andragogy and its principles (1984) are necessary aspects to consider, so that an optimal rapport may be established among all parts engaged in educational encounters. Within this framework, Borsari and his collaborators (2016) found that an exposure of agriculture students to extension education in rural Panama provided excellent learning opportunities for all students and farmer participants, when these learners were engaged in hands-on activities and sharing of their knowledge.

For these reasons, Ebun-Cole (1992) added that a more positive attitude about agriculture should be encouraged in Sierra Leone since the years of primary education, because food production is relevant to the livelihood of a broad majority of people. To accomplish this, curricula in teachers' preparation could consider selected agriculture topics and offer these in their education study programs. Also, the school garden, farm and similar open spaces become distinctive environments where education becomes experiential, ecocentric and potentially effective to educate students about climate change and remediation strategies to regain planetary health (Borsari & Kunnas, 2023; Borsari & Vidrine, 2022). Too often the school farm, or garden in Sierra Leone has been used as a labor space to punish delinquent students (Swanson & Tucker, 1978), instead of becoming an asset for hands-on learning activities where most curriculum subjects connect to climate change education.

Whether these learning opportunities in agriculture are offered through on-farm practicum courses, visits, or experiential workshops, the scientific application of tested agroecological practices could be offered for any level of instruction and thus, benefit all learners (Table 13.1).

Table 13.1 Selected agroecological practices whose application advantages have been amply documented

Practice	Description	Benefits	Reference
Agroforestry	The practice of including trees in crop or animal production agroecosystems	Agronomic, biological Environmental	Borsari (2022) Gliessman et al.(2023)
Double-digging method	Technique used to increase soil drainage and aeration, while loosening two layers of soil, and the addition of organic matter	Agronomic, biological	Jeavons (1982)
Intercropping	Multiple cropping practice involving growing two, or more crops in proximity	Agronomic	Borsari and Vidrine (2000) Gliessman et al.(2023)
Cover cropping	Plants that are grown to cover the soil rather than for the purpose of being harvested	Agronomic	Borsari and Vidrine (2000) Gliessman et al.(2023)
Compost	Organic matter that has been decomposed and recycled as a fertilizer, or soil amendment	Agronomic, environmental	Borsari and Vidrine (2000) Minnich and Hunt (1979)
Biochar	Charcoal produced from plant biomass and stored in the soil to sequester carbon dioxide from the atmosphere	Agronomic, biological Environmental	Agegnehu et al., (2017)

Field Observations

Study programs as the one in agriculture at SJVI have a duration of 3 academic years and these have been attracting students from every region of the country, for the last 37 years (Fig. 13.2). The vocational emphasis of the curriculum is equally split between morning classroom instruction time (50%) and afternoon work in the 20 ha. farm of SJVI (50%), under the guidance of three agriculture teachers, in collaboration with the farm manager. The food grown at the farm was used to offer lunch at a nominal cost to all students at SJVI (about 200) and teachers (15–18 employees).

Major, staple cultivated crops were rice, palm-oil (*Elaeis guineensis* Jacq.), sweet potato (*Ipomoea batatas* L.), cassava (*Manihot esculenta*). However, a broad diversity of vegetable species was cultivated also (from beans to onions, eggplants, peppers, tomatoes) and fruits, to expose students to these foods thus, aiming at improving/diversifying their daily nutrition. Although the student population at SJVI was male only, a higher number of women has always been observed working the fields, in all thirty farms that were visited by the author in Sierra Leone, between 1987 and 1989.

In 1988 an alley cropping system was established on a sloping field (1.5 ha) at the SJVI farm that demonstrated an agroforestry approach to soil conservation and fertility management, with rows of the nitrogen-fixing tree *Leucaena leucocephala* (Borsari, 1996). This effort was also a tangible pedagogical approach to educate students in climate change abatement through carbon sequestration by trees

Fig. 13.2 The author and some of his second-year students' cohort at the SJVI farm (November 1987)

and an enhancement of biodiversity that was achieved with an intercropping of vegetable species between the tree rows. Another notable achievement in the same year consisted in the establishment of a first agriculture cooperative in the village of Worebana (between Lunsar and Makeni), with five graduates who had been granted permission to farm rice on a 25 ha. boliland (seasonal swamp), by the local community council and chief. This effort aimed at strengthening the idea of cooperation in agriculture that would have employed more SJVI students from other programs (e.g.: masonry, metalwork, carpentry). The expectation consisted in establishing a thriving local economy thus, demonstrating sustainable development in a small rural, community of the Northern Province.

The Case of Rice Intensification (SRI) in Sierra Leone

Rice is a keystone staple crop in Sierra Leone, with 85% of farmers growing it during the rainy season, to fulfill an annual consumption demand of about 76 kg per person (International Trade Administration, 2021). Therefore, a special emphasis was given to this iconic plant species in this study, for the role that rice plays in the sustenance of the country's food system and overall economy.

According to Yamah (2002), SRI is an agroecological rice production system because it utilizes efficiently land and water resources of inland valley swamps, while needing less seed to achieve outstanding yields. Generous applications of organic matter in the form of compost, or manure, constitute the ecological, on-farm input, which supports the crop's nutritional needs and health (Toungos, 2018).

According to a World Bank report (2014), 10,865 farmers in Sierra Leone adopted the introduced cultivation approach, demonstrating the feasibility of scaling up the System of Rice Intensification in west Africa, through an agroecological method. However, during the 2015 Ebola crisis SRI promotion efforts had to be interrupted to give priority to local and national efforts of controlling the propagation of the deadly virus. More recently, due to an increased isolation caused by the Covid-19 pandemic, labor demands of local food production rose, slowing economic growth in the farming sector of Sierra Leone. For 2022, as the effects of the post-COVID-19 pandemic attenuate, it is predicted that inflation will go back to 10.5% due to recovery difficulties affecting the domestic food production system (World Bank, 2020). Despite this major public health disaster, an evaluation of SRI in Sierra Leone substantiated that this intensification approach was economically viable, producing rice yields that were three times (6.2 t/ha) higher than the average yield (2.0 t/ha), obtained with conventional farming practices (Harding et al., 2017). Videos of SRI in Sierra Leone are available from the SRI-Rice channel Africa playlist (https://www.youtube.com/watch?v=_vwuDGPtlKw&list=PL452A22F2821BB152). These include a video of Gerald Aruna's work with SRI in Mendesora village and chief Pa Foday Kanu, who was presented the Annual Chairperson's Award in 2019, for his leadership in implementing SRI cultivation practices in this community.

Discussion: Climate Change Education Begins with Agroecology Education

Climate change has negative impacts on food production systems with potential to jeopardize livelihoods and quality of life, especially in Africa. To address these challenges, there is an urgent need of transformational change in education to occur, as Borsari and Mora (2020) presented to fulfill the requirements of sustainable development goal (SDG 4), of the United Nations' Agenda 2030. The needed reform in education should consider establishing stronger connections to food production systems across curricula, to enhance public awareness about foods and agriculture, including implications for climate change adaptations (Borsari et al., 2014). The need for improving connectivity between education and food production was substantiated through a recent report by Leippert and team (2020), which was inspired by the idea that transformation in agriculture will only happen through a coordinated strategy among all stakeholder levels and thus, engage the broadest set of agencies like FAO, research institutes (FIBL, Biodiversity, ISRA) and community supported organizations. Initial attempts of combining evidence from a broad range of backgrounds and perspectives have been showing that agroecology can be an excellent global approach to farming, including Africa (IPES-Food, 2016). However, there is still a missing link in this emerging paradigm, and this consists in connecting solidly to every institution of education and make persuasive the case to all teachers that agroecology could become a keystone feature of every curriculum to improve nutrition, environmental education and quality of life. This will be the necessary, additional step to maximize awareness in support of climate change education across societal strata in Sierra Leone (Fig. 13.3).

Food production and security remain relevant and very personal for most people in Sierra Leone thus, an inclusion of agroecology in all curricula can make a pivotal contribution to spur interest and actions in climate change mitigation (Borsari, 2012). Experiences in urban gardening that include agroecological practices have been developed with success in many African cities and engaged mainly the participation

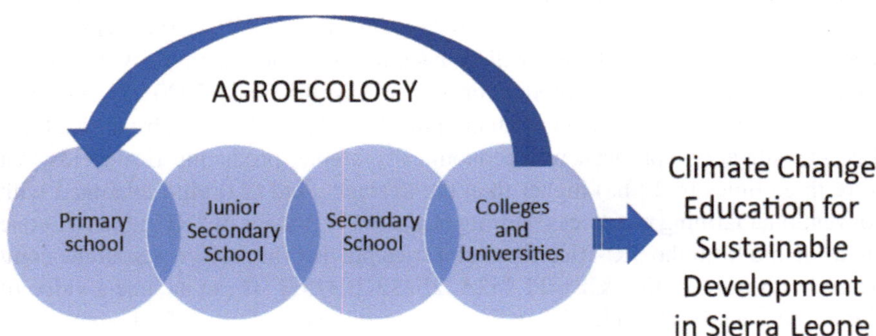

Fig. 13.3 An inclusion of agroecology in every curriculum in Sierra Leone fosters climate change awareness and sustainable development

of women (Prainet al., 2010), indicating the importance of resolving inequalities, while expanding education opportunities across genders. Finally, even the smallest garden space on a primary school compound has potential to inspire children (under the guidance of well prepared and dedicated teachers), to understand the ecological role of plant communities in sequestering carbon from the atmosphere into the soil (Borsari & Vidrine, 2022), while yielding nutritious food.

Conclusion and Recommendations

The impacts of climate change are an undeniable reality that demands immediate attention for preserving ecosystems and human communities, from regional to planetary levels, to impede uncontrollable escalations of bad weather patterns and irreparable consequences to the biosphere (IPCC, 2022). Adapting to climate change in view of the 2030 Agenda for Sustainable Development provides a guiding framework for stabilizing the climate phenomenon before this may spiral out of control. Education can contribute significantly to mitigate climate change when learning veers toward ecocentrism (Borsari, 2012), while becoming accessible to every citizen. A focus on women's education is of paramount importance because these continue playing a remarkable role at every level of the African food system, from production to preparation (Gnisci, 2016). Therefore, women have potential when adequately educated, to ensure securing food for their families and communities while enhancing resiliency, despite uncertainties and traumas inflicted by climate change events (Muhanguzi, 2019). In Sierra Leone women have already been engaged in adopting the knowledge and technology of the SRI cultivation method in comparable numbers to those of their male counterparts (Aruna, 2013), through extension education that preceded the spreading of Covid-19. The feasibility of SRI cultivation in Sierra Leone proves to be an effective agroecological practice that could become employable in climate change education and curricula to more countries in Africa.

From this research a holistic model has been conceived to demonstrate the nexus between agroecology curricula and climate change education (Fig. 13.4).

The interdisciplinary and systemic nature of agroecology is key for a true, transformational pursuit of sustainability in agriculture and education at the same time. Paradoxically, these traits can be also the challenges for each other, unless multidisciplinary research, policy revisions and periodic assessments that are aimed at maintaining these two concepts in balance are conducted.

Climate change education is intertwined to curricula in agroecology, and both serve as foundation items for every other study, or extension program. This biunivocal relationship between the two central constructs of the model is regulated by four education factors: education reform, instruction, evaluation and teacher preparation. A reform of education is the necessary process within this adaptive model (Borsari & Mora, 2020). This will gear instruction toward agroecology, while ensuring quality teachers' preparation, including efforts of implementing climate change education in curricula. Periodic evaluations are aimed at adapting, or creating specific curricula

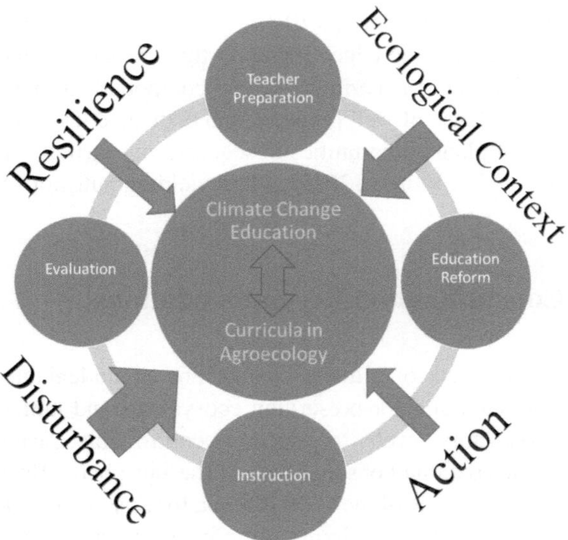

Fig. 13.4 Holistic model for a climate change education through an employment of curricula in agroecology

in climate change with agroecology as a focus. The proposed model also shows four external forces that regulate the equilibrium between curricula in agroecology and climate change education. Ecological context refers to the knowledge of biotic assets, abiotic factors, and resources (including indigenous knowledge) of growing, or collecting food from the wilderness, in specific agroecosystems, or environments. Thus, a commitment to agrobiodiversity conservation becomes an indicator of paramount importance when considering any ecological context for food production purposes (Borsari, 2022) because agrobiodiversity will make any environment most resilient to disturbances, caused by unexpected weather events. The fourth force is action intended as hands-on, restoration works and implications for human and planetary health, that teachers can employ to optimize instruction and learning about climate change (Borsari & Vidrine, 2022). Therefore, the motivation, dedication and preparedness of teachers are keystone attributes to enhance the benefits provided by the model. However, preparing teachers in Sierra Leone remains a very challenging endeavor that continues to be affected by multiple and complex factors thus requiring monetary support from international agencies and commitment from local leaders and stakeholders (World Bank, 2007). These strategies are pivotal to transform education in Sierra Leone and, according to Wright (2018), these conjunct efforts should yield significant improvements in favor of an equitable, transformative, education in this west African country.

In sum, feasible recommendations that emerged from this work consist in:

- Recognizing that the knowledge base of fostering agroecology to build ecosystems' resilience is employable as a viable climate change adaptation strategy, whose application is feasible and much needed for education purposes, at every level of instruction.

- Expanding education opportunities for girls and women in Sierra Leone and worldwide because they are primary food growers, caregivers, and thoughtful stewards of ecosystems (World Bank, 2020).
- Amplifying the scaling-up of agroecology needs on the success achieved already with the SRI method, to be addressed through an improved access to knowledge and understanding of systemic approaches that should be fostered across multiple sectors of local economies, stakeholders, and scales.
- Investing most generously in schoolteachers' preparation with agroecology as a priority need for their training, so that they may become effective leaders to enhance an education for climate change in Sierra Leone.

Further comparative research on the multidimensional effects of agroecology is needed, in conjunction with evaluation studies that will substantiate further the validity of the suggested educational approaches here proposed.

Acknowledgements This work was made possible through the financial assistance of Cooperazione Internazionale (COOPI), Italian Non-Governmental Organization. The author dedicates this work to Mr. Gerald Aruna and to all the dedicated faculty at SJVI in Lunsar, Sierra Leone. Many thanks also to the anonymous reviewers whose assistance improved the quality of this manuscript.

References

Adekunle Adeyeye, B., & Ighorojeh, B. (2019). Education and gender disparity in west African countries: The Nigeria scenario. *European Scientific Journal, ESJ, 15*(28), 43. https://doi.org/10.19044/esj.2019.v15n28p43

Adger, W. N., Huq, S., Brown, K., Conway, D., & Hulme, M. (2003). Adaptation to climate change in the developing world. *Progress in Development Studies, 3*(3), 179–195. https://doi.org/10.1191/1464993403ps060oa

Agegnehu, G., Srivastava, A. K., & Birda, M. I. (2017). The role of biochar and biochar-compost in improving soil quality and crop performance: A review. *Applied Soil Ecology, 119*, 156–170. https://www.sciencedirect.com/science/article/pii/S0929139316304954?via%3Dihub

Aijuka, J., Bassey, M., Belay, M., Bassermann, L., Goïta, M., Herre, R., Koch, J., Maina, A., Mkindi, A., Senzia, D. A., Tanzmann, S., Urhahn, J., Wamunyima, M., & Wangre, A. (2021). A Sting in the AGRA Tale: Independent expert evaluations confirm that the Alliance for a Green Revolution has failed. https://afsafrica.org/a-sting-in-the-agra-tale-independent-expert-evaluations-confirm-that-the-alliance-for-a-green-revolution-has-failed/

Altieri, M., Nicholls, C., Henao, A., & Lana, M. (2015). Agroecology and the design of climate change-resilient farming systems. *Agronomy for Sustainable Development, 35*(3), 869–890. https://doi.org/10.1007/s13593-015-0285-2

Aruna, G. (2013). Facilitators guide for conducting farmer field schools on SRI. https://sriwestafrica.files.wordpress.com/2014/06/sierra-leone-sri-facilitators-guide.pdf

Bayar, M., & Aral, M. M. (2019). An analysis of large-scale forced migration in Africa. *International Journal of Environmental Research and Public Health, 16*(21), 4210. https://doi.org/10.3390/ijerph16214210

Berton, G. (1999). Una *voce lontana. Diario dalla Sierra Leone. Editrice Artística Bassano.* Castelfranco Veneto, Italy

Borsari, B., Kunnas, J. (2023). Historical memory and eco-centric education: Looking at the past to move forward with the 2030 agenda for sustainable development. In W. Leal Filho, A. M.

Azul, F. Doni, & A. L. Salvia (Eds.), *Handbook of sustainability science in the future*. Springer. https://doi.org/10.1007/978-3-031-04560-8_40

Borsari, B. (2023). From agroecology to food systems sustainability: An evolutionary path shifting toward sustainable agriculture and development. In W. Leal Filho, A. M. Azul, F. Doni, & A. L. Salvia (Eds.), *Handbook of sustainability science in the future*. Springer. https://doi.org/10.1007/978-3-031-04560-8_8

Borsari, B. (2022). Agrobiodiversity as necessary standard for the design and management of sustainable farming systems. *Planta Tropika, 10*(1). https://doi.org/10.18196/pt.v10i1.14105

Borsari, B., Vidrine& , M. F. (2022). Planetary health begins on campus: Enhancing students' well-being and health through prairie habitat restoration. In W. Leal Filho (Ed.), *Handbook of human and planetary health*. Climate Change Management. Springer. https://doi.org/10.1007/978-3-031-09879-6_14

Borsari, B., & Mora, C. (2020). Education reform through a systems approach for sustainable development. In W. Leal Filho, A. Azul, L. Brandli, P. Özuyar, & T. Wall (Eds.), *Quality education. Encyclopedia of the UN sustainable development goals*. Springer.

Borsari, B., Espinosa-Tasón, J., & Hassán, J. (2014). Agroecology and a vision for sustainable agriculture in spite of global climate change. *Investigación y pensamiento critico, 2*(5), 63–79. https://revistas.usma.ac.pa/ojs/index.php/ipc/article/view/36

Borsari, B., Peréz de Gracia, N., & Castro Peralta, J. (2016). Students' engagement in an extension program in agroecology for subsistence farmers at the Universidad Católica Santa Maria La Antigua (USMA), Panamá. In W. Leal Filho & P. Pace (Eds.), *Teaching education for sustainable development at University level* (pp. 147–161). Springer International Publishing. https://doi.org/10.1007/978-3-319-32928-4_10

Borsari, B. (1999). Teaching agriculture in tropical Africa: Understanding the local culture for the design of a sustainable curriculum. *Journal of Sustainable Development in Africa, 1*(2), 1–7.

Borsari, B., & Vidrine, M. F. (2000). An evaluation tool for improving undergraduate curricula in agriculture. Sustainable agriculture-fertile ground for growth in the agricultural sciences. *Journal of College Science Teaching, 29*(4), 235–240

Borsari, B. (2012). Curriculum framework for sustainability education. *Academic Exchange Quarterly, 16*(1), 74–78.

Borsari, B. (August, 1996). Practical utilization of *Leucaena leucocephala* in Sierra Leone: a successful example of sustainable farming. (Abstract). In *11th international conference of organic agriculture (IFOAM)*, Copenhagen Denmark (pp. 11–15, p. 39).

Bunce, M., Rosendo, S., & Brown, K. (2010). Perceptions of climate change, multiple stressors and livelihoods on marginal African coasts. *Environment, Development and Sustainability, 12*(3), 407–440. https://doi.org/10.1007/s10668-009-9203-6

Burrows, K., & Kinney, P. L. (2016). Exploring the climate change, migration and conflict nexus. *International Journal of Environmental Research and Public Health, 13*(4), 443. https://doi.org/10.3390/ijerph13040443

Coates, S. J., Enbiale, W., Davis, M. D., & Andersen, L. K. (2020). The effects of climate change on human health in Africa, a dermatologic perspective: A report from the international society of dermatology climate change committee. *International Journal of Dermatology, 59*(3), 265–278.

Creswell, J. W., & Creswell, J. D. (2018). *Research design: Qualitative, quantitative, and mixed methods approaches* (5th ed.). Sage Publications Inc.

Dube, T., Moyo, P., Ncube, M., & Nyathi, D. (2016). The impact of climate change on agro-ecological based livelihoods in Africa: A review. *Journal of Sustainable Development, 9*(1), 256–267.

Ebel, R., Ahmed, S., Valley, W., Jordan, N., Grossman, J., Byker Shanks, C., Stein, M., Rogers, M., & Dring, C. (2020). Co-design of adaptable learning outcomes for sustainable food systems undergraduate education. *Frontiers in Sustainable Food Systems, 29*. https://doi.org/10.3389/fsufs.2020.568743

Ebun-Cole, W. A. (1992). Adult Learning principles for the improvement of agricultural extension in Sierra Leone. *Convergence, 25*(3), 53–65.

Foley, J. A., Ramankutty, N., Brauman, K. A., Cassidy, E. S., Gerber, J. S., Johnston, M., Mueller, N. D., O'Connell, C., Ray, D. K., West, P. C., Balzer, C., Bennett, E. M., Carpenter, S. R., Hill, J., Monfreda, C., Polasky, S., Rockström, J., Sheehan, J., Siebert, S., … Zaks, D. P. M. (2011). Solutions for a cultivated planet. *Nature*. https://doi.org/10.1038/nature10452

Glesne, C., & Peshkin. A. (1992). *Becoming qualitative researchers: An introduction*. White Plains, NY: Longman Publishing Group.

Gliessman, S. R., Méndez, E. V., Izzo, V. M., Engels, E. W., & Gerlicz, A. (2023). *Agroecology. Leading the transformation to a just and sustainable food system* (4th Ed.). Boca Raton, FL: CRC Press. https://doi.org/10.1201/9781003304043

Gnisci, D. (2016). Women's roles in the west African food system: Implications and prospects for food security and resilience. West African papers, No. 3, OECD Publishing, Paris. https://doi.org/10.1787/5jlpl4mh1hxn-en

Harding, S. S., Mahmood, N., Chernor Sullayk, J. M. K., & Toure, A. (2017). Assessing the suitability and profitability of the system of rice intensification (SRI) methodology under farmers' circumstances in Sierra Leone. *Agronomie Africaine*, *29*(1), 41–52 https://www.ajol.info/index.php/aga/article/view/163168

Human Development Report. (2009). 2007/2008 human development report. Sierra Leone. https://web.archive.org/web/20090429194950/http://hdrstats.undp.org/en/countries/data_sheets/cty_ds_SLE.html

Ickowitz, A., Powell, B., Rowland, D., Jones, A., & Sunderland, T. (2019). Agricultural intensification, dietary diversity, and markets in the global food security narrative. *Global Food Security, 20*, 9–16. https://doi.org/10.1016/j.gfs.2018.11.002

International Trade Administration. (2021). Sierra Leone—country commercial guide. Agriculture Sector. https://www.trade.gov/country-commercial-guides/sierra-leone-agriculture-sector

IPCC. (2022). Climate Change 2022: Impacts, adaptation and vulnerability. Retrieved September 18, 2022, from https://www.ipcc.ch/report/sixth-assessment-report-working-group-ii/

IPCC. (2019). Summary for policymakers. In Climate change and land: an IPCC special report on climate change, desertification, land degradation, sustainable land management, food security, and greenhouse gas fluxes in terrestrial ecosystems. Retrieved May 16, 2022, from https://www.ipcc.ch/report/srccl/

IPES-Food. (2016). From uniformity to diversity: A paradigm shift from industrial agriculture to diversified agroecological systems. *International Panel of Experts on Sustainable Food Systems*. Retrieved September 19, 2022, from https://www.researchgate.net/publication/303737887_From_Uniformity_to_Diversity_A_paradigm_shift_from_industrial_agriculture_to_diversified_agroecological_systemsEnter_title

Jeavons, J. (1982). *How to Grow more vegetables than you evert thought possible on less land than you can imagine*. Ten Speed Press.

Jones-Parry, R. (Ed.). (2006). *Commonwealth education partnerships 2007*. Nexus Strategic Partnerships Ltd. ISBN 978-0-9549629-1-3

Kellaghan, T., & Stufflebean, D. L. (Eds.). (2003). *International Handbook of educational evaluation*. Klüver Academic Publisher.

Kerr, R. B. (2012). Lessons from the old Green Revolution for the new: Social, environmental and nutritional issues for agricultural change in Africa. *Progress in Development Studies, 12*(2–3), 213–229. https://doi.org/10.1177/146499341101200308

Knowles, M. S. (1984). The adult learner. A neglected species (3rd ed.). Gulf Publishing Co., Houston, Texas.

König, D. (2008). Linking agriculture to tourism in sierra Leone—a preliminary research. Master's thesis GRIN Verlag. Retrieved May 26, 2022, from https://www.grin.com/document/91629

Leippert, F., Darmaun, M., Bernoux, M., & Mpheshea, M. (2020). *The potential of agroecology to build climate-resilient livelihoods and food systems*. FAO and Biovision. https://doi.org/10.4060/cb0438en

Mai, T. T. (2019). Project information document—sierra Leone free education project—P167897. Washington, D.C. Retrieved June 10, 2022, from http://documents.worldbank.org/curated/en/711051560267527870/Project-Information-Document-Sierra-Leone-Free-Education-Project-P167897

Martínez-Torres, M. E., & Rosset, P. M. (2010). La Vía Campesina: The birth and evolution of a transnational social movement. *The Journal of Peasant Studies, 37*(1), 149–175. https://doi.org/10.1080/03066150903498804

McDermott, P., & Allen, N. (2015). Successes and challenges of implementing a teacher education project in rural Sierra Leone. *International Journal of Educational Research, 71*, 16–25. https://doi.org/10.1016/j.ijer.2015.02.001

Miles, R. (2015). Complexity, representation and practice: Case study as method and methodology. *Issues in Educational Research, 25*(3), 309–318. Retrieved September 12, 2022, from https://search.ebscohost.com/login.aspx?

Minnich, J., & Hunt, M. (1979). *The Rodale guide to composting*. Rodale Press.

Muhanguzi, F. K. (2019). Women and girls' education in Africa. In O. Yacob-Haliso & T. Falola (Eds.) *The Palgrave handbook of African women's studies*. Palgrave Macmillan. https://doi.org/10.1007/978-3-319-77030-7_34-1

National Electoral Commission. (2018). https://necsl.org/index_files/2017_Press_Releases/Completed_RegistrationPressRelease2017.pdf

Njala University College—NUC. (2007). Sierra Leone: Sierra Leone Encyclopedia.https://web.archive.org/web/20070311011418/http://www.daco-sl.org/encyclopedia/1_gov/1_7njala.htm

Onwueme, I., Borsari, B., & Leal Fihlo, W. (2008). An analysis of some paradoxes in alternative agriculture and a vision of sustainability for future food systems. *International Journal of Agricultural Resources, Governance and Ecology, 7*(3), 199–210.

Patel, R. (2012). The long green revolution. *The Journal of Peasant Studies, 40*(1), 1–63. https://doi.org/10.1080/03066150.2012.719224

Patten, M. L. (2004). *Understanding research methods. An overview of the essentials* (4th ed.) Pyrkzak Publishing.

Prain, G., Karanja, N., & Lee-Smith, D. (Eds.). (2010). *African urban harvest agriculture in the cities of Cameroon, Kenya and Uganda*. Springer. https://idl-bnc-idrc.dspacedirect.org/bitstream/handle/10625/45136/IDL-45136.pdf?sequence=1&isAllowed=y

Randell, H., & Gray, C. (2019). Climate change and educational attainment in the global tropics. *PNAS, 116*(18), 8840–8845. https://doi.org/10.1073/pnas.1817480116

Rasmussen, L. V., Coolsaet, B., Martin, A., Mertz, O., Pascual, U., Corbera, E., Dawson, N., Fisher, J. A., Franks, P., & Ryan, C. M. (2018). Social-ecological outcomes of agricultural intensification. *Nature Sustainability*. https://doi.org/10.1038/s41893-018-007

Rockström, J., & Gaffney, O. (2021). *Breaking boundaries: The science of our planet*. Johan Dorling Kindersley Publisher. https://www.penguinrandomhouse.com/books/659581/breaking-boundaries-by-johan-rockstrom-and-owen-gaffney/

Rosenzweig, C., & Parry, M. L. (1994). Potential impact of climate change on world food supply. *Nature, 367*, 133–138.

Rossini, G. (2007). *La Guerra e' il mio pane. La guerra civile in Sierra Leone 1991–2001. Meroni Editrice*. Albese con Cassano, Como, Italy

Statistics Sierra Leone. (2017). Sierra Leone 2015 population and housing census—national analytical report. Freetown. https://www.statistics.sl/images/StatisticsSL/Documents/Census/2015/2015_census_national_analytical_report.pdf

Swanson, B. E., & Tucker, S. W. (1978). Land lab experiences in Sierra Leone and Illinois. *Agricultural Education, 50*(9), 199–203.

Thompson, C. B. (2012). Alliance for a Green Revolution in Africa (AGRA): Advancing the theft of African genetic wealth. *Rev African Pol Econ, 39*(132), 345–350. https://doi.org/10.1080/03056244.2012.688647

Toungos, M. D. (2018). System of rice intensification: A review. *International Journal of Innovative Agriculture & Biology Research, 6*(2), 27–38.

United Nations. (2015). General assembly resolution 70/1. Transforming our world: the 2030 Agenda for Sustainable Development. A/RES/70/1. https://www.un.org/en/development/desa/population/migration/generalassembly/docs/globalcompact/A_RES_70_1_E.pdf

University of Makeni. (2018). *University of Makeni prospectus 2018–2019*. Makeni, Sierra Leone. https://unimak.edu.sl/wp-content/uploads/2021/06/FINAL-PROSPECTUS-2018-2019.pdf

van Oort, P. A. J., & Zwart, S. J. (2018). Impacts of climate change on rice production in Africa and causes of simulated yield changes. *Global Change Biology, 24*(3), 1029–1045. https://doi.org/10.1111/gcb.13967

Wang, L. (2007). *Education in Sierra Leone: Present challenges, future opportunities* (p. 2). World Bank Publications. ISBN 0-8213-6868-0

Wezel, A., Bellon, S., Doré, T., Francis, C., Vallod, D., & David, C. (2009). Agroecology as a science, a movement and a practice. A review. *Agronomy for Sustainable Development, 29*(4), 503–515. https://doi.org/10.1051/agro/2009004

World Bank. (2007). *Education in Sierra Leone, education in Sierra Leone*. World Bank. https://doi.org/10.1596/978-0-8213-6868-8

World Bank. (2014). *Agricultural program helps rice crops and food security grow in Sierra Leone*, Washington, DC. https://www.worldbank.org/en/news/feature/2014/06/18/agricultural-program-helps-rice-crops-food-security-grow-sierra-leone

World Bank. (June, 2020). *Sierra Leone economic update. The power of investing in girls* (3rd ed.). World Bank Group. https://documents.worldbank.org/en/publication/documents-reports/documentdetail/131511593700755950/sierra-leone-economic-update-2020-the-power-of-investing-in-girls

Wren Media. (2007). Settling for a future in Sierra Leone. New Agriculturist. https://web.archive.org/web/20200804200645/http://www.new-ag.info/en/focus/focusItem.php?a=291

Wright, C. (2018). A comprehensive situation analysis of teachers and the teaching profession in Sierra Leone: Final report presented to the Teaching Service Commission. https://tsc.gov.sl/wp-content/uploads/2020/11/18-448-Sierra-Leone-Teaching-report-web.pdf

Yamah, A. (2002). The practice of the system of rice intensification in Sierra Leone. Paper presented at the international conference on assessments of the system of rice intensification (SRI), April 1–4, in Sanya, China (pp. 3–8).

United Nations. (2015). [illegible] Agenda for sustainable development [illegible]

University of [illegible]

[illegible] R. [illegible] Z. [illegible] and causes of [illegible] Africa [illegible]

Wang, L. [illegible] World Bank Publications. [illegible]

Wessel, A. [illegible]

World Bank. (2021). [illegible] World Bank Report [illegible]

World Bank. [illegible] Washington [illegible]

World Bank. [illegible]

Wren-Lewis [illegible]

Wright, C. (2019). [illegible]

Yaseli [illegible]

Chapter 14
Indigenizing Education for Climate Management in Cameroon

Ernest L. Molua, Francis E. Ndip, Marco Alberto Nanfouet, Lionel P. Kemeni Kambiet, and Sophie E. Etomes

Abstract Climate change is associated with food and nutrition insecurity, with the potential to further exacerbate water, health, and energy insecurity. The education of farmers is crucial for measures beyond food security. Farmers need education to remain abreast of technological innovations that affect agricultural operations. However, farmers encounter unique obstacles and must receive education and training to be successful. In this chapter we examine the impact of indigenizing climate change education on the perceptions of small holder farmers in the humid tropical agroecologies of Cameroon, whose forests are vital to climate change mitigation in the Congo basin. Farmers were studied together with their households in southern regions of Cameroon. The field study shows positive perception about a changing climate, and that farming households are employing coping and adaptive measures based on local and indigenous knowledge. Farmers diversify and reduce risk with off-farm income in the face of climate-related challenges. As a means of bolstering resilience and mitigating vulnerability, farmers are incorporating native tree species into agroforestry. The findings are critical for shaping climate change policy on awareness-building, education, and training for effective adaptation.

E. L. Molua (✉)
Department of Agricultural Economics and Agribusiness, University of Buea, Buea, Cameroon
e-mail: emolua@yahoo.com; emolua@cidrcam.org

Centre for Independent Development Research, Buea, Cameroon

F. E. Ndip
Centre for Development Research (ZEF), Bonn, Germany

M. A. Nanfouet
Faculty of Agriculture, Institute for Food and Resource Economics, University of Bonn, Bonn, Germany

L. P. K. Kambiet
Mohamed VI Polytechnic University, Ben Guerir, Morocco

S. E. Etomes
Department of Educational Foundation and Administration, Faculty of Education, University of Buea, Buea, Cameroon
e-mail: sophie.ekume@ubuea.cm

M. F. Mbah et al. (eds.), *Practices, Perceptions and Prospects for Climate Change Education in Africa*, https://doi.org/10.1007/978-3-031-84081-4_14

Keywords Indigenizing education · Climate change · Adaptation · Innovation · Technology adoption · Indigenous people and local communities

Introduction

Climate is a natural amenity with significant global impact in the modern economy (Arora-Jonsson & Gurung, 2023; Halsnæs & Kærgård, 2023; Mbeva et al., 2023). In several national and international forums, policy directives concerning climate change education for adaptation and mitigation are therefore emerging. As a result, educational actors are in a prime position to reprioritize the role of training for effective use of natural resources, including the climate as a global good and to continue expanding their capacities in this area. According to estimates from Sharaky (2014), Lebert (2015), and Tumushabe (2018), 30% of the world's recognized mineral reserves are located in the continent of Africa. Despite this abundance and the promise of a sustainable future, the negative consequences of climate change pose a danger to some of the advancements being made in the continent's pursuit of the SDGs (Ogwu, 2019; Tumushabe, 2018). Studies recommend education as a critical tool for altering people's attitudes and encouraging positive behavioral behaviors to adapt to climate change consequences and lessen global warming as the climate crisis develops (Simpson et al., 2021; Lehtonen et al., 2019; Cordero et al., 2020; Anderson, 2012). In order to achieve a long-term solution, numerous countries, particularly in the industrialized world, are including education (Lehtonen et al., 2019).

Contrarily, it might be argued that nothing is being done to address environmental challenges through its educational institutions, despite the fact that Africa has been establishing policies for sustainability (Franco et al., 2019; Ochieng & Koske, 2013). The practices, perceptions, and potential for climate change education in Africa must therefore be recorded (Anderson, 2012). Building the capacity of the present and future generations to solve what has been deemed the most serious global issue requires incorporating the topic of climate change throughout formal and informal sectors of education (Molthan-Hill et al., 2021; Cordero et al., 2020; Lehtonen et al., 2019). According to Feinstein and Mach (2020), the distinct and profound impact of education is found in its ability to provide adaptive learning support, encompassing curricular, pedagogical, and technical resources. These resources equip people with the necessary skills for intricate adaptive decision-making and aid in reinforcing their learning while engaged in practical tasks.

However, the inadequacies in educational frameworks designed to support a future that is climate resilient can also be revealed by gaining insight into current practices and perceptions in Africa in relation to climate change (Simpson et al., 2021). The indigenization of education to address the issue of climate change may also be a crucial tactic. This opens up discussion about opportunities and potential directions for climate change education. To address SDG13 and target 3, which emphasize the need to "improve education, awareness-raising and human and institutional capacity on climate change mitigation, adaptation, impact reduction, and early warning," policymakers, governments, and other stakeholders must develop pathways and put into practice effective strategies.

Most farmers believe that supernatural forces are causing climate change (Salite, 2019; Haluza-DeLay, 2014). According to Alotaibi et al. (2020), climate change poses a serious danger to agricultural production, food security, and the management of natural resources. For effective climate change adaptation and food security, smallholder farmers' opinions of conservation agriculture are crucial (Huffman, 2001; Nyanga et al., 2011). For instance, Nyanga et al. (2011) note that actors involved in promoting conservation agriculture have frequently overlooked smallholder farmers' perceptions of climate change and conservation agriculture as an adaptation strategy and that adoption of conservation agriculture was significantly correlated with perceptions regarding floods and droughts. Education and science-based communication can both address this (Huffman, 2001). Huffman (2001) presents a summary of the key contributions and effects of education in agriculture.

"Science-based climate information is the foundation of resilience building, a cornerstone of climate change adaptation, as well as an oasis for sustainable livelihoods for achieving the goals of the Africa Agenda 2063 for development," says the African Union Commission's Commissioner for Rural Economy and Agriculture (African Union, 2015, p. 1). The scarcity of accurate and timely climate information is a contributing factor to Africa's low adoption and usage of climate information services in development planning and practice. In order to enhance adoption rates, conservation agriculture programs should, according to Nyanga et al. (2011), not only concentrate on technical solutions but also take into account social factors like perceptions, which are crucial to conservation agriculture. It is crucial to include climate change communication in order to encourage the interchange of climatic data that can help smallholder farmers relate to conservation agriculture as an adaptation approach.

There are historic reports that indict the lack of capacity for African agriculture to play its quintessential role for the economy (Holmén & Hyden, 2011; Voortman et al., 2003; Wiggins, 2014). While agriculture is important for food security, raw material supply, income and job creation opportunities, the overreliance of developing country and African tropical agriculture on rain and direct external climatic factors exposes the sector to weather variability and climate change. Coupled with challenges to access information and inadequate training, overreliance on very simple technologies and an ineffective institutional framework particularly the extension service impedes the capacity of the agricultural sector from responding, coping, and adapting to climate variability and change. Ample reports on the dearth of educational capacity in a continent where farming has been left to rural vulnerable masses, makes it problematic for an important sector as agriculture to fulfil its mandate for households, communities, region, and continent. According to Kydd et al. (2004, p. 1) "there is widespread concern at continuing and deepening poverty and food insecurity in sub-Saharan Africa and the lack of broad-based economic growth….Policy needs to focus more on agriculture, and recognize and address the diversity of institutional, trade, technological and governance challenges to poverty-reducing growth in Africa."

With these challenges omnipresent, the survival of local peoples and indigenous communities (IPLCs) hinges on digging deep into their inherent capabilities acquired locally over the years. In fact, increasing evidence suggests that indigenous knowledge (IK) plays a crucial role in adapting to climate change (Naess, 2013; Nyong et al., 2007; Reyes-García et al., 2024). The collection of studies reviewed in Nyadzi et al. (2021), Mbah et al. (2021) and Petzold et al. (2020) indicate that the indigenous knowledge acquired through long-term and careful observation is highly effective in mitigating climate change. It is demonstrated that indigenous practices for land and resource management not only enhance biodiversity, surpassing even the outcomes of protected areas, but also aid in curbing deforestation, carbon emissions, and the occurrence of wildfires. Where IPLCs have limited access to weather data, whether farmers, fishers, herders, or hunters, they have traditionally relied on natural signs, such as the first rainfall, the growth of specific plants, or the arrival of certain bird species, to determine the appropriate timing for activities such as planting, harvesting, or other tasks. However, as a result of climate change, numerous biological patterns have undergone significant alterations.

In our case-study in Cameroon's Southwest and Littoral regions, IK has been developed over many generations via routine daily activities and an intimate knowledge of the surrounding ecosystems. Indigenous education that integrates a wealth of traditional knowledge which forms the basis of indigenous peoples' identities, cultural heritage, civilizations, livelihoods, and coping mechanisms across many generations will be crucial in solving the climate change crisis. This strategy is appealing because it centres adaptation strategies on indigenous knowledge. Insofar as farmers are familiar with such methods, it facilitates their transfer to contemporary methods produced through formal education-based training. It could be possible that successful indigenous techniques are taken up by non-IPLCs in an attempt to gain from some perceived benefits. The transfer of such knowledge could be done formally by education institutions and training centres, properly packaged in climate change education by public or private agricultural extension services.

The principal research question we thus seek to address is what role does indigenous education play in climate change perception and management? We explore this question by examining the relationship between education and the adoption of measures to adapt to perceived climate change by households in rural Cameroon whose primary occupation is agriculture. The hypothesis is that famers' educational status significantly explains their farm returns following adoption and uptake of autonomous measures to adapt to a changing climate. Some previous expositions have delved into the issue and established both correlation and causality on the effects of education and climate change adaptation (Asare-Nuamah et al., 2022; Sibanda & Manik, 2022; Paltasingh & Goyari, 2018; Ochieng & Koske, 2013; Anderson, 2012; Nyanga et al., 2011). Climate change education, according to Anderson (2012), is likely to be more effective. Also, formally preparing the future generation of people who will have to deal with climate concerns is achieved through indigenous climate change education. Because of this, it has the dual benefit of formally assisting both the present and coming generations in adjusting to climate change. With five distinct ecological zones and a range of meteorological conditions, Cameroon serves as an essential test bed for the phenomenon of climate change. Cameroon is a desirable location for this kind of project since its geography is indicative of the many circumstances seen throughout the continent of Africa.

To properly address the context of this chapter, we present the remainder of the key information in four sections. These include the theoretical nexus of education and adoption of climate change adaptation options, then the methodology of the empirical research, the field observations with hypothesis testing and our concluding comments complete with suggestions to enhance the capacity of indigenising education to promote climate change education.

Theoretical Nexus: Education, Adoption and Climate Change Adaptation

Agriculture and Climate Change Adaptation

Agriculture's inherent biophysical relationship with climate puts the sector in the frontline to bear the brunt of climate change. The reliance of society's survival for food and by-products makes the agricultural sector an important conduit for national and international development. This means that changes in global climate patterns is a development related emergency requiring commensurate response guided by empirical analysis that untangles the key components.

The nature of climate change, however, will define the micro and macro-level actions required for response. By characterisation, climate change is the change in global or regional climate patterns, attributed largely to the increased levels of atmospheric carbon dioxide produced by the use of fossil fuels.[1] Clearing land and forests especially for agriculture, also releases carbon dioxide. As emissions continue to rise, the Earth is now warmer (IPCC, 2021). The consequences of climate change now include, among others, intense droughts, water scarcity, severe fires, rising sea levels, flooding, melting polar ice, catastrophic storms, and declining biodiversity. In addition, climate change affects health, food supply, housing, safety and work.

The unfolding change in climate requires not only climate change adaptation; responding to climate change involves a two-pronged approach: (a) reducing emissions of and stabilizing the levels of heat-trapping greenhouse gases in the atmosphere ("mitigation"); and (b) adapting to the unfolding climate change. Mitigation—reducing climate change—involves reducing the flow of heat-trapping greenhouse gases into the atmosphere, either by reducing sources of these gases (for example, the

[1] Climate change refers to long-term shifts in temperatures and weather patterns. These shifts may be natural, such as through variations in the solar cycle. But since the 1800s, human activities have been the main driver of climate change, primarily due to burning fossil fuels like coal, oil and gas. Burning fossil fuels generates greenhouse gas emissions that act like a blanket wrapped around the Earth, trapping the sun's heat and raising temperatures. Examples of greenhouse gas emissions that are causing climate change include carbon dioxide and methane.

burning of fossil fuels for electricity, heat, or transport) or enhancing the "sinks" that accumulate and store these gases (such as the oceans, forests, and soil).[2] Adaptation (adapting to life in a changing climate) involves adjusting to actual or expected future climate. The goal is to reduce our risks from the harmful effects of climate change (like sea-level rise, more intense extreme weather events, or food insecurity). It also includes making the most of any potential beneficial opportunities associated with climate change (for example, longer growing seasons or increased yields in some regions).

Reviews have shown that throughout history, people and societies have adjusted to and coped with changes in climate and extremes with varying degrees of success (Berrang-Ford et al., 2015; McNamara & Buggy, 2017; Owen, 2020). Climate change (drought in particular) has been at least partly responsible for the rise and fall of civilizations. Adapting to climate consequences protects people, homes, businesses, livelihoods, infrastructure, and natural ecosystems. It covers current impacts and those likely in the future. There is ample evidence that humans have been adapting to their environments throughout history by developing practices, cultures and livelihoods suited to local conditions as well behavioural shifts such as individuals using less water, farmers planting different crops and more households and businesses buying flood insurance.[3] However, adaptation is required everywhere, but must be prioritized for the most vulnerable people with the fewest resources to cope with climate hazards (Filho et al., 2022; Schlingmann et al., 2021).

Indigenizing Education for Climate Action

One form of empowering societies is indigenizing education for climate action. It is well established that education has a crucial role in tackling the problem of climate change. Climate change education facilitates comprehension and resolution of the consequences of the climate issue, equipping individuals with the information, abilities, principles, and mind-sets necessary to actively contribute to transformative actions. Furthermore, education has the potential to foster a shift in individuals' attitudes and behaviour, while also equipping them with the necessary knowledge to make well-informed choices (Feola et al., 2015; Masud et al., 2016). According to Feinstein and Mach (2020), education supports climate change adaptation through three distinct but overlapping pathways, each offering concrete policy options: education infrastructure, general education, and adaptation learning support.

[2] The goal of mitigation is to avoid significant human interference with Earth's climate, stabilize greenhouse gas levels in a timeframe sufficient to allow ecosystems to adapt naturally to climate change, ensure that food production is not threatened, and to enable economic development to proceed in a sustainable manner.

[3] Adaptation measures may be planned in advance or put in place spontaneously in response to a local pressure. They include large-scale infrastructure changes—such as building defences to protect against sea-level rise or improving the quality of road surfaces to withstand hotter temperatures.

Education can be a powerful tool in enabling effective adaptation to climate change (Bangay & Blum, 2010; Frankenberg et al., 2013; Lutz et al., 2014; UNICEF, 2012). Feinstein and Mach (2020) delineate three distinct yet intersecting policy applications. Firstly, safeguarding and implementing educational infrastructure, encompassing the social and material resources that uphold education, can aid in reducing susceptibility and fostering adaptability. Secondly, enhancing overall education, as evaluated through reading proficiency, school attendance, and overall academic performance, can enhance the ability to adapt. Lastly, the implementation of research-based adaptation learning assistance can expedite social and policy change by prioritizing the acquisition of knowledge prior to and throughout the process of making adaptive decisions. Overall, investments in education can significantly contribute to the formation of climate resilience and the promotion of climate mitigation and adaptation. Climate change education enhances individuals' understanding of climate hazards and provides them with increased resources and strategies to mitigate these hazards and cope with associated disruptions. Furthermore, integrating climate education into mainstream schooling will have a transformative impact on people's attitudes and actions towards climate change. This will also contribute to the promotion of enhanced readiness and adaptability to climate-induced disruptions across entire populations.

The global architecture acknowledges the significance of education and training in tackling climate change. For instance, the Kyoto Protocol, the Paris Agreement of the United Nations Framework Convention on Climate Change (UNFCCC) and the related agenda (United Nations, 1992, 2015) urge governments to provide education, empowerment, and engagement to all stakeholders and significant groups about climate change policies and activities. The UNFCCC mandates Parties of the Convention to conduct educational and public awareness initiatives about climate change. Additionally, they are required to facilitate public involvement in programs and provide access to information on the subject. By doing this, they will effectively communicate and improve awareness and ability to develop innovative strategies for adapting to climate change.

The education materials could comprise of local and indigenous knowledge from IPLCs who have long observed climatic and environmental changes and devised adaptive strategies. Filho et al. (2023) documents communities across the global south that use Indigenous and Local Knowledge (ILK) to predict weather events and climate hazards, as well as ILK for adaptation as IPLCs experience severe impacts of the climate crisis. An important approach of education is thus by learning from IPLCs and attempting to transfer that knowledge to non-indigenous communities, as well as integrating these efforts into policy frameworks. IPLCs, nonetheless, continue to remain susceptible against the negative effects of climate change since they depend on nature for their existence and economic sustenance.

Broadly, indigenization or naturalizing ILK systems and making them evident to transform spaces, places involves bringing IK and its approaches together with Western knowledge systems. Indigenization is seen as a collaborative process of naturalizing indigenous intent, interactions, and processes and making them evident to transform spaces, places, and hearts (Louie et al., 2017; Samek, 2021). In the context of climate change, indigenization benefits not only indigenous community members but seeks relevant programs and support services, as well as fundamental shift in the ways that institutions function (Cameron et al., 2021; Johnson et al., 2022). Through indigenous perspectives, values, and cultural understandings in policies and daily practices, IK can serve as conduit for adaptation and mitigation of climate change.

Whether indigenous or not, however, some have argued that climate change should be included in all school curricula and should play a central role in updated Nationally Determined Contributions (NDCs). In other words, compulsory climate education is needed. Countries like Mexico have included compulsory environmental education in its constitution as an initial measure in a holistic strategy, recognizing that significant changes can only be accomplished via the acquisition of information, consciousness, and collaborative efforts (Juarez, 2014; Mendoza-Zuany, 2019). Environmental education is thus a crucial means of achieving Sustainable Development Goals, serving as a vital instrument in combating the climate catastrophe and fostering a significant cultural shift towards ensuring the long-term viability of our planet (Acosta Castellanos & Queiruga-Dios, 2022; Agbedahin, 2019; Bonnett, 2018).

Hosen et al. (2020) aptly conceptualised the place of traditional ecological knowledge (TEK) as key component for adaptation to climate change, whereby ILK produced based on detailed observation of local ecosystems are generated and transmitted over many centuries (Fig. 14.1). Indigenous people conduct a careful observation of exogenously driven elements to direct their subsistence activities across time and space, including agriculture, hunting, fishing, and gathering. The feasibility of their actions and success embodies the TEK components, resilience theory and socio-ecological frameworks to explore communities' adaptation to climate change in coupled socio-ecological systems. Their framework identifies categories as (1) local knowledge of the environment, (2) land and resource management, (3) social networks and institutions and (4) worldview and belief systems. The model surmises that TEK is the root of resilience or the pillar of their capacity to adapt to environmental change and uncertainties based on an in-depth understanding of their environment.

Education and Innovation Diffusion Models for Climate Change Adaptation

How proven ILK and TEK are taken up by other sectors of the economy will depend the innovative capacity of adopters. Education can facilitate this capacity. While it is important for climate change education, education in general is the foundation of human capital development and remains important in classical theory of production (Becker, 2009; Hanushek & Woessmann, 2023; Wößmann, 2003). Education which births knowledge is important for innovation and technological development and it is required to push outwards the development frontier. The path through which

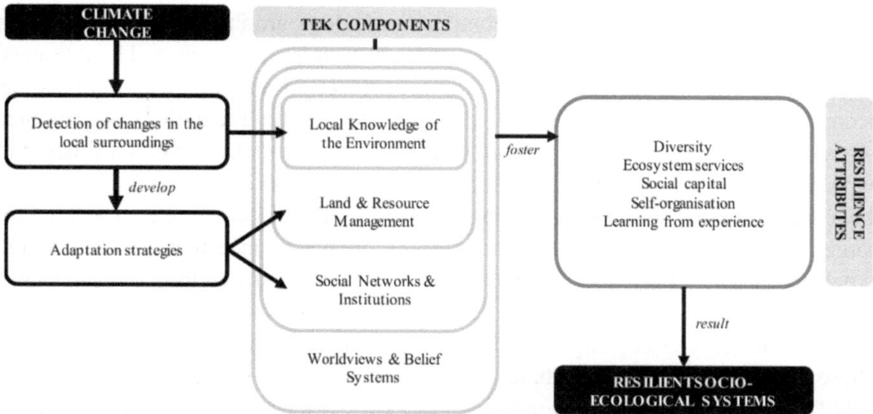

Fig. 14.1 Traditional ecological knowledge as key component for adaptation to climate change (Hosen et al., 2020)

education pushes the production frontier passes through technological innovation (Lin, 1991). Innovation whether related to new idea, process, or product is thus critical advancing production outcomes.[4]

However, innovation must be accepted and then adopted, for it to play an important role in the production process. Straub (2009) describes adoption as when an individual integrates a new innovation into their life and diffusion as "the collective adoption process over time." Straub (2009) then notes that adoption-diffusion theories, "refer to the process involving the spread of a new idea over time." Based on this, there may be three innovation theories that may define the role of climate change education awareness and adoption of adaptation measures; these are the Rogers' Innovation Diffusion Theory, Hall's Concerns-Based Adoption Model, and the Technology Acceptance Model. The mechanics of these three models suggest that technology adoption is a complex, inherently social, developmental process; individuals construct unique yet malleable perceptions of technology that influence their adoption decisions. In other words, successfully facilitating technology adoption must address cognitive, emotional, and contextual concerns.

Recent empirical accounts of innovation diffusion abound (e.g. Asare-Nuamah et al., 2022; Ayisi et al., 2022). For instance, Asare-Nuamah et al (2022) examined the adoption of innovations in mango production, which is susceptible to climate change risk and showed that mango farmers associated post-harvest losses, frequent droughts, rising temperature, declining rainfall, and pests and diseases to climate change. Their results further reveal that farmers' adoption of innovation is largely

[4] The first systematic effort by an economist to analyse the process of innovation was undertaken by Joseph Schumpeter in the first half of the twentieth century (Godin, 2017), where he identified three stages of the process: invention, innovation and diffusion. For Schumpeter, invention is the first demonstration of an idea; innovation is the first commercial application of an invention in the market; and diffusion is the spreading of the technology or process throughout the market (Godin, 2012).

determined by their contextual and compositional demographic characteristics such as age, education, size of plantation and years of farming experience. The influence of socioeconomic factors to promote the acceptance of adaptive technologies has been the preoccupation of Davis (1989) in proposing the Technology Acceptance Model (TAM). The TAM asserts that it is in fact a potential adopter's attitude and expectations of the innovation that affects the chances for its adoption. Davis believed that ease of use has a direct impact on perceived usefulness as, the easier an adopter perceives an innovation to be able to use, the greater chance they will use it and experience higher productivity thus proving to be useful to the adopter (Davis, 1989).

A top-down approach to promote ILK and TEK is increasingly promoted by some government agencies seeking to respect global climate change adaptation mandates. These efforts, no matter the debate on their reliability, are within the realm of Hall (1979) Concerns-Based Adoption Model (CBAM) which stemmed from the need for a model to address the traditional top-down approach to change. CBAM approaches innovation adoption from the perspective of those impacted by the adoption of the innovation and also charged with implementing the subsequent change. The idea is that by addressing the concerns during the adoption process, the challenges experienced during the change process will be lessened. Extant studies of climate change have looked at the implementation of change (Alotaibi et al., 2020; Choi, 2009). According to Choi (2009) the following are major issues that should be considered for efficient and effective technology transfer: conceptions of technology, technological activity and transfer, communication channels, factors affecting transfer, and models of transfer.

Overall, while indigenous knowledge offers invaluable insights for how to approach climate change, its uptake and dissemination remains a challenge. Scientists, governments, and communities are increasingly recognizing that taking action to adapt to climate change is urgent and an increasing number of people are looking to indigenous knowledge for insight. Gaps therefore exist on the experiences of different communities attempting to exploit the potential of disseminating ILK and TEK using modern education practices. This chapter thus attempts to fill this gap by studying the relationship between education and the adoption of measures to adapt to perceived climate change by households in rural Cameroon.

Materials and Method

Ethical Clearance

Mindful that Research Integrity embodies a range of good research practice and conduct which can include intellectual honesty, accuracy, fairness, intellectual property, and protection of human and animal subjects involved in the conduct of research, we sought ethical clearance for this research. We applied both to the Ethics Commission of the Ministry of Scientific Research and the University of Buea Ethics

Committee (UBEC) for approval. At the time of commencement of the research, the UBEC granted Ethical waiver. During the field research, the survey manager presented the research permit to respective region, district and village leaders. In addition, we sought verbal agreement from all respondents prior to the interviews to guarantee their willingness to participate. To protect confidentiality, respondents' names and personal information are kept anonymous.

Research Location

This research is undertaken in the Southwest and Littoral Regions of Cameroon, which are in the humid rainforest ecological regions of the country. These administrative regions are made up of Divisions and sub-Divisions (Fig. 14.2). Two Divisions are simply randomly selected and farming communities further sampled for the study. In the South west region, Fako Division was selected and studied. In the Littoral region, we studied the Moungo Division. Fako Division lies westwards of the region spreading from Muyuka on the east to Limbe on the west, and Tiko on the South, while bounded by the Atlantic Ocean and the Mount Cameroon range on its northern flank. The Moungo Division spans 3,947 km^2 between latitude 4.15° to 4.95° north of the equator and longitude 9.8° and 10.65° east of the prime meridian. It is bordered north westward and south eastward by the Nkam and Sanaga maritime Divisions, respectively. While the Wouri river forms a natural border to the northeast and the Atlantic Ocean to the southwest, providing access to the coast. The Moungo is composed of the municipalities of Mbanga, Njombe, Loum, Manjo, Souza with the capital being Nkongsamba while Fako is split into Limbe, Tiko, Muea, Mile 16 and the regional capital Buea.

Both Divisions are characterised by a diverse geography which displays coastal plains, river valleys, hills, and mountain and dotted a display of rivers, streams, springs, and swamps. The coastal plains and fertile river valleys support the cultivation of crops such as cocoa, coffee, palm oil, rubber, plantains, bananas, maize, and cassava. Traditionally, the interplay has encouraged the development of agricultural activities and settlement of diverse ethnicities within these areas. According to Beckline et al. (2022), Fako Division is home to 400,000 inhabitants, 70% of whom depend on agriculture either directly or indirectly for their livelihood. They participate by either selling their labour to large scale farmers and agricultural companies like the Cameroon Tea Estate and Cameroon Development Corporation or engaging in subsistence production of a variety of crops including vegetables, cereals, plantain, oil palm, root, and tubers. The case is not any different in the Moungo. More than 452,722 people live in this area and the local community is either gainfully employed in individual or agricultural firms like *Plantation du Haut Penja* (PHP) or self-employed in family holdings of less than 2 ha producing spices, fruits vegetables, banana, cocoa, coffee, and tubers. The availability of road infrastructure and the proximity of these areas to urban centers like Douala further enhanced the development of commercial agricultural activities. This region account for a bulk of agricultural

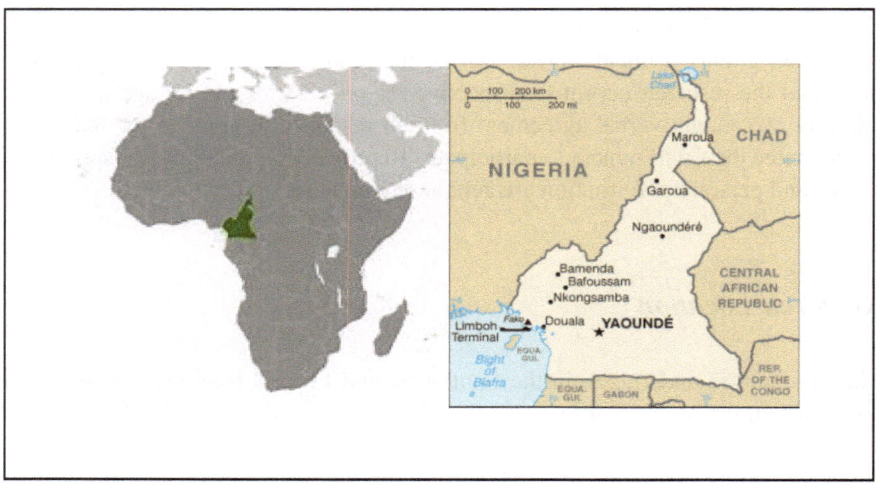

Fig. 14.2 Map of Cameroon showing study areas

exports and is critical for national food security, national unemployment reduction, and foreign exchange earnings.

Nature and Source of Data

The research employs a multi-stage sampling technic to sample farming households from the study area between the periods of July to September 2022. Buea municipality was purposively selected in the first stage based on the intensity of agricultural activities and bio-cultural diversity. Following the same selection mechanism of stage one, six villages were selected in the second stage from a pool of 18 villages which cuts across. A total of 20 questionnaires were randomly administered by well-trained enumerators to indigenous farming household heads in each village, making a total of 120 households. Prior to administration, the questionnaires were validated through a pilot survey in June 2022. The questionnaire elicited information on household demographics, farming systems, the resource mobilisation and use, the returns to the farming system, the perception to climate change, adaptive response to climate variation, indigenous knowledge as well as access to capacity development for climate change adaptation. After cleaning the data, 99 households were retained for further examination. The data is supplemented by expert elicitations from professionals and policy actors including extension officers, lecturers, researchers, traditional authorities, and the Ministry of Agriculture. This process involves personal interviews which are recorded using digital recording devices. The outcome of these interviews is then transcribed word verbatim, and the information generated used to strengthen further analysis.

Data Analysis

In our analysis we used mixed methods of qualitative and quantitative data analysis of the answers of the interview questions. This allows us to find out respondents' perception of climate variability, their behavioural responses and their coping mechanism or adaptation strategy. Descriptive statistical techniques give insights on detailed description of their self-characterization, leading us to key information on farmer socioeconomic characteristics, farm types, indigenous processes, and farm production. Recordings allowed us to reproduce the narratives and opinion of key informants and other frontline actors in the study area, capturing some important recurring themes to contextualize their account on climate and adaptation. The continuous data is then subjected to correlational analysis and simple regression using Ordinary Least Squares (OLS) technique to test the suggested hypothesis.

Results and Discussion

We present the results with the associated discussion in four sub categories. First, we describe the farmers we encountered. Second, we present their response to climate variation and change. Third, we present information on the role of indigenous and local knowledge in contributing to farmers' response to climate change. Finally, we test whether farmers' educational status and perception influence their farm performance. Experienced farmers with formal education manage larger farms, with modern farming technologies typical among experienced farmers. Farmers diversify and reduce risk with off-farm income in the face of local climate variations. Farmers suffer from floods, pest infestations and depleted groundwater as temperatures rise and precipitation patterns fluctuate. Native trees are increasingly used in farms for agroforestry, as well intercropping especially vegetables and legumes to enhance resilience and reduce vulnerability of their farming effort. Overall, the level of experience that farmers have is positively connected with their income. This level of experience measured by a variety of parameters, including age and level of education have a positive correlation with farm revenue.

Description of Local and Indigenous Households

The interviews were collated and analysed, with the first effort directed to summarising the socioeconomic and ecological characteristics of the households and homesteads. The households are relatively large with an average size of about four adults and six children. The average age of the household members is around 29 years thus suggesting that households have relatively younger demography. Household members are relatively literate with an average education of over nine years which goes beyond primary schooling. This indicates that most of the farmers have at least primary education level, with the First School Leaving Certificate, which gives them the ability to read and write, and interact with the Extension Service Officials. This provides the base for farmers to further their education or participate in activities related to climate change to improve on their knowledge. According to Christoffels and De Groot (2004) and Purcell-Gates (2004), basic reading and writing abilities translate to cognitive power which is relevant for engagement and comprehension of educative materials of various types.

Over fifty percent of households sampled have access to either formal or informal extension services via participating in village programmes organised by the Ministry of Agriculture and Rural Development (MINADER), NGOs, and INGOs undertaken in rural areas. Extension is one of the main conduits of agricultural information. With over half of households having access to extension implies that they have access to agricultural information. Net farm income from agriculture is relatively large. On average households earn over 3 million FCFA ($6000) annually from all their agricultural activities. These figures may seem high by the standards of peasant agriculture, however, a detailed review indicates that these observations are symptomatic to the living conditions of rural households in SSA (Azzarri & Signorelli, 2020; Ekeocha & Iheonu, 2021; Ibisomi & De Wet, 2014). For example, the distribution of the annual income of $6000 for a household of 4 persons and 6 children indicates this is an insignificant amount unable to guarantee average living conditions.

We triangulated our field observations with key-informant discussions with some Agricultural Officials, who provided further insights on the state of farming in the study region:

> Farmers have formal education, manage small and fragmented parcels of cultivatable land no more than one hectare, and have more than ten years of agricultural experience. Farmers that obtained additional land to augment their crops acquired less than one hectare. Experienced farmers are more likely to have access to more arable land and to use contemporary farming practices. Smallholder farmers were involved in non-farming activities. This suggests that, in addition to agriculture, they had other businesses that provided revenue. Off-farm income implies revenue diversification and risk smoothing across many enterprises. Off-farm activities lessen income uncertainty since the income can be utilized to cover any shortages in farm income and to acquire recommended modern farm inputs to reduce labour drudgery (Extension Official 1, Southwest Regional Delegation of Agriculture and rural development).

For the treatment variables of interest, over 23% of households have attended a training related to climate change. Given that formal trainings regarding climate change are usually scarce, this figure suggests that many households now take climate change seriously by attending trainings that can help them cope against it. In terms of climate education, over 76% of the sampled households report that they were at least taught about studies in geography that included concepts on climate while in school. Further information on the recent phenomena of climate change are received over the radio and other media. These observations may pass as suggestive evidence that most formal education trainings now consider climate change as a serious issue that warrants it being taught.

According extension officers, farmers are demonstrate good awareness about weather and climate variation:

> When we meet farmers, they always talk about the weather and its changing patterns. Our farmers are conscious of climate change. This is significant since the ability to manage with climate change is determined by one's level of understanding of the issue. Farmers who understand climate change, its causes, and consequences are more likely to implement adaptation and mitigation methods to cope with the negative effects of climate change. However, our farmers lack adequate adaptive capacity to climate change and global warming. They require support with information, equipment, and infrastructure. Climate change education is critical for preparing for the effects of climate change and teaching them how to adapt effectively. (Extension Official 2, Littoral Regional Delegation of Agriculture and rural development).

Considering that education enhances farm productivity (Ninh, 2021; Paltasingh & Goyari, 2018), providing in-service education for farmers will improve on adaptation mechanisms for climate change which Dhakal et al. (2022) found that it mitigates the risk associated with climate change and increase resilience.

Climate Variation, Change and Farmers' Response

Over 90% of farmers perceived changing weather patterns that align with scientifically observed climatic factors, as indicated by location-specific indicators. The most commonly reported observations pertain to changes in the atmospheric system. This include subtle observations of changes in precipitation patterns especially erratic rainfall, temperature, flood wind direction, fog, weather predictability, and seasonality. These observations are linked to variations in river dynamics, a decrease in natural springs, a reduction in soil moisture, an increase in soil erosion, and a decline in groundwater quality. Farmers have been very keen on these changes since they directly affect their agricultural production decisions by altering their production calendars.

Exchange with meteorological officials report that:

> There is discernable evidence in their records that changes are occurring in the local climate. Changes in mean temperature and seasonal mean temperature are the most commonly reported indicators of climate change. Farmers are confronting devastating new difficulties as temperatures rise and precipitation patterns shift, including severe drought, floods, and pest infestations, as well as increased desertification and depleted groundwater sources. Crop production variations, particularly for arable seasonal crops, are among the most commonly reported indicators of climate consequences. Climate change and its consequences disproportionately affect rural populations who face social and economic inequality. These effects are most obvious in people who rely on their immediate surroundings for sustenance and livelihood. (Meteorology official)

A varied array of practices are used by farmers to cope with climate vagaries. Farmers report changing their crop types, while others altered their use of herbicides and fertilizers. Farmers also make other adaptations, such as diversifying crop types and varieties and implementing soil and water conservation strategies. Some report employing water collection technologies to cope with the negative effects of climate change. However, not all adaptations are based on local knowledge. Responses such as introducing chemical fertilizers and pesticides, adopting hybrid varieties, or switching to off-farm jobs are implemented based on extension agency recommendations or by imitating nearby farmers. Conversation with an Extension official revealed certain methods that may be classified as:

> Fast-growing legume varieties, drought-resistant crops, improved water management, expanded peri-urban farming, the return to native plants or types, and agroforestry. (Extension Official 3, Southwest Regional Delegation of Agriculture and rural development)

Agroforestry is the integration of farms into natural ecosystems to promote reforestation and soil health, as well as the use of forest-grown goods such as cacao, mushrooms, and acai to support their lives. Over 11% of farmers use agroforestry as a climate change adaptation strategy. This is relatively low. A plausible explanation for this low number is that agroforestry may be very costly for the farmers. Identifying the most important trees to incorporate into the farm as well as the cost of planting them may discourage farmers. Hence, only a few privileged well-to-do farmers may engage in such practices. Moreover, most farmers cultivate food crops which may deter them from engaging in agroforestry.

About 33% of the farmers practice intercropping. This is expected given that most smallholder farmers have a dual motive of production. Implicit in this dual production motive is the cultivation of different crop types which may serve as important climate mitigation strategies (Ndip et al., 2023). Crop rotation is practiced by over 44% of the sampled farming households. This may be facilitated through land fragmentation which is a quintessential feature of African agriculture. Many smallholders have different plots of lands on which they may choose to plant different types of crops. Mulching is the most widely used adaption measure with over 58% of the sampled households using it. This has been an age-old practice used by smallholder farmers to adapt to climate change, especially in extremely dry periods. Mulching is a natural conservation technic which is cost effective but labour demanding (Moine & Ferry, 2018). The technique traps soil water through organic matter covering on the soil thus making it more available for crop reabsorption.

Overall, most of the local practices used to manage the environment and provide food security are broadly conservation agricultural practices. The implementation of these conservation technologies does not, however, preclude farmers' socio-psychological capacity and education. An effective foundation for facilitating a technology transfer process is as espoused in Choi's (2009) well-developed model of technology transfer. Some researchers have found that innovative characteristics indirectly influence behavioral intention (Valizadeh et al., 2020). Drought frequency and perceptions of climate change have an impact on smallholder farmers' adoption of innovation (Alotaibi et al., 2020; Asare-Nuamah et al., 2022; Ayisi et al., 2022).

ILK and Role of Education in Farmers' Response

On exploring through the face-to-face structured interviews with each of the famers we established an enhanced and positive perception of the roles of education in climate change. Given their farming activities, farmers indicate the need for climate related capacity to boost their farm productivity. They suggest possible areas for which capacity building is needed to include: training on how to reduce the effects of climate change, as well as training on how to mitigate or prevent climate change from happening.

When probed on whether they have ever attended any training programme on climate change, few reported affirmatively. For those that had attended training programmes on climate change in the last five years, information obtained from such training covered broad areas related to soil management, crop-plant management, and water management. For the educated farmers, during studies in school, college or university, they indicate exposure to information on climate variability and climate change. At school they particularly received information on the nature of the climate system, importance of forest and the effect of deforestation; on water and natural resource management; on the source and causes of pollution; as well as information on the effect of their activities on climate change and global warming.

Besides the structured trainings participated in, farmers reported being aware of some other training programs or educational scheme or capacity building program for climate resilient farming organized by some organizations. It is reported by local experts that, "these programmes included seminars and workshops by the Regional Delegations of Agriculture, Forestry and the Environment." Further interrogation of key informants revealed that the diverse themes covered in the training include the effects of deforestation and on how to increase food availability through afforestation, coping with climate variability, application of agroforestry, organic farming and regenerative agriculture.

The goal of this capacity building schemes was to provide knowledge to improve crop yields, ensure food security, improve human and livestock health, consolidate farm incomes and promote growth of farmer-owned enterprises. However, whether these capacity building efforts were enough to give relevant information or training on adaptation to climate change was acknowledged by few farmers as being adequate. The majority who reported the insufficiency based their views on the lack of enough time to prepare and attend the trainings.

The farmers generally perceived climate to be changing, and were enthusiastic to improve on their coping and adaptive capacity. Most of the farmers expected more information on water and irrigation management; farm pest management, weather prediction and adaptation to fluctuating weather parameters, and harvest and post-harvest management practices to enhance their farming practice. Their expectations for the content of organized trainings included desire for more information on modern farming techniques, weather forecast, irrigation techniques and household-based coping strategies. With respect to the best method for delivering climate change content, farmers identified class-based, regular workshops, seminars and farmer-field school.

According to academics in local colleges of agriculture, farmers are enthusiastic in sharing their experiences on sustainable agriculture and are similarly motivated to learn from scientists:

> Recently, I have been educating farmers and conducting cooperative participatory research alongside them. We work with both crop and livestock farmers. After holding several meetings, we were able to pinpoint a major information gap about climate change and its potential impacts on agriculture and other sectors. Sustainable agriculture and climate change are the topics covered in the curriculum, particularly the role of agroforestry, which is presented to farmers through musical performances, hands-on activities, small group conversations, and dramatic presentations. The researchers and farmers worked together to develop the training program's initial curriculum. We suggested topics for the trainings, but the farmers gave us advice based on their knowledge and experience. The farmers' approval of the final package of adaptation technique was boosted when they participated as frontline participants in group discussions, theater, and music. The farmers, however, at the end of our training sessions always request for the need of financial assistance to help them invest the new ideas in their farms. (Lecturer, Faculty of Agriculture and Veterinary Medicine)

Farmers in both regions employ local-indigenous knowledge in adapting their farming practices to changing climate. The local and indigenous knowledge in adapting farming practice to changing climate related to mulching, green-manuring, water capturing, improved crop-rotation practice, staking and wind braking, as well as water collection. Knowledge given by parents, grandparents or ancestors to help overcome conditions of difficult environment or climate change included farm preparation methods, crop management methods, water capture, and postharvest management methods. Indigenous techniques exploited are related to mulching to reduce weed, conserve soil moisture and fertilize the soil, green manuring, wood ash application, integrate crop with planted trees to provide shade during the dry season, rain prediction and rainwater harvesting.

According to researchers from the national Institute for Agronomic Research and Development (IRAD), they are observing farmers either taking up new practices or recalling old practices to guarantee stability of households' food and income security:

> Use of native trees in farms for agroforestry, intercropping legumes with tubers, rotating crops, changing crop choices especially vegetables and legumes, growing more dry condition tolerant foods such as cassava tubers at the expense of other carbohydrate-rich tubers like yams and potatoes, nurturing seedlings of native trees of 'forgotten' foods e.g., jackfruit, wild plums, cashew, sour-sop; collecting water from rooftops, sharing farm harvests among households, use of local chickens, goats and pigs to assure continued practice of livestock farming, new techniques of river fishing along the river Wouri and river Mungo, new techniques for hunting, and selling commodities to new markets for traditional food products. (Research Official, IRAD, Ekona, South West Region).

A larger proportion of farmers reported that these techniques have been helpful over the years. Given the benefits of such information, farmers reported that they would like to get more information on ILK employed by other communities. Such case studies could be provided during training on how to cope with changing climate.

In their view indigenous and local knowledge was beneficial for climate change adaptation as well as for increasing productivity and household income.

We further sought to find out if there were opportunities or challenges for trainers or educators to work with indigenous farmers to frame the content of capacity building program. These opportunities included researchers obtaining local information, students carrying out internships, graduates and undergraduate students participating in farm interviews and questionnaire administration, national festivals and competitions. However, the existing challenges included poor communication, information blackout, shortage of resource-persons or trainers and low financial motivation, amongst others. Some of the farmers especially the educated ones reported they would be interested to collaborate with educators in developing training content for climate change education for Cameroon.

Additional information from local experts revealed that, "some activities are undertaken in formal institutions (schools, colleges and universities) that support farmers with relevant skills, knowledge or information include University-based institutions such as the Faculty of Agriculture and Veterinary Medicine (FAVM) through its outreach programmes of FAVM. Others include technical and vocational training centers and outposts of IRAD. Apart from other education or training providers, the University is thought of as being helpful in building the capacity for climate resilient farming, via increase awareness, developing of standards for effectiveness training, and it trails ground for developing new techniques."

Famers' Educational Status, Climate Perception and Farm Returns: Hypothesis Testing

We subjected the quantitative information collected on social and economic factors that could explain education, perception and adaptation to climate change to correlational analysis to understand the principal drivers. Rather than in tabular form we summarize the outcome of the analysis in a figure. In Fig. 14.3, the correlates of perception of climate change and farm income from an OLS regression are presented. Experience measured as age, level of education and household size have a positive correlation with farmers' income. However, the results reveal that perception of climate change is negatively correlated with farm income. In other words, the higher the income or wealthier the household the less they report changes in climate. On the contrary, the lower the income and closer the household is to rudimentary peasant economy that's tied to nature, the more they report weather or climate variability and change.

Focusing on perception, the fact that low-income farmers perceive climate change may warrant that they employ necessary adaptation measures to counter the effects. Hence, perception may not affect their farm income. This corroborates Methorst et al. (2016, 2017) who highlight that perception is about recognizing and not seeing.

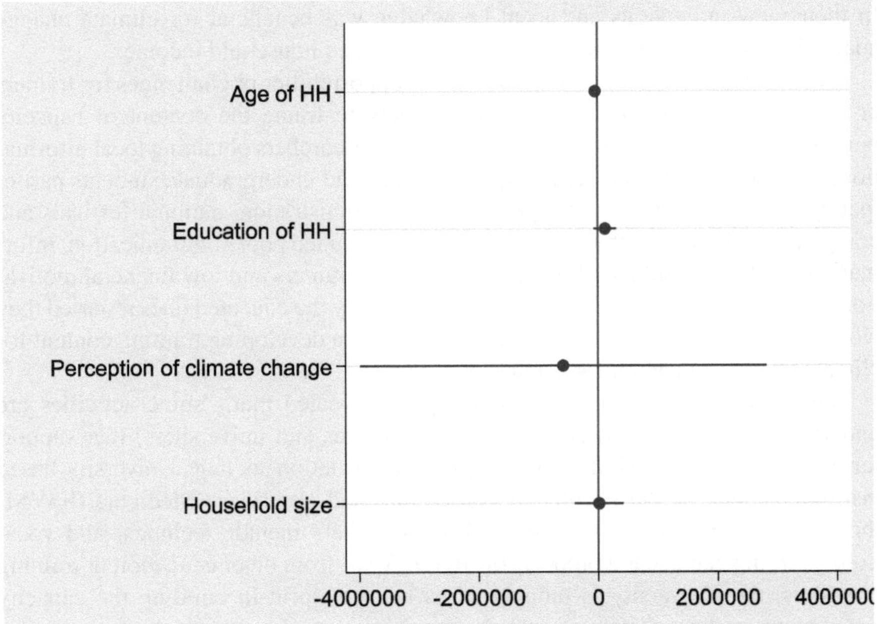

Fig. 14.3 Correlation of social factors and perception of climate change with farm income (Francs CFA)

Implying farmers who perceived risk as well as opportunities within their environment generally adopt methods to adapt to changes. In our case, most farmers (94%) have perceived climate change in one form or the other which explains the adoption of adaptation strategies. This observation further bolsters the position of Foley (2017) on behavioral entropy and decision making under uncertainty which turns to seek loss minimization rather than profit optimization. Methorst et al. (2016) highlight the inherent realities of decision-making in smallholdings which is guided by emotions rather than data and hence more intuitive rather than objective as proposed by conventional economic modelling. These predispositions could additionally be supported by the concepts of risk aversion highlighted in Thaler and Sunstein (2008) and Tversky and Kahneman (1992) which explain how human fear of risk may convert their decision towards a risk minimization behaviour rather than a profit maximizing one under uncertainty. Given that loses are more painful than gains, it is possible that every household adopts a strategy to climate change which maintains their income at a reference point rather than double their income above each reference point. Figure 14.3 summarizes the OLS estimates of perception of climate change and farm income.

In the context of the effect of education on choice of adaptation strategy, education provides positive correlates, but no relationship is observed between education and farm income. This is not a surprising observation, since these are smallholder farmers using simple technology over the year to continuously turn out multiple farm products used primarily for household consumption and a smaller share of marketable-surplus for generating income to purchase other household goods. Basic education can still guarantee these levels of subsistence living, more especially if these folks remain in the rural areas and exploit readily available resources of land and family labour, and even lower transaction costs when commodities are sold at the farm gates. However, the argument that education may promote adoption of modern techniques and the possibility of higher returns still remains valid, if cost of production and commodity prices are properly managed.

Overall, our findings contribute to the discourse that climate change is a complex and evolving phenomenon, with location specific experiences thus requiring local adaptation choices. Local adaptation evokes the need to dig deep into historic ILK which IPLCs have used over years to overcome the challenges of physical weather. The path forward is designing and deploying resilient adaptation strategies, techniques, and technologies. To achieve this, however, may require to design and redesign policies which meet current and near future realities. Public and private collaborations are thus needed with the government, public institutions and policy makers jointly collaborating with farmers from strategy conception to implementation. Such initiatives might prove futile if policy makers fail to consider that perception is not about seeing but rather recognising. In this case different people may see the same thing but recognise different things based on their contexts and their operating realities. More studies delving into an in depth understanding of farmers behaviours might prove more productive and effective for future strategies. No matter the strategy, education and training is required as the pivot of any approach to promote timely responsible response to the diktats of climate variation and change.

Conclusion

Climate change and its impacts, especially on agriculture have been well established. Consequently, mitigating the impacts of climate change has remained an important policy question both at the local and international policy arena. In our study areas, ILK including nature-based solutions which is transmitted through generation remains crucial in enhancing the ability of farming populations to withstand the impacts of climate variation and change. The challenge however, is extending these proven methods to non-indigenous communities and other practicing farmers. Hence, while a plethora of adaptation measures are in existence, teaching farmers about such measures may be more beneficial. Education is thus crucial to promote climate action. It would help people understand and address the impacts of the climate crisis, empowering them with the knowledge, skills, values and attitudes needed to act as agents of change. Teaching however does not guarantee learning on the farmer's part. Rural farmers in Cameroon and in many parts of Africa are unique as they are older with varying motivations and possess relatively less cognitive capacity to understand some inherent scientific concepts of climate change. But they perceive, understand climate change and its implication of their livelihood. Teaching them may require using relatable tools and techniques. Indigenising climate change education to contexts whereby farmers are formally taught about climate change using relatable figures may offer even more benefits. Despite these theoretical benefits, there is paucity of empirical evidence, which can be the preoccupation of future research.

Based on the results of the analyses we recommend that:

1. Farmers should be trained on the different drivers of climate change adaptation such as Agroforestry, mulching, crop rotation, and intercropping. Training should incorporate indigenous knowledge of farmers.
2. Farmers are also encouraged to enroll for formal education as it plays a critical role in opportunity and risk sense making which partly determine adaptation success and failures.
3. Farmers should also be encouraged through the provision of extension officers who operate regular visits for follow ups (knowledge assessment), capacity enforcement and external motivation for adopting new technics.
4. Farmers, especially young farmers, should be trained in formal educational systems like in colleges of agricultural sciences. Enhancing general education will necessitate modifications to education curriculum and infrastructure, as both general education and education infrastructure are key components for supporting adaptive learning.

Farmers' level of climate change education has a positive impact on climate change adaptation which is mostly acquired through formal education. This has a multiplier effect on yields. However, indigenous knowledge of climate change also influences climate change mitigation and adaptation. This would mean that the well-established and proven practices of IPLCs should be taken into account when making policy decisions and should be incorporated into adaptation frameworks, with the recognition of their rights. Therefore, providing continuous in-service training for farmers is relevant for sustainability of farm output. Farmers education regardless of the nature (formal or informal) and setting is an indispensable policy tool for enhancing farmers adaptive capacity and resilience. Education forms the bases for sense and decision making under intuition. Farmer perceptions of are only honed when they understand the differences between risks and opportunities (in the context of climate change or adaptation respectively). An effective way to ensure maximum efficiency of knowledge transfer is undeniably the indigenisation of climate change knowledge to capture contextual realities.

References

Acosta Castellanos, P. M., & Queiruga-Dios, A. (2022). From environmental education to education for sustainable development in higher education: A systematic review. *International Journal of Sustainability in Higher Education, 23*(3), 622–644.

African Union. (2015). The Africa we want: A shared strategic framework for inclusive growth and sustainable development & a global strategy to optimize the use of Africa's Resources for the benefit of all Africans. https://au.int/sites/default/files/documents/33126-doc-01_background_note.pdf

Agbedahin, A. V. (2019). Sustainable development, education for sustainable development, and the 2030 Agenda for sustainable development: emergence, efficacy, eminence, and future. *Sustainable Development, 27*(4), 669–680.

Alotaibi, B. A., Kassem, H. S., Nayak, R. K., & Muddassir, M. (2020). Farmers' beliefs and concerns about climate change: An assessment from southern Saudi Arabia. *Agriculture, 10*(7), 253.

Anderson, A. (2012). Climate change education for mitigation and adaptation. *Journal of Education for Sustainable Development, 6*(2), 191–206.

Arora-Jonsson, S., & Gurung, J. (2023). Changing business as usual in global climate and development action: Making space for social justice in carbon markets. *World Development Perspectives, 29*, 100474.

Asare-Nuamah, P., Antwi-Agyei, P., Dick-Sagoe, C., & Adeosun, O. T. (2022). Climate change perception and the adoption of innovation among mango plantation farmers in the Yilo Krobo municipality. *Ghana. Environmental Development, 44*, 100761.

Ayisi, D. N., Kozári, J., & Krisztina, T. (2022). Do smallholder farmers belong to the same adopter category? An assessment of smallholder farmers' innovation adopter categories in Ghana. *Heliyon, 8*(8), e10421.

Azzarri, C., & Signorelli, S. (2020). Climate and poverty in Africa South of the Sahara. *World Development, 125*, 104691.

Bangay, C., & Blum, N. (2010). Education responses to climate change and quality: Two parts of the same agenda? *International Journal of Educational Development, 30*, 359–368.

Becker, G. S. (2009). *Human capital: A theoretical and empirical analysis, with special reference to education.* University of Chicago Press.

Beckline, M., Sun, Z., Ntoko, V., Ngwesse, D., Manan, A., Hu, Y., Che., M. & Foncha, J. (2022). Rural livelihoods and forest incomes in the Etinde Community Forest of Southwest Cameroon. *Open Access Library Journal, 9*(7), 1–15.

Berrang-Ford, L., Pearce, T., & Ford, J. D. (2015). Systematic review approaches for climate change adaptation research. *Regional Environmental Change, 15*, 755–769.

Bonnett, M. (2018). Sustainable development, environmental education, and the significance of being in place. In *Education for sustainable development* (pp. 82–103). Routledge.

Cameron, L., Mauro, I., & Settee, K. (2021). "A return to and of the land": Indigenous knowledge and climate change initiatives across the Canadian prairies. *Journal of Ethnobiology, 41*(3), 368–388.

Choi, H. J. (2009). Technology transfer issues and a new technology transfer model. *Journal of Technology Studies, 35*(1), 49–57.

Christoffels, I. K., & De Groot, A. M. (2004). Components of simultaneous interpreting: Comparing interpreting with shadowing and paraphrasing. *Bilingualism: Language and Cognition, 7*(3), 227–240.

Cordero, E. C., Centeno, D., & Todd, A. M. (2020). The role of climate change education on individual lifetime carbon emissions. *PLoS ONE, 15*(2), e0206266.

Davis, F. D. (1989). Technology acceptance model: TAM. *Al-Suqri, MN, Al-Aufi, AS: Information Seeking Behavior and Technology Adoption, 205*, 219.

Dhakal, C., Khadka, S., Park, C., & Escalante, C. L. (2022). Climate change adaptation and its impacts on farm income and downside risk exposure. *Resources, Environment and Sustainability, 10*. https://doi.org/10.1016/j.resenv.2022.100082

Ekeocha, D. O., & Iheonu, C. O. (2021). Household-level poverty, consumption poverty thresholds, income inequality and quality of lives in sub-Saharan Africa. *African Development Review, 33*(2), 234–248.

Feola, G., Lerner, A. M., Jain, M., Montefrio, M. J. F., & Nicholas, K. A. (2015). Researching farmer behaviour in climate change adaptation and sustainable agriculture: Lessons learned from five case studies. *Journal of Rural Studies, 39*, 74–84.

Feinstein, N. W., & Mach, K. J. (2020). Three roles for education in climate change adaptation. *Climate Policy, 20*(3), 317–322.

Filho, W. L., Wolf, F., Totin, E., Zvobgo, L., Simpson, N. P., Musiyiwa, K., Kalangu, J. W., Sanni, M., Adelekan, I., Efitre, J., & Donkor, F. K. (2023). Is indigenous knowledge serving climate adaptation? Evidence from various African regions. *Development Policy Review, 41*(2), e12664.

Filho, L. W., Barbir, J., Gwenzi, J., Ayal, D., Simpson, N. P., Adeleke, L., Tilahun, B., Chirisa, I., Gbedemah, S. F., Nzengya, D. M., & Sharifi, A. (2022). The role of indigenous knowledge in climate change adaptation in Africa. *Environmental Science & Policy, 136*, 250–260.

Foley, D. K. (2017). *Crisis and theoretical methods: Equilibrium and disequilibrium once again* (No. 1703).

Franco, I., Saito, O., Vaughter, P., Whereat, J., Kanie, N., & Takemoto, K. (2019). Higher education for sustainable development: Actioning the global goals in policy, curriculum and practice. *Sustainability Science, 14*(6), 1621–1642.

Frankenberg, E., Sikoki, B., Sumantri, C., Suriastini, W., & Thomas, D. (2013). Education, vulnerability, and resilience after a natural disaster. *Ecology and Society, 18*, 16–28.

Godin, B., 2017. Models of innovation: The history of an idea. MIT press.

Godin, B. (2012). "Innovation studies": The invention of a specialty. *Minerva, 50*(4), 397–421.

Haluza-DeLay, R. (2014). Religion and climate change: Varieties in viewpoints and practices. *Wiley Interdisciplinary Reviews: Climate Change, 5*(2), 261–279.

Hall, G. E. (1979). The concerns-based approach to facilitating change. *Educational Horizons, 57*(4), 202–208.

Halsnæs, K., & Kærgård, N. (2023). Climate Problems: Market and Ethics. *Market, ethics and religion: The market and its limitations* (pp. 285–295). Springer International Publishing.

Hanushek, E. A., & Woessmann, L. (2023). *The knowledge capital of nations: Education and the economics of growth.* MIT Press.

Holmén, H., & Hyden, G. (2011). African agriculture: From crisis to development? African smallholders. *Food crops, markets and policy*, 23–44.

Hosen, N., Nakamura, H., & Hamzah, A. (2020). Adaptation to climate change: Does traditional ecological knowledge hold the key? *Sustainability, 12*(2), 676.

Huffman, W. E. (2001). Human capital: Education and agriculture. *Handbook of Agricultural Economics, 1*, 333–381.

Ibisomi, L. D., & De Wet, N. (2014). The dynamics of household structure in sub-Saharan Africa. In *Continuity and change in sub-Saharan African demography* (pp. 197–215). Routledge.

IPCC. (2021). *Climate Change 2021: The Physical Science Basis. Contribution of Working Group I to the Sixth Assessment Report of the Intergovernmental Panel on Climate Change* [V. Masson-Delmotte, P. Zhai, A. Pirani, S.L. Connors, C. Péan, S. Berger, N. Caud, Y. Chen, L. Goldfarb, M. I. Gomis, M. Huang, K. Leitzell, E. Lonnoy, J. B. R. Matthews, T. K. Maycock, T. Waterfield, O. Yelekçi, R. Yu, B. Zhou].

Johnson, D. E., Parsons, M., & Fisher, K. (2022). Indigenous climate change adaptation: New directions for emerging scholarship. *Environment and Planning E: Nature and Space, 5*(3), 1541–1578.

Juarez, F. J. C. (2014). The Earth Charter as an environmental policy instrument in Mexico: a soft law or hard policy perspective. In *The earth charter, ecological integrity and social movements* (pp. 270–283). Routledge.

Kydd, J., Dorward, A., Morrison, J., & Cadisch, G. (2004). Agricultural development and pro-poor economic growth in sub-Saharan Africa: Potential and policy. *Oxford Development Studies, 32*(1), 37–57.

Lebert, T. (2015). Africa: A continent of wealth, a continent of poverty. *New Internationalist, 24.*

Lehtonen, A., Salonen, A. O., & Cantell, H. (2019). Climate change education: A new approach for a world of wicked problems. In *Sustainability, human well-being, and the future of education* (pp. 339–374). Palgrave Macmillan.

Lin, J. Y. (1991). Education and innovation adoption in agriculture: Evidence from hybrid rice in China. *American Journal of Agricultural Economics, 73*(3), 713–723.

Louie, D. W., Poitras-Pratt, Y., Hanson, A. J., & Ottmann, J. (2017). Applying indigenizing principles of decolonizing methodologies in university classrooms. *Canadian Journal of Higher Education, 47*(3), 16–33.

Lutz, W., Muttarak, R., & Striessnig, E. (2014). Universal education is key to enhanced climate adaptation. *Science, 346*, 1061–1062.

Masud, M. M., Al-Amin, A. Q., Junsheng, H., Ahmed, F., Yahaya, S. R., Akhtar, R., & Banna, H. (2016). Climate change issue and theory of planned behaviour: Relationship by empirical evidence. *Journal of Cleaner Production, 113*, 613–623.

Mbah, M., Ajaps, S., & Molthan-Hill, P. (2021). A systematic review of the deployment of indigenous knowledge systems towards climate change adaptation in developing world contexts: Implications for climate change education. *Sustainability, 13*(9), 4811.

Mbeva, K., Makomere, R., Atela, J., Chengo, V., & Tonui, C. (2023). The evolving geopolitics of climate change. *Africa's right to development in a climate-constrained world* (pp. 85–126). Springer International Publishing.

McNamara, K. E., & Buggy, L. (2017). Community-based climate change adaptation: A review of academic literature. *Local Environment, 22*(4), 443–460.

Mendoza-Zuany, R. G. (2019). Challenges in tackling environmental concerns in indigenous education in Mexico. *Southern African Journal of Environmental Education, 35.*

Methorst, R., Roep, D., Verhees, F., & Verstegen, J. (2016). Drivers for differences in dairy farmers' perceptions of farm development strategies in an area with nature and landscape as protected public goods. *Local Economy, 31*(5), 554–571.

Methorst, R. R., Roep, D. D., Verhees, F. F., & Verstegen, J. J. (2017). Differences in farmers' perception of opportunities for farm development. *NJAS-Wageningen Journal of Life Sciences, 81*, 9–18.

Moine, B., & Ferry, X. (2018). Plasticulture: Economy of resources. In *XXI international congress on plastics in agriculture: Agriculture, plastics and environment 1252* (pp. 121–130).

Molthan-Hill, P., Blaj-Ward, L., Mbah, M. F., & Ledley, T. S. (2021). Climate change education at universities: relevance and strategies for every discipline. In M. Lackner, B. Sajjadi, & W.-Y. Chen (Eds.), *Handbook of climate change mitigation and adaptation* (pp. 1–64). Springer.

Naess, L. O. (2013). The role of local knowledge in adaptation to climate change. *Wiley Interdisciplinary Reviews: Climate Change, 4*(2), 99–106.

Ndip, F. E., Molua, E. L., Mvodo, M. E. S., Nkendah, R., Choumbou, R. F. D., Tabetando, R., & Akem, N. F. (2023). Farmland fragmentation, crop diversification and incomes in Cameroon, a Congo Basin country. *Land Use Policy, 130*, 106663.

Ninh, L. K. (2021). Economic role of education in agriculture: Evidence from rural Vietnam. *Journal of Economics and Development, 23*(1), 47–58. https://doi.org/10.1108/JED-05-2020-0052

Nyadzi, E., Ajayi, O. C., & Ludwig, F. (2021). Indigenous knowledge and climate change adaptation in Africa: A systematic review. *CAB Reviews: Perspectives in Agriculture, Veterinary Science, Nutrition and Natural Resources, 2021.*

Nyanga, P. H., Johnsen, F. H., & Aune, J. B. (2011). Smallholder farmers' perceptions of climate change and conservation agriculture: Evidence from Zambia. *Journal of Sustainable Development, 4*(4).

Nyong, A., Adesina, F., & Osman Elasha, B. (2007). The value of indigenous knowledge in climate change mitigation and adaptation strategies in the African Sahel. *Mitigation and Adaptation Strategies for Global Change, 12*, 787–797.

Ogwu, M. C. (2019). Towards sustainable development in Africa: The challenge of urbanization and climate change adaptation. In *The geography of climate change adaptation in Urban Africa* (pp. 29–55). Palgrave Macmillan, Cham.

Ochieng, M. A., & Koske, J. (2013). The level of climate change awareness and perception among primary school teachers in Kisumu municipality, Kenya. *International Journal of Humanities and Social Science, 3*(21), 174–179.

Owen, G. (2020). What makes climate change adaptation effective? A systematic review of the literature. *Global Environmental Change, 62*, 102071.

Paltasingh, K. R., & Goyari, P. (2018). Impact of farmer education on farm productivity under varying technologies: Case of paddy growers in India. *Agricultural Economics, 6*, 7. https://doi.org/10.1186/s40100-018-0101-9

Petzold, J., Andrews, N., Ford, J. D., Hedemann, C., & Postigo, J. C. (2020). Indigenous knowledge on climate change adaptation: A global evidence map of academic literature. *Environmental Research Letters, 15*(11), 113007.

Purcell-Gates, V. (2004). Family literacy as the site for emerging knowledge of written language. In *Handbook of family literacy* (pp. 101–116). Routledge.

Reyes-García, V., García-Del-Amo, D., Porcuna-Ferrer, A., Schlingmann, A., Abazeri, M., Attoh, E. M., da Cunha, V., Ávila, J., Ayanlade, A., Babai, D., Benyei, P., & Calvet-Mir, L. (2024). Local studies provide a global perspective of the impacts of climate change on Indigenous Peoples and local communities. *Sustainable Earth Reviews, 7*(1), 1.

Salite, D. (2019). Explaining the uncertainty: Understanding small-scale farmers' cultural beliefs and reasoning of drought causes in Gaza Province, Southern Mozambique. *Agriculture and Human Values, 36*(3), 427–441.

Samek, T. (2021). Indigenous-engaged education: a Canadian viewpoint. In *Social justice design and implementation in library and information science* (pp. 218–235). Routledge.

Schlingmann, A., Graham, S., Benyei, P., Corbera, E., Sanesteban, I. M., Marelle, A., Soleymani-Fard, R., & Reyes-García, V. (2021). Global patterns of adaptation to climate change by Indigenous Peoples and local communities. A systematic review. *Current Opinion in Environmental Sustainability, 51*, 55–64.

Sharaky, A. M. (2014) *Mineral resources and exploration in Africa.* Department of Natural Resources, Institute of African Research and Studies, Cairo University, Egypt.

Sibanda, A., & Manik, S. (2022). Reflecting on climate change education (CCE) initiatives for mitigation and adaptation in South Africa. *Environmental Education Research*, 1–18.

Simpson, N. P., Andrews, T. M., Krönke, M., Lennard, C., Odoulami, R. C., Ouweneel, B., Steynor, A., & Trisos, C. H. (2021). Climate change literacy in Africa. *Nature Climate Change, 11*(11), 937–944.

Straub, E. T. (2009). Understanding technology adoption: Theory and future directions for informal learning. *Review of Educational Research, 79*(2), 625–649.

Thaler, R., & Sunstein, C. (2008). Nudge: Improving decisions about health, wealth and happiness. In *Amsterdam Law Forum; HeinOnline: Online* (p. 89).

Tumushabe, J. (2018). Climate change, food security and sustainable development in Africa. In *The Palgrave handbook of African politics, governance and development* (pp. 853–868). New York: Palgrave Macmillan.

Tversky, A., & Kahneman, D. (1992). Advances in prospect theory: Cumulative representation of uncertainty. *Journal of Risk and Uncertainty, 5*, 297–323.

UNICEF. (2012). *Climate change adaptation and disaster risk reduction in the education sector.* UNICEF Division of Communication.

United Nations. (2015). *Paris agreement on climate change.* United Nations. https://unfccc.int/process-and-meetings/the-paris-agreement/the-paris-agreemen

United Nations. (1992). U.N. treaty collection, Chapter XXVII Environment, 7.1. *Kyoto protocol to the United Nations framework convention on climate change.* United Nations. Retrieved December 11, 1997, from https://treaties.un.org/Pages/ViewDetails.aspx?src=TREATY&mtdsg_no=XXVII-7-a&chapter=27&clang=_en

Valizadeh, N., Rezaei-Moghaddam, K., & Hayati, D. (2020). Analyzing Iranian farmers' behavioral intention towards acceptance of drip irrigation using extended technology acceptance model. *Journal of Agricultural Science and Technology, 22*(5), 1177–1190.

Voortman, R. L., Sonneveld, B. G., & Keyzer, M. A. (2003). African land ecology: Opportunities and constraints for agricultural development. *Ambio: A Journal of the Human environment, 32*(5), 367–373.

Wiggins, S. (2014). African agricultural development: Lessons and challenges. *Journal of Agricultural Economics, 65*(3), 529–556.

Wößmann, L. (2003). Specifying human capital. *Journal of Economic Surveys, 17*(3), 239–270.

Woolcock, M. ([illegible]). Social capital [illegible]. Theory and Society, 27([illegible]), 151–208.

Chapter 15
Challenges and Opportunities for Climate Change Literacy in Botswana: A Critical Review

Thato Majola, Neo Scholarstic Ntirase, and Kgosietsile Velempini

Abstract One of the most urgent global issues, climate change, has long-term effects on countries' sustainable development. It affects everything from rising sea levels and catastrophic flooding to changing weather patterns that threaten food security. Even though they do not bear the same responsibility as nations in the global north, nations like Botswana are susceptible to effects of climate change. The purpose of this work is to critically examine key challenges and opportunities of climate change literacy in Botswana. This chapter followed a qualitative research design and relies on extensive review of journal articles and official documents in Botswana about climate change literacy. Our work presented emerging findings such as lack of climate change literacy in school curriculum, lack of professional development for educators and lack of resources to teach environmental education. We recommend an employment of local and indigenous knowledge into climate change literacy to build on the rich experiential wisdom passed down by our ancestors.

Keywords Climate change literacy · Environmental education · Teacher education · Local knowledge · Botswana

Introduction

Due to changing weather patterns, one of the most dangerous global issues is climate change, which has long-term repercussions for nations' sustainable development. Climate change leads to extreme rainfall that causes flooding and rising sea levels that threaten food security (Apollo & Mbah, 2021; IPCC, 2022). Botswana is one of the nations in sub-Saharan Africa region that bears the burden of changing weather patterns (Akinyemi & Abiodun, 2019; Kolawole et al., 2014). Contemporary research

T. Majola (✉) · N. S. Ntirase
University of Botswana, Environmental Education Unit, Gaborone, Botswana
e-mail: 201200740@ub.ac.bw

K. Velempini
Department of Environmental Sciences, University of North Carolina, Wilmington, USA

M. F. Mbah et al. (eds.), *Practices, Perceptions and Prospects for Climate Change Education in Africa*, https://doi.org/10.1007/978-3-031-84081-4_15

warns that such extreme climate-related events could be worse than predicted in the near future (Battisti & Naylor, 2009; IPCC, 2022; Kemp et al., 2022; Spratt & Dunlop, 2019). Global warming, primarily attributed to carbon-intensive industrialization, is the primary cause of climate variability (Hu et al., 2006; Shi et al., 2019). According to NASA (2022), while climate change is a long-term change in the typical weather patterns that have come to define Earth's local, regional, and global climates, global warming is the long-term warming of Earth's surface observed since the pre-industrial period because of anthropogenic factors and particularly the burning of fossil fuels. Despite not bearing as much of the blame for climate change as nations in the global north, nations in the global south, like Botswana, are still susceptible to its effects (Alenda-Demoutiez, 2022). Indeed, this claim is supported by data from the Intergovernmental Panel on Climate Change (IPCC), with recent decadal analyses that indicate increased warming trends in the continent of Africa over the last 50–100 years.

Research suggests that the average national climate change literacy rate in Africa is 37%, which is lower than in Europe and North America, where rates are over 80% (Davids, 2021; Kastner et al., 2012; Simpson et al., 2021a, 2021b). Climate change literacy differs substantially across African countries, as well as within countries. For instance, the climate change literacy rate is 66% in Mauritius and 62% in Uganda, 25% in Mozambique, 23% in Tunisia and 69% in Botswana (Simpson et al., 2021a, 2021b, Trisos et al., 2022). According to UNESCO (2004), literacy is "the ability to identify, understand, interpret, create, communicate, compute and use printed and written materials" (p. 13), that relate to various contexts and enabling individuals to achieve their goals to develop knowledge and potential to participate in the wider society's decision-making processes. Johnston (2019) stated that literacy can also be defined as competence or knowledge in a specific area, whereas climate change literacy can be referred to as competence or knowledge about climate change, its impacts, and its solutions. Selormey et al. (2019) stated that

> A quarter of Africans (28%) are fully "climate change literate," meaning they are aware of the phenomenon, understand its negative effects, and know that human activity is at least partially to blame. The only nation where the majority is aware of climate change is Mauritius, with a literacy rate of 57%. In six nations (Liberia, Mozambique, Namibia, Niger, South Africa, and Tunisia), less than one in five people are knowledgeable about climate change. (p. 3)

Drawing on Alenda-Demoutiez (2022), UNESCO (2004), Johnston (2019), and Selormey et al. (2019), climate change literacy is defined in this chapter as the knowledge or ability to understand climate change with regard to its causes, impacts, adaptations, mitigation, and coping strategies. It is critical for producing citizens who can understand and interpret the causes and effects of climate change, as well as coping and mitigation strategies. Citizens should be able to apply their local knowledge and competencies in their roles as active community members. To address the effects of climate change, it is important to consider knowledge creation and skill development through education. Climate change literacy, both formal and informal, is important in enabling the population in actively acquiring skills to adapt and respond to climate risks and shocks. Specifically, education becomes a key for promoting positive behavioral attitudes (Bangay & Blum, 2010). Climate change knowledge and understanding require policy orientation and pedagogies that address community attitudes and values (Ketlhoilwe, 2019). Local communities could be provided with knowledge, skills, and opportunities for value addition and sustainable development through environmental education.

The purpose of this chapter is to critically examine challenges and opportunities for climate change literacy in Botswana, which is an upper-middle income country with a stable socio-economic environment. The chapter contributes to the review of 2018 Botswana's Climate Change strategy and 2021 Botswana Climate Change policy. According to the Republic of Botswana (2018), the concept of climate change is still relatively new in Botswana and the know-how in terms of incorporating climate change responses into governance is still in the process of its development. The sector of education offers untapped opportunities for efficient climate change adaptation and mitigation through the generation of knowledge and skills. Republic of Botswana (2021) stated, "Climate change decisions shall be informed by research and as a cross sectoral discipline, its understanding is imperative to effectively adapt to the varied circumstances and mitigate potential impacts" (p. 33). This chapter contributes to promoting effective climate change literacy in Africa. It adheres to SDG target 13.3, which says "Improve education, awareness-raising, and human and institutional capacity on climate change mitigation, adaptation, impact reduction, and early warning" (United Nations, undated, p. 1). Alenda-Demoutiez (2022) buttressed as follows: "Climate change literacy is a crucial component to develop strategies, carry out policies, and alter behaviors in order to achieve SDG 13 and respond to the urgency of the situation" (p. 1). Therefore, in order to get communities and governments to work together in the fight against climate change-related events and uncertainties, climate change literacy is crucial.

Even though it is obvious that climate change literacy is vital in addressing the impacts of climate change, there is scarce literature about this topic in Botswana. Botswana's climate is known as semiarid to arid and has high intra and inter-seasonal rainfall variability. As a result of climate change, there have been notable shifts in Botswana's climate. Intra-seasonal rainfall patterns, which describe the distribution of rainfall within a single season, and inter-seasonal variability, referring to changes in rainfall between different years, have become more pronounced. These changes exacerbate drought conditions, significantly impacting farmers who experience reduced agricultural output. This heightens vulnerability of farmers to the effects of climate change and consequently suffers reduced agricultural production, whereas biodiversity is impacted through wildlife migrations that also become negatively affected and less predictable (Akinyemi, 2017). Schools in Botswana, particularly in rural and remote areas, experience water scarcity, which affects teaching and learning due to reduced proper sanitation and lack of proper hygiene. Consequently, Hambira, Saarinen, and Moses (2020) called for an active learning procedure that concentrates on the effects of climate change in Botswana, arguing that this can result in better strategies for enhancing adaptive capacity.

Climate Change and the Global Community

The global community and its wide range of socioeconomic activities, such as health, livelihood, and food security, are seriously challenged by climate change (Amjath-Babu et al., 2016; Mitchell & Van Aalst, 2008; Romieu et al., 2010). The economic status of people in both developed and underdeveloped countries is impacted by climatic changes. However, underdeveloped countries are more impacted. Various international treaties have been signed by countries to address the negative effects of climate change in response to the challenges it poses (Ostrom, 2017). The United Nations Framework Convention on Climate Change (UNFCCC), which was created during the Earth Summit in 1992 in Brazil to stop harmful human interferences with the climate system, was the first agreement. Following in 2008, the Kyoto Protocol began discussions to strengthen the international response to climate change, and its implementation period ended in 2012. While total greenhouse gas emissions from nations that met the Kyoto targets were significantly reduced, Rosen (2015) argued that the protocol's binding emission targets and condensed timeframe for action made it the incorrect answer at the time. The Paris Agreement on Climate Change (Paris Accords or the Paris Climate Accords) was later adopted by 196 Parties at COP 21 in Paris in order to speed up and intensify the actions taken as a result of the previous convention. The Paris Agreement aimed at advancing bold initiatives to combat climate change and encourage financial assistance for developing nations to adapt to its effects. All nations came to an agreement on a Climate Pact at the Conference of Parties (COP26) conference in Glasgow, United Kingdom, in November 2021. The agreement would have prevented average global temperatures from rising by more than 1.5 °C. The Glasgow Climate Pact prioritized collaboration (working together to achieve goals), finance (enabling countries to meet their climate goals), adaptation (helping those already impacted by climate change) and mitigation (reducing emissions) (United Nations Framework Convention on Climate Change, 2021). Although it is too soon to judge the effectiveness of the new climate change agreement from the COP 2021, it is crucial to have a discussion about general inequality issues like carbon caps, climate finance, and the degree of policy ambition between developed and developing countries. The development of sustainable climate change mitigation strategies has become difficult because of these disparities and power dynamics in the climate issue. Consequently, putting a focus on climate change literacy, particularly through education, presents a significant opportunity to add climate change to the development agendas of both national and sub national levels of governments. It is possible that both men and women can become more knowledgeable about climate change through education (Momsen, 2009). According to Bissinger and Bogner (2018), people with a high school diploma are 19% more likely to be knowledgeable about climate change than people with no formal education. University graduates are 36% more likely to be knowledgeable about climate change than non-graduates. Ketlhoilwe (2019) wrote as follows:

> The acquisition of climate change knowledge and competencies should be oriented towards giving more attention to individual role to solve environmental problems. It should emphasize both behaviorist and cognitivist-oriented outcome to promote a shift from community educator oriented to learner pedagogy in its approach promoting interest in development of climate change adaptation and mitigation strategies for sustainable development and resilience. (p. 11)

In retrospect, understanding the causes of climate change and its effects on the global community are essential components of climate change literacy (Educational International, 2021; Hoath & Dave, 2022). People who lack literacy will be less able to adapt to the detrimental effects of climate change, including both potential opportunities as well as anticipated negative economic and environmental effects. Climate change literacy strengthens more informed responses to climate change (Joseph, 2021; Park et al., 2012). For instance, Austria started the project k.i.d.Z.21-to address issues with teen climate change awareness and literacy. In the eight participating schools, the k.i.d.Z-21-Austria project (2015–16 pilot project) is an essential part of daily life and encourages students to discuss climate change with their families, friends and local community members (Williams et al., 2017).

Climate Change Literacy in Africa

Due to its high reliance on rainfed agriculture and its limited capacity for adaptation, Africa experiences severe negative direct effects from climate change. The Intergovernmental Panel on Climate Change (IPCC) Representative Concentration Pathway 8.5 indicates that warming scenarios will have negative effects on crop production and food security (United Nations, 2022). "The number of undernourished people in the drought-prone African countries has increased by 45.6% since 2012" (United Nations, 2022, p. 1) More informed climate change adaptation may be possible throughout Africa with the application of indigenous knowledge practices and increased climate change literacy (Nakashima et al., 2018; Simpson et al., 2021a, 2021b). The majority of Africans are aware of climate change and concur that various actions should be taken into practice to create more awareness about it (Lee et al., 2015). However, very few believe that common people can stop it, and even fewer comprehend its anthropogenic factors (Cohen, 1995). Food insecurity is one of the most pervasive and harmful outcomes of climate change in the East Africa region, with predictions of more frequent emergencies and famine. According to the United Nations Food and Agriculture Organization (FAO), there is a straightforward relationship between climate change and food insecurity in East Africa as a result of shifted growing seasons, which are interspersed with more frequent droughts and floods that destroy food crops (FAO, 2021). Africans with more money and mobility, as well as those who live in cities, are said to be more knowledgeable about climate change. Climate change literacy is weakened by poverty. There is a distinction based on gender. The literacy rates for women in countries about climate change are, on average, 12.8% lower than those for men. Climate change literacy is also impacted by the environment. Increased climate change literacy is linked to historical precipitation trends and perceived drought experiences (Simpson et al., 2021a, 2021b). However, until recently, little was understood about how climate change literacy rates vary among African populations, what causes variation, and what predicts changes in these literacy rates.

Botswana's Endeavor in Climate Change Literacy

The Minister of Tertiary Education, Research, Science, and Technology of Botswana stated as follows in his official opening remarks at the 2020 Regional Policy dialogue for CAP ESD: "there is need to come up with adaptive strategies to climate change" (Botswana Government, 2020, p. 5). In order to instill Education for Sustainable Development in students from an early age, the Minister noted that these strategies "need to be integrated into our education policies and programmes for teacher education as well as the curriculum at all levels of education" (p. 5). It is crucial that the Botswana education system reconsiders its pedagogical methods and improves teachers' capabilities in incorporating climate change into the curriculum. Also, Alenda-Demoutiez (2022) stated that "Different aspects, such as education…may impact interpretation of climate change and thus climate change literacy. Education in particular has been shown to play an important role in the comprehension and necessity to undertake adaptation strategies among African farmers" (p. 4). The action areas of Education for Sustainable Development (ESD) are supported by Botswana's educational system (Ketlhoilwe & Jeremiah, 2010). Enhancing ESD in the educational systems of Botswana has become possible mainly due to the Sustainability Starts with Teachers (SST) consultation workshop held in 2019 and the subsequent capacity building for teacher educators (UNESCO, 2020).

The ongoing effects of climate literacy in Botswana have led to the adoption of a national response strategy to address current and potential climate change impacts (United Nations Development Programme, 2022). The policy can make it easier to integrate climate change literacy into the curriculum. The goal of Botswana's climate change strategy is to make the nation resistant to the rising temperatures and shifting rainfall patterns that endanger food security. The policy opens up opportunities for investigation and the use of local and indigenous knowledge in combating climate change. According to Darst and Dawson (2019), framing climate change within an underpinning that includes public health, social justice, and national climate security in addition to values influencing behaviors and lifestyles, would be helpful. To maintain national climate change resilience to shocks and stresses, the policy is intended to promote compatible development based on SDGs (e.g., SDG4 on Quality Education, SDG7 on Affordable and Clean Energy, and SDG13 on Climate Action). Although the 1994 RNPE was without a focus on climate change education, it was the first initiative in Botswana to emphasize the inclusion of environmental education in the school curriculum. However, Botswana's recently adopted outcome-based education (OBE) strategy can make it possible for students to respond to climate change issues (Botswana Government, 2015). OBE was designed to make it easier to introduce multiple pathways in schools. This has led up to the swift implementation of educational reforms under the Education and Training Sector Strategic Plan (ETSSP) (Botswana Government, 2015), with a view to promoting a competence-based curriculum that gives students the chance to incorporate climate change topics into every study program.

Methodology

This chapter employed a qualitative research design (Creswell & Creswell, 2022; Patton, 2002; Rubin & Rubin, 2012) and relied on extensive review of journal articles ($n = 19$) and official documents ($n = 10$) from Botswana about climate change literacy. Essentially, the existing and online policies, strategies, plans and government, private and NGO sector programs on climate change literacy are reviewed to explore key challenges and opportunities of climate change literacy in Botswana. The journal articles and official documents served as secondary sources and were searched through key word searches (e.g. climate change literacy in Botswana, climate change education in Botswana) from the internet (e.g. https://scholar.google.com/; https://www.researchgate.net/) and ad hoc readings (Onwuegbuzie & Frels, 2016). The official documents enabled a further critical review of climate change literacy in Botswana. Qualitative research studies are designed in such a way that they enable the researcher means of understanding a phenomenon under investigation (Denzin & Lincoln, 2008). Purposive sampling (Patton, 2002) was followed in identifying the above secondary sources. We engaged in a process of line-by-line manual analysis and reading and re-reading the documents at length, while writing down examples and concepts that relate to climate change literacy in Botswana (Charmaz, 2008). We also followed steps suggested by Rubin and Rubin (2012), which included manually coding excerpts (examples, concepts, themes). We note that retrieving secondary sources and tracking analytical insights that start to emerge is part of qualitative analysis (Patton, 2015). Then, through reading official documents and identifying data pertaining to the purpose of the research, relevant descriptors in particular span were identified on climate change literacy.

Results and Discussion

The findings from the secondary sources that have been examined are presented in this section, along with suggestions for future actions regarding the difficulties and possibilities for climate change literacy in Botswana.

Challenges for Climate Change Literacy

Lack of Climate Change Education in the School Curriculum

Some schools in Botswana have been undertaking initiatives through various school activities such as clean up campaigns (e.g. rubbish collection in Kgatleng district) and water harvesting in Shoshong in the Central district (Silo, 2009; Selabe & Minyoi, 2018). It is through these activities that teaching and learning in Botswana has continued to give prominence to conservation and environmental management discourse that reveal practical applications of the 1994 RNPE on environmental education (Ketlhoilwe, 2007). Various activities take the form of clean up campaigns and water conservation activities such as using watering cans/buckets to conserve water in gardens and households. According to Silo, Velempini and Ketlhoilwe (2022), some schools in Botswana have upscaled their responses to climate change education. For instance, a school in southern Botswana named Phuthisutlha Primary School built an underground rainwater storage facility to support a vegetable garden. However, due to lack of resources to support teaching and learning, these initiatives and responses have not been fully utilized. These are noble efforts that focus on responding to climate related challenges like water scarcity. Yet, these have not been making way into the mainstream curriculum in a whole school approach to prepare learners to become competent participants in environmental education (Silo, 2017). Currently, environmental education remains integrated mainly in social studies and science subjects. Teachers' professional development and neglect by educational authorities constitute additional barriers to an enhancement of environmental education in alignment with learning about climate change in Botswana. Often teachers end up neglecting the environmental module of the course syllabus as they believe there are more important topics they should be focusing on. This claim is supported by research from Velempini (2017) and Ketlhoilwe (2003), which pointed out that teachers at the secondary level rarely include environmental education because students are not tested on it to evaluate their preparedness for continuing their education. Even though teachers are required to incorporate environmental education topics into their lessons, they occasionally find it challenging to do so because they were not given the proper training. According to Verma and Dhull (2017), there is much discussion surrounding the inclusion of environmental education as a separate subject in the curriculum. This makes it simpler for teachers to give the subject a distinct focus and identity. They claimed that the socioeconomic makeup of a nation and its specific conditions affect how successfully environmental education is incorporated into the school curriculum. According to Verma and Dhull (2017), "Bolstad (2005) found that schools are likely to find space for Environmental Education if it can be associated with existing subjects in the curricula rather than creating a new subject" (p. 1551). However, Johnson-Pynn and Johnson (2005) reported that the integration of environmental education into various subjects poses restrictions and difficulties for educational systems.

Lack of Professional Development for Educators

Ketlhoilwe (2007) argued that there has been a standardization of environmental education in the existing school culture through activities with learners that are limited on instruction about how to manage environmental challenges such as water scarcity and waste. In essence, teachers are not prepared to teach climate change education. Teachers find themselves stuck with the structured way of operating schools where learners participate in limited school environmental activities constrained by minimal, or even non-existing educational resources, large class sizes, and lack of external support. Silo (2011) focused research on active learner participation in Botswana schools and reported that attempts by teachers to abide by policy imperatives through prescription of rules, and ascribing roles to learners in environmental management activities, create tensions among learners. Consequently, goals for children engagements become unclear. This gives rise to an elusive learning context because the purpose for students' participation is not becoming clear in terms of how their engagements respond to environmental challenges such as water scarcity and pollution from waste (Silo, 2011). Clearly, the capacity building of teachers to enable them to teach for practical experiences to climate change related challenges that young people face daily is constrained. Pre-service teachers are inadequately trained to teach environmental education which climate change literacy is infused (Velempini, 2017). They are not equipped with relevant methodologies to teach climate change in schools. This is a problem because if teachers are not trained to teach content for environmental education, they cannot teach it effectively in addition to the problem of environmental education not having a fully-fledged syllabus.

Lack of Resources to Teach Environmental Education

Clearly, a lot of effort is being made in schools to address the challenges posed by climate change since the 1994 introduction of the RNPE. However, funding for teaching climate change education in Botswana does not appear to have been given top priority. Various degrees of insufficient support were reported by teachers (Ketlhoilwe, 2007; Silo, 2011). Teachers' standardization of the strategies based on policy (mis)interpretation serves as an illustration of this (Ketlhoilwe, 2007). The 1994 RNPE in Botswana introduced environmental education into the formal school curriculum, but policy implementation at the school level appears to have been based on teachers' ideas of how environmental education should be conceptualized and applied without ever consulting the students, who are supposed to be the policy's intended audience (Silo, 2011). Truly, this impediment hinders effective and meaningful implementation of climate change literacy. Instead, funds should be allocated specifically to support the teaching and learning of environmental education, which includes a focus on climate change. Ketlhoilwe and Velempini (2021) emphasized that school trips to outdoor places discussed in classroom promote active learning in environments that cannot be replicated in the classroom. For instance, when teaching about greenhouse gases, students should visit places where greenhouse gases are being produced, to learn how they end up being produced and which industrial activity results to a production of these gases. Such learning presents first-hand experiences framed within a constructivist learning approach. In most cases hindrances to implementing this type of education is due to lack of money for excursions carrying out programs that advocate for environmental education and sustainable management of natural resources. For instance, a teacher might want to take students on an excursion at a coal mine, which causes air pollution by emission of particular matter and gases including Methane (CH_4) and Sulphur dioxide (SO_2), but due to shortage of funds the trip fails which results failure to implement project based learning in climate change literacy. Velempini (2017) explained that schools face a lack of resources (scarcity of transport for outdoor experiential activities). Thus, the need to have policy that advocates for funds specifically for resources needed to teach environmental education in remediation of transport constraints is a contributory factor to limited fieldwork activities (Ketlhoilwe, 2003).

Opportunities for Climate Change Literacy

Teacher Training on Climate Change Literacy

Project-based learning has the potential to be a viable approach to transformative pedagogies, which has opened opportunities for capacity building in Botswana's teacher education through the Sustainability Starts with Teachers program, a UNESCO initiative. Eventually opportunities cascade to schools through teachers who have completed teacher-education programs (Ketlhoilwe & Silo, 2016). Therefore, environmental education should be made a major field of studies in teacher training colleges. This will allow teachers who are interested in environmental education to take it as a teaching subject. This will also improve in-service training as teachers will continue their training in environmental education. Ketlhoilwe (2003) also highlights the importance of enhanced teacher education (in-service and pre-service) as a recommendation to improve environmental education. In order to strengthen the emphasis on sustainability and climate change education, the University of Botswana reviewed the environmental education teacher preparation programs (Ketlhoilwe et al., 2020). Moreover, to enhance teacher preparation in the area of climate change literacy, the University of Botswana has since introduced the Master of Education and Doctor of Philosophy degrees in Environmental Education, which includes elements of climate change literacy.

Use of Project Based Learning in Climate Change Teaching

Engaging in ESD change projects in teacher education institutions to promote environmental and sustainability education through a whole school approach of change projects that are experiential and could offer insights for solutions to climate change has been one catalyst for new forms of learning in Botswana (Ketlhoilwe & Silo, 2016; UNESCO, 2020). Project-based learning is a viable method for implementing transformative pedagogies for education for sustainable development in teacher preparation, according to the results of change projects. To encourage effective sustainability among student-teachers in teacher education institutions in Botswana, place-based learning is advised. The change projects approach has offered opportunities for young student teachers in participatory approaches involving genuine community problems with real consequences in these institutions through whole institutions planning, creating and reflection on institutional capacity to embed sustainability practices in the curriculum.

Teacher trainers at Serowe College of Education in Botswana have come up with an innovative solution: creating teaching materials from recycled rubbish. Teacher trainers and student teachers collect, sort and clean waste material. The latter is used to create different learning tools like musical instruments, masks and toys. In this manner, student teachers are given the opportunity to explore their creative talents during this process (UNESCO, 2020).

In Molepolole College of Education, change project also focused on improved use of old cooking oil for soap (Ketlhoilwe & Silo, 2016).The teacher trainer observed that there is used oil from the college kitchen that is meant to be collected every Wednesday, but this is not done and consequently the kitchen area has been dirty, and the used oil polluted the environment. This led to the idea that the used oil can make soap. The plan of making soap included inviting community women to the college to demonstrate soap making. This Community of Practice (CoP) included local women, students, and science and arts lecturers. In the process, some ingredients like wooden charcoal for cosmetic therapy were used as well. There was an assessment of significant learning and the integration of cultural knowledge and science in the project through soap making. This project creates opportunities as well for college and community collaborations through experiential activities for unemployed youths in areas around the college. One of the teachers for this Change Project in Botswana said,

> The process of developing a Change Project has helped me to develop an in-depth understanding of the significance of sustainability. I have gained insights into how to develop an inclusive working Change Project ideal for transforming community activities that contribute to greenhouse gas emissions. (UNESCO, 2020, p 1.)

Consequently, the above sustainability and climate change literacy related initiatives motivated capacity building in young student teachers geared towards implementation of their teaching strategies in schools where they are going to be teaching. To strengthen their own agency for more adaptable approaches to climate change education, teachers and teacher educators should be given the opportunity to engage in place-based learning experiences.

The Application of Local Knowledge in Climate Change Literacy

Local knowledge plays an important role in gradually transforming human behavior towards long-term sustainability. Local knowledge on climate change is taught from birth in villages in Botswana. For example, the cultural way of life of Botswana is inclusive of climate change. In *Setswana* culture there were some trees like *mosetlha* (*Peltophorum africanum*), or *mohalatsamaru* (*Asparagus aspergillum* (Campbell & Hitchcock, 1985), which were not supposed to be burnt as they produced gases that were not good for the environment. This cultural practice not only underscores the intrinsic connection between traditional knowledge and environmental stewardship but also highlights indigenous strategies for environmental conservation and sustainability. Thus, there is an opportunity to increase climate change literacy by promoting indigenous and local knowledge. As a result, it is critical to conduct more research on indigenous knowledge as a means of promoting climate change literacy in Botswana. It is important that local cultural education for sustainability can be taught to youth to explore different livelihood pathways.

Conclusion and Recommendations

This chapter examined challenges and opportunities of climate change literacy in upper-middle income Botswana, a nation with a stable socio-economic environment. Our work presented emerging findings such as:

- lack of climate change literacy in school curriculum;
- lack of professional development for educators;
- lack of resources to teach environmental education;
- Opportunities for climate change literacy.

We suggest an employment of local and indigenous knowledge in climate change literacy. Our ancestors passed their rich knowledge and wisdom of stewardship from generation to generation through millennia. Their understanding of nature and its resources enhanced their resilience to adverse weather events and should remain embedded in our culture now, more than ever. It is necessary to involve all students and teachers in societal transformation and take into account the learning implications of environmental challenges like climate change in order to effectively implement climate change literacy in schools (Lotz-Sisitka & Lupele, 2017).

Climate change issues like water scarcity and pollution can be linked to changes in pedagogical processes and consequently affecting local communities (Silo, 2011). The Sustainability Starts With Teachers' Change Projects, which focus on building capacity, have significantly improved responses to climate change literacy (UNESCO, 2020). In light of this, this chapter urges all interested parties, including local people, policy and decision makers, curriculum developers, and researchers, to set aside time to collaborate with teachers in order to implement climate change literacy projects at the school and community levels. Future policy-making procedures and related teacher education initiatives should pay more attention to the ongoing, complex climate change issues that have an impact on both local communities and schools.

References

Akinyemi, F. O. (2017). Climate change and variability in Semi arid Palapye, eastern Botswana: An assessment from smallholder farmers' perspective. *Weather, Climate and Society, 9*(43), 349–365. https://doi.org/10.1175/WCAS-D-16-0040.1

Akinyemi, F. O., & Abiodun, B. J. (2019). Potential impacts of global warming levels 1.5 °C and above on climate extremes in Botswana. *Climatic Change, 154*, 387–400. https://doi.org/10.1007/s10584-019-02446-1

Alenda-Demoutiez, J. (2022). Climate change literacy in Africa: The main role of experiences. *International Journal of Environmental Studies, 79*(6), 981–997. https://doi.org/10.1080/00207233.2021.1987059

Amjath-Babu, T. S., Krupnik, T. J., Aravindakshan, S., Arshad, M., & Kaechele, H. (2016). Climate change and indicators of probable shifts in the consumption portfolios of dryland farmers in Sub-Saharan Africa: Implications for policy. *Ecological Indicators, 67*, 830–838. https://doi.org/10.1016/j.ecolind.2016.03.030

Apollo, A., & Mbah, M. F. (2021). Challenges and opportunities for climate change education (Cce) in East Africa: A critical review. *Climate, 9*(6), 1–16. https://doi.org/10.3390/cli9060093

Bangay, C., & Blum, N. (2010). Education responses to climate change and quality: Two parts of the same agenda? *International Journal of Educational Development, 30*, 359–368. https://doi.org/10.1016/j.ijedudev.2009.11.011

Battisti, D. S., & Naylor, R. (2009). Historical warnings of future food insecurity with unprecedented seasonal heat. *Science, 323*(5911), 240–244. https://doi.org/10.1126/science.1164363

Bissinger, K., & Bogner, F. X. (2018). Environmental literacy in practice: Education on tropical rainforests and climate change. *Environment, Development and Sustainability, 20*(5), 2079–2094. https://doi.org/10.1007/s10668-017-9978-9

Bolstad, R. (2005). School-based curriculum development: Is it coming back into fashion? *Curriculum Matters, 1*. https://doi.org/10.18296/cm.0068

Botswana Government. (2015). Education & training sector strategic plan (ETSSP 2015–2020). https://www.gov.bw/sites/default/files/2020-03/ETSSP%20Final%20Document.pdf

Botswana Government. (2020). Official opening remarks by Dr. Douglas Letsholathebe Hon. Minister. Ministry of Tertiary Education, Research, Science and Technology—Botswana at the Policy Dialogue as part of the Sustainability Starts with Teachers, a Capacity Building Programme for Teacher Educators on ESD (CAP ESD) . Republic of Botswana.

Campbell, A., & Hitchcock, R. (1985). Some Setswana names of woody plants. *Botswana Notes and Records, 17*, 117–129. https://www.jstor.org/stable/40979742

Charmaz, K. (2008). Grounded theory as an emergent method. In S. N. Hesse-Biber & P. Leavy (Eds.), *Handbook of emergent methods* (pp. 155–170). The Guilford Press. http://www.sxf.uevora.pt/wp-content/uploads/2013/03/Charmaz_2008-b.pdf

Cohen, J. E. (1995). How many people can the earth support? *The Sciences*, *35*(6), 18–23. https://lab.rockefeller.edu/cohenje/assets/file/229bCohenHowManyPeopleCanEarthSupportNewEthics4PublicHealthOUP1999.pdf

Creswell, J. W., & Creswell, J. D. (2022). *Research design. Qualitative, quantitative, and mixed methods approaches*. Sage Publications.

Darst, R. G., & Dawson, J. I. (2019). Putting meat on the (Classroom) table: Problems of denial and communication. In T. Lloro-Bidart & V. S. Banschbach (Eds.), *Animals in environmental education*. Palgrave Studies in Education and the Environment. Palgrave Macmillan. https://doi.org/10.1007/978-3-319-98479-7_12

Davids, N. (2021). Education, gender and poverty affect climate change literacy in Africa. https://www.news.uct.ac.za/article/-2021-10-08-education-gender-and-poverty-drive-climate-change-literacy-in-africa

Denzin, N. K., & Lincoln, Y. S. (2008). Introduction: The discipline and practice of qualitative research. In N. K. Denzin & Y. S. Lincoln (Eds.), *Strategies of qualitative inquiry* (pp. 1–43). Sage Publications Inc.

Educational International. (2021). COP26 event roundtable: Why climate literacy and civic skill building will solve the climate crisis—answers from civil society, educators, and governments. https://www.ei-ie.org/en/item/25495:cop26-event-roundtable-why-climate-literacy-and-civic-skill-building-will-solve-the-climate-crisis-answers-from-civil-society-educators-and-governments

FAO. (2021). *The impact of disasters and crises on agriculture and food security: 2021*, Rome. https://doi.org/10.4060/cb3673en

Hambira, W. L., Saarinen, J., & Moses, O. (2020). Climate change policy in a world of uncertainty: Changing environment, knowledge, and tourism in Botswana. *African Geographical Review, 39*(3), 252–266. https://doi.org/10.1080/19376812.2020.1719366

Hoath, L., & Dave, H. (2022). Implementing a sustainability and climate change strategy, *103*(384). https://fed.education/wp-content/uploads/2022/04/ssr-march-2022-072-075-hoath.pdf

Hu, C., Chen, L., Zhang, C., Qi, Y., &, Yin, R. (2006). Emission mitigation of CO_2 in steel industry: Current status and future scenarios. *Journal of Iron and Steel Research, International, 13*(6), 38–52. https://doi.org/10.1016/S1006-706X(06)60107-6

IPCC. (2022). *Climate change 2022: Impacts, adaptation and vulnerability. Contribution of working group II to the sixth assessment report of the intergovernmental panel on climate change* (H.-O. Pörtner, D.C. Roberts, M. Tignor, E.S. Poloczanska, K. Mintenbeck, A. Alegría, M. Craig, S. Langsdorf, S. Löschke, V. Möller, A. Okem, & B. Rama (Eds.)] (3056 pp.). Cambridge University Press. https://report.ipcc.ch/ar6/wg2/IPCC_AR6_WGII_FullReport.pdf

Johnson-Pynn, J. S., & Johnson, L. R. (2005). Successes and challenges in east African conservation education. *The Journal of Environmental Education, 36*(2), 25–39. https://doi.org/10.3200/JOEE.36.2.25-39

Johnston, J. D. (2019). Climate change literacy to combat climate change and its impacts. In W. Leal Filho, A. Azul, L. Brandli, P. Özuyar, & T. Wall (Eds.), *Climate action*. Encyclopedia of the UN Sustainable Development Goals. Springer. https://doi.org/10.1007/978-3-319-71063-1_31-1

Joseph, S. (2021). A brief introduction to COP26 and the fight for global quality climate education. https://populationeducation.org/a-brief-introduction-to-cop26-and-the-fight-for-global-quality-climate-education/

Kastner, T., Rivas, M. J. I., Koch, W., & Nonhebel, S. (2012). Global changes in diets and the consequences for land requirements for food. *Proceedings of the National Academy of Sciences, 109*(18), 6868–6872. https://doi.org/10.1073/pnas.1117054109

Kemp, L., Xu, C., Depledge, J., Ebi, K. L., Gibbins, G., Kohler, T. A., Rockstrom, J., Scheffer, M., Schellnhuber, H. J., Steffen, W., & Lenton, T. M. (2022). Climate endgame: Exploring catastrophic climate change scenarios. *PNAS. Perspective, 119*(34). https://doi.org/10.1073/pnas.2108146119

Ketlhoilwe, M. J. (2003). Environmental education policy implementation in Botswana: The role of secondary education officers and school heads. *Southern African Journal of Environmental Education, 20*, 75–84. https://www.ajol.info/index.php/sajee/article/view/122667

Ketlhoilwe, M. J. (2007). Environmental education policy implementation challenges in Botswana schools. *The Southern African Journal of Environmental Education, 24*, 171–184. https://www.ajol.info/index.php/sajee/article/view/122752

Ketlhoilwe, M. J. (2019). Building community resilience through environmental education: A local response to climate change. Building sustainability through education. *IGI Global*. https://www.igi-global.com/book/building-sustainability-through-environmental-education/209765

Ketlhoilwe, M. J., & Jeremiah, K. (2010). Mainstreaming environment and sustainability in institutions of higher education: The case of the University of Botswana. *International Journal of Scientific Research in Education, 3*(1), 1–9.

Ketlhoilwe, M. J., & Silo, N. (2016). Change project-based learning in teacher education in Botswana. *Southern African Journal of Environmental Education, 32*, 1–16. https://www.ajol.info/index.php/sajee/article/view/152737

Ketlhoilwe, M. J., & Velempini, K. (2021). Wilding educational policy: The case of Botswana. *Policy Futures in Education, 19*(3), 358–371. https://doi.org/10.1177/1478210320986350

Ketlhoilwe, M. J., Silo, N., Velempini, K. (2020). Enhancing the roles responsibilities of higher education institutions in implementing the sustainable development goals. In G. Nhamo & V. Mjimba (Eds.), *Sustainable development goals and institutions of higher education*. Sustainable Development Goals Series. Springer.

Kolawole, D. O., Wolski, P., Ngwenya, B., Mmopelwa, G., & Thakadu, O. (2014). Responding to climate change through joint partnership: Insights from the Okavango Delta of Botswana. *World Journal of Science, Technology and Sustainable Development, 11*(3), 170–181. https://doi.org/10.1108/WJSTSD-06-2014-0010

Lee, T. M., Markowitz, E. M., Howe, P. D., Ko, C. Y., & Leiserowitz, A. A. (2015). Predictors of public climate change awareness and risk perception around the world. *Nature Climate Change, 5*(11), 1014–1020. https://doi.org/10.1038/nclimate2728

Lotz-Sisitka, H., & Lupele, J. (2017). ESD, learning and quality education in Africa: Learning today for tomorrow. In H. Lotz-Sisitka, O. Shumba, J. Lupele, & D. Wilmot (Eds.), *Schooling for sustainable development in Africa. Schooling for sustainable development*. Springer. https://doi.org/10.1007/978-3-319-45989-9_1

Mitchell, T., & van Aalst, M. (2008). Convergence of disaster risk reduction and climate change adaptation. *A Review for DFID, 44*, 1–22. https://www.preventionweb.net/files/7853_ConvergenceofDRRandCCA1.pdf

Momsen, J. (2009). *Gender and development* (3rd ed.). Routledge.

Nakashima, D., Krupnik, I., & Rubis, J.T (2018). *Indigenous knowledge for climate change assessment and adaptation*. Local & indigenous knowledge 2. Cambridge University Press and UNESCO. https://unesdoc.unesco.org/ark:/48223/pf0000265504?posInSet=1&queryId=4cfd60f1-decd-40b8-bf8c-c24194eb62e4

NASA. (2022). Global warming vs climate change. https://climate.nasa.gov/global-warming-vs-climate-change/

Onwuegbuzie, A. J., & Frels, N. (2016). *Seven steps to a comprehensive literature review: A multimodal and cultural approach*. Sage.

Ostrom, E. (2017). Polycentric systems for coping with collective action and global environmental change. In *Global justice* (pp. 423–430). Routledge.

Park, S. E., Marshall, N. A., Jakku, E., Dowd, A. M., Howden, S. M., Mendham, E., & Fleming, A. (2012). Informing adaptation responses to climate change through theories of transformation. *Global Environmental Change, 22*(1), 115–126. https://doi.org/10.1016/j.gloenvcha.2011.10.003

Patton, M. Q. (2002). *Qualitative research and evaluation methods.* SAGE Publications.

Patton, M. Q. (2015). *Qualitative research & evaluation methods: Integrating theory and practice* (4th ed.). Sage.

Republic of Botswana. (2018). A national climate change strategy for Botswana. Final strategy. https://drmims.sadc.int/sites/default/files/document/2020-03/2018_Botswana%20Climate%20Change%20Strategy.pdf

Republic of Botswana. (2021). *The Botswana climate change policy.* Ministry of Environment, Natural Resources, Conservation and Tourism, Gaborone, Botswana.

Romieu, E., Welle, T., Schneiderbauer, S., Pelling, M., & Vinchon, C. (2010). Vulnerability assessment within climate change and natural hazard contexts: Revealing gaps and synergies through coastal applications. *Sustainability Science, 5*, 159–170. https://doi.org/10.1007/s11625-010-0112-2

Rosen, A. M. (2015). The wrong solution at the right time: The failure of the Kyoto protocol on climate change. *Politics and Policy, 43*(1), 30–58. https://doi.org/10.1111/polp.12105

Rubin, H. J., & Rubin, I. S. (2012). *Qualitative interviewing: The art of hearing data* (3rd ed.). Sage Publications.

Selabe, M., & Minyoi, M. K. (2018). Perceived barriers to rainwater harvesting as a source of water supply in Botswana: A case of Gaborone city. *Botswana Journal of Technology, 23*(1), 86–96. https://journals.ub.bw/index.php/bjt/article/view/1493

Selormey, E., Dome, M. Z., Osse, L., & Logan, C. (2019). Change ahead: Experience and awareness of climate change in Africa. Africa portal. https://www.africaportal.org/publications/change-ahead-experience-and-awareness-climate-change-africa/

Shi, Y., Han, B., Zafar, M. W., & Wei, Z. (2019). Uncovering the driving forces of carbon dioxide emissions in Chinese manufacturing industry: An intersectoral analysis. *Environ Science and Pollution Research, 26*, 31434–31448. https://doi.org/10.1007/s11356-019-06303-7

Silo, N. (2009). Exploring learner participation in waste-management activities in a rural Botswana Primary School. *Southern African Journal of Environmental Education, 26*. https://www.ajol.info/index.php/sajee/article/view/122820

Silo, N. (2011). Children's participation in waste management activities as a place-based approach to environmental education. *Children, Youth and Environments, 21*(1), 128–148. https://doi.org/10.7721/chilyoutenvi.21.1.0128

Silo, N. (2017). Integrating learners' voices into school environmental management practices through dialogue. In H. Lotz-Sisitka, O. Shumba, J. Lupele, & D. Wilmot (Eds.), *Schooling for sustainable development in Africa.* Schooling for Sustainable Development. Springer. https://doi.org/10.1007/978-3-319-45989-9_12

Silo, N., Velempini, K., & Ketlhoilwe, M. J. (2022). Botswana teachers educators' and schools' responses to climate change. In H. Lotz-Sisitka & E. Rosenberg (Eds.), Education in times of climate change. NORRAG special issue 07. https://www.norrag.org/app/uploads/2022/05/Norrag_NSI_07_EN_ToC.pdf

Simpson, N. P., Andrews, T. M., Krönke, M., Lennard, C., Odoulami, R. C., Ouweneel, B., Steynor, A., & Trisos, C. H. (2021a). Climate change literacy in Africa. *Nature Climate Change, 11*(11), 937–944. https://doi.org/10.1038/s41558-021-01171-x

Simpson, N. P., Trisos, C., Kronke, M., & Andrews, T. M., (2021). Africa's first continent-wide survey of climate change literacy finds education is key. https://theconversation.com/africas-first-continent-wide-survey-of-climate-change-literacy-finds-education-is-key-169426

Spratt, D., & Dunlop, I. (2019). *Existential climate-related security risk: A scenario approach.* Policy Paper. Breakthrough—National Centre for Climate Restoration, Melbourne, Australia. http://mycoasts.org/commons/library/2019_Spratt_Dunlop.pdf

Trisos, C. H., Adelekan, I. O., Totin, E., Ayanlade, A., Efitre, J., Gemeda, A., Kalaba, K., Lennard, C., Masao, C., Mgaya, Y., Ngaruiya, G., Olago, D., Simpson, N. P., & Zakieldeen, S. (2022). Africa. In *Climate change 2022: Impacts, adaptation and vulnerability. Contribution of working Group II to the sixth assessment report of the intergovernmental panel on climate change* [H.-O. Pörtner, D. C. Roberts, M. Tignor, E. S. Poloczanska, K. Mintenbeck, A. Alegría, M. Craig, S. Langsdorf, S. Löschke, V. Möller, A. Okem, & B. Rama (Eds.)] (pp. 1285–1455). Cambridge University Press. https://www.ipcc.ch/report/ar6/wg2/downloads/report/IPCC_AR6_WGII_Chapter09.pdf

UNESCO. (2004). Literacy. https://uis.unesco.org/node/3079547

UNESCO. (2020). Botswana change projects. https://sustainabilityteachers.org/change-projects/botswana/

United Nations Development Programme. (2022). Botswana climate change policy. https://www.undp.org/botswana/publications/country-programme-document-botswana-2022-2026

United Nations. (2022). Africa's practical realities for COP 27 and Beyond: The food, energy and climate nexus. https://www.un.org/osaa/news/africa%E2%80%99s-practical-realities-cop-27-and-beyond-food-energy-and-climate-nexus

United Nations Framework Convention on Climate Change. (2021, November 13). Glasgow climate pact (UNFCCC/PA/CMA/2021/L.16). In Conference of the parties serving as the meeting of the parties to the paris agreement, Third session, Glasgow, 31 October–12 November 2021. https://unfccc.int/sites/default/files/resource/cma2021_L16_adv.pdf

Velempini, K. (2017). Infusion or confusion: A meta–analysis of environmental education in the 21st century curriculum of Botswana. *Africa Education Review, 14*(1), 42–57. https://doi.org/10.1080/18146627.2016.1224560

Verma, G., & Dhull, P. (2017). Environmental education as a subject in schools. *International Journal of Advanced Research (IJAR), 5*(8), 1547–1552. https://doi.org/10.21474/IJAR01/5214

Williams, S., McEwen, L. J., & Quinn, N. (2017). As the climate changes: Intergenerational action-based learning in relation to flood education. *The Journal of Environmental Education, 48*(3), 154–171. https://doi.org/10.1080/00958964.2016.1256261

Part IV
Prospects for the Future

Chapter 16
Climate Change Education in West Africa: Prospects and Problems

Edidiong Samuel Akpabio, Kemi Funlayo Akeju, and Moses Metumara Duruji

Abstract In recent years, the challenges of climate change have accentuated with human factors playing a significant role in its sustenance and spread. This climatic instability which has affected the socio-political dynamics of nations in West Africa come along with threats of political instability. It is due to these myriads of unpalatable experiences that global agencies and policymakers have advocated the imperative to engage climate change education as a remediation tool for the rising challenges of climate change in the sub-continent. It is however important to sound a caveat that as beneficial as climate change education (CCE) can be when adopted, it is laden with prospects and problems for implementation. This represents the motivation for this study which examines the prospects and problems of adopting climate change education in West Africa. In a bid to actualize this quest, we explored the frameworks for climate change education in Africa and engaged in a cross-continental appraisal of climate change education across the globe. This allowed us to identify strategies that can be adopted and replicated in West-Africa while unearthing loopholes to be plugged. The chapter adopted a qualitative research methodology and utilized secondary sources of information among other sources to interrogate the embedded issues in the discourse. The findings suggest that the West Africa region can defeat climate change by enhancing capacities in the area of climate education and ensuring that CCE is made a compulsory subject to be taught in all schools. It was observed that although some schools particularly at the tertiary level teach climate change, much emphasis is always on the scientific dimensions to the neglect of political, social and economic components. This indicated that beyond educating the populace on climate change, emphasis should also be on the approaches or methodologies adopted.

E. S. Akpabio (✉)
Department of Political Science, Trinity University, Yaba, Nigeria
e-mail: edidiong.akpabio@trinityuniversity.edu.ng

K. F. Akeju
Economics Department, Ekiti State University, Ado Ekiti, Nigeria
e-mail: kemi.akeju@eksu.edu.ng

M. M. Duruji
Political Science Department, Covenant University, Ota, Nigeria

M. F. Mbah et al. (eds.), *Practices, Perceptions and Prospects for Climate Change Education in Africa*, https://doi.org/10.1007/978-3-031-84081-4_16

Introduction

Climate change has remained an issue of concern across the globe (Charles et al., 2023). However, Africa can be described as a continent requiring more attention, given the fact that seven out of the ten most vulnerable countries to climate change in the world are domiciled in the continent (Simpson et al., 2021; UN, 2023; UNICEF, 2023). This implies that concerted action to mitigate the effects of climate change is required which is a far cry from the current situation where issues of climate change are handled with levity and hardly given priority attention. It is worrisome to observe that apart from insecurities challenges facing the sub-continent, West African states are continually confronted with threats of climate change which often act as propellants to the spate of insecurities they experience especially with respect to climatically induced conflicts. This is a pointer that the impacts of climate change are already been felt in West Africa as she is currently contending with its negative effects (Sylla et al., 2016). One major sector affected by climate change in West Africa is the agricultural sector with the concomitant consequences on food insecurity, a clear indicator of a weak adaptive capacity in the sub-region (Sorgho et al., 2020a, 2020b). Despite its negative consequences, it is pertinent to state that as frightening as climate change and its accompanying aftermaths are (Simpson et al., 2021; Busby, 2018), it is not insurmountable as it can be addressed in the classrooms and beyond.

Across the globe, climate education is believed to have potentials for combating climate change(Fernandez et al., 2014).This is a truism in all respects as education is a major agent in addressing climate change in the contemporary world (Wade, 1999). Hence, the clamor for the inclusion of climate education in current schools' curricula. It is expedient at this juncture to make it explicitly clear that while awareness on climate change is one thing, having a clear-cut understanding of what it represents and steps to be taken in combating it represent another. Hence, to buttress the importance of climate change education, stakeholders such as international governmental organizations, civil society organizations among others have called on west African political leaders at various fora to ensure that everyone is educated on issues of the climate. It is, however, worrisome to observe that as vital as education is to the lingering issue of climate change, it has been plagued by an avalanche of challenges hence impeding its applicability in addressing the raving and raging climate debacle confronting the international political system (Pruneau et al., 2010). For instance, the West African sub-region has not done well with respect to climate change education because of the impediments and daunting challenges it is plagued with (UNEP, 2023). Hence, there is need to give climate education more attention in a bid to achieve large scale education of West Africans on the dangers inherent in climate change and how they can combat it through engaging in climate friendly practices. The underlying assignment in this chapter is to interrogate climate change education in West Africa in its entirety. We also unearth the problems confronting the sub-region with respect to the entrenchment of climate change education while identifying the prospects embedded in it.

Conceptual Appraisal and Theoretical Discourses

The term "climate change" is defined as any form of changes in the climate over a period of time which could evolve as a result of natural variability or human induced actions (Pielke, 2004; IPCC, 2007). It connotes a periodic modification to the climate as a result of disorder in atmospheric conditions led by biological and geographical factors which could result in an extended change in the typical meteorological conditions such as rainfall, wind, global warming, among other manifestations (Mahato, 2014; Noor et al., 2020). This indicates that these climatic conditions are expected to change over time into extreme conditions because the biologically and geographically induced causal-factors persist. For decades, options for meeting the challenges of climate change have been clearly identified as the mitigation of future effect which can be done through having a clear understanding of its causal effects and the adaptation to changes which is expected to be addressed through the knowledge of factors that make the society and its environment vulnerable to the effect of climate change (Pielke, 2004).

The place of education is critical when dealing with climatic change challenges because it provides information, knowledge and understanding of the impacts and strategies of its adaptation and mitigation to the people (Cordero et al., 2020). Climate Change Education is the utilization of teaching and learning methods to provide adequate information and knowledge to people to boost their understanding of changes in climatic conditions while helping to develop effective adaptation and mitigation responses to climate changes (Mochizuki & Bryan 2015; Sibanda & Manik 2022). Climate change education can help to create awareness on the benefits of environmental conservation vis-a-vis the impacts of the failure to do so on sustainable development. In the same vein, education literature about the environment provides perspectives on the association between behaviorial change and education (Molthan-Hil et al., 2019), showcasing the capability of learning to imbibe changes as reflected by the constructivism theory. Currently, the youth are offered knowledge and information about the world's climatic conditions now and the appropriate mitigation strategies that must be implemented, as they will be the ones to deal with the impacts of climate change (Apollo & Mbah, 2021).

The UNESCO Climate Change Education for Sustainable Development programme (CCESD) which was established in 2010 works with countries national governments towards including information on climate change in educational system with the aim of promoting people's understanding through innovative learning approaches (Mochizuki & Bryan, 2015). It strives to improve education and make it an indispensable component of the global climate change strategy and attempts to promote climate literacy particularly in young people, by assisting individuals in comprehending the effects of global warming today (Amanchukwu et al., 2015; Kolenatý et al., 2022). Climate change issues are recognized as a productive setting for both learning and instruction as situated at the interface of science and society (Szczepankiewicz et al., 2021), as such, the primary actors in confronting the problems associated with climate change are educators and students.

Educational knowledge on climatic changes is anchored on learning that promotes sustainability while of recent, the gamification theory has been considered as a useful model for enhancing learning and conformity of households and communities to understanding the procedures for mitigation and climate change resilience (Douglas & Brauer, 2021; Mazur-Stommen & Farley, 2016; Fernández Galeote et al. 2021).

Gamification approach to learning has been identified as the application of suitable game principles to non-game contexts such as climate change mitigation and adaptation methods (Csoknyai et al., 2019). The principle of gamification fosters an atmosphere where people are genuinely driven to interact with content pertaining to the behavioral change that is desired (Wee & Choong 2019). Games are considered promising for engaging people with climate change as it provides a structured and replicable approach which can influence behavior, increase system knowledge, awareness, impact norms, facilitates discussions and increase legitimacy of decision. Long games have the ability to foster increased player connection in the field of adapting to climate change, challenging players' preconceived notions, altering their behavior, and sparking action (Flood et al., 2018). Application of games when played across several sessions and are integrated with other teaching techniques, are more successful at promoting cognitive learning (Wouters et al., 2013).

Review of Theoretical Literature

The globe is currently affected by climate change, which is a complicated issue that will continue to evolve (IPCC 2014). Research has revealed that gamification fosters greater motivation through discovery learning, pictorial reference, and social engagement, giving relatedness that improves participants' cognition and inspires behaviour that elicits a reaction to climate change (HSU, 2022; Koivisto & Hamari 2019). Borrowing from social constructivist theories, where ideas are developed through interaction, games frequently offer interactive places where reality may be experienced and modified thus making games an efficient tool for transforming reality and creating interactive environments where it can be experienced (Monroe et al., 2019a, 2019b).By utilizing people's talents and skills, gamification may take climate change communication to a new level (Rajanen & Rajanen, 2019; Brannon et al., 2022). Applications for climate change gamification, for instance, might emphasize the urgency of the issue, raise knowledge of it more quickly, and give users a greater sense of control over the future by changing their behavior now.

The importance of students' knowledge and literacy in science have been emphasized, (Hapgood & Palincsar 2006). Previous research has emphasized the need to confront the issues related to climate change (Boon, 2010; Sharma, 2012). The difficulties that the nature of climate change education presents to students have been established in the literature (Bodzin et al., 2014), and the teachers' lack of training and preparation on the study guide and adequate curricula on climate change (Hestness et al., 2014; Seneviratne et al. 2022), have been identified as impediments to climate change education. These limitations are also readily observed in teachers' content knowledge (Seow & Ho 2016), thus signifying that existing literature acknowledged the vital need for the scientific instruction research community to adopt more engaging and effective teaching strategies on climate change (McGinnis et al., 2016; Sharma, 2012).

Emphasis has been placed on the role of environmental education in providing to various groups different approaches and tools that can enhance support for their environmentally related attitudes, awareness, values, perceptions, skills and knowledge (Armstrong & Krasny, 2020). Literature has also identified the best strategies for effective climate change education as the use of active and engaging teaching methods addressing misunderstandings and executing climate change projects in school and communities (Monroe et.al., 2019). On the strength of the identified lapses observed from the lenses of literature although from other climes of the world, we have launched this inquiry to identify challenges militating against climate change education that are specific to the West African sub-region. We are optimistic that the identified gaps will help in no mean measure to entrench CCE in the sub-continent.

Climate Change Education: Processes and Regional Experiences

The agreement of stakeholders, including State parties in 1992 establishing the UNFCCC is the fulcrum under which climate change education is anchored. The United UNFCCC Article 6 demonstrated the need for climate change-action in manifold sectors, such as education, training and public awareness (UNFCCC 1992). In the UNFCCC Articles 4, the imperative for climate change education was stressed that countries should promote and collaborate in education, training, and public awareness to encourage an increased degree of involvement in climate change adaptation and mitigation. As such, the UNFCCC prioritized six major spheres of action such as, education, public engagement, public awareness, public access to information, and international cooperation.

Collectively, the six areas for intervention would be supported largely at different levels of government and entrusted with guiding and supporting academic standards and administration, public services and products in the semi and informal teaching and learning process, and also encouraging voluntary action (Carter, 2019). Nonetheless, the more than 20-year absence of significant advancement on these duties needs to be recognized as a key factor in creating the frustrations that is felt in the implementation and even discourse on seeing the realization of Article 12 of the Paris Accord (UNFCCC 2020).

Cross-National Comparisons of Climate Change Education

Climate change which is often referred to as the long-term, sustained shift in weather conditions in factors like temperature, precipitation, atmospheric pressure, and wind systems. Global warming, which results from greenhouse gas (GHG) emissions into the atmosphere, is linked to climate change. This emission progressively depletes the ozone layer thus causing changes in global temperature. The impact of this sustained changes in climatic condition has caused untold hardship to the planet, earth and life (Duruji & Duruji-Moses, 2018). Studies have confirmed over and again that it is the activities of man mainly through industrialization and deforestation that is responsible for the emission of GHG responsible for climate change (Duruji 2018).

The adoption and utilization of multilateral platforms and channels on a global scale to find solution to climate change beginning from the Earth Summit in Rio de Janeiro 1992 to the biennial conference of the parties under the aegis of UNFCCC have come out with resolutions, chatters, accords and agreements which recognize that the solution lies in a global concerted effort and responsibility at national, local and individual levels to address the issues associated with climate change (Duruji & Duruji-Moses, 2017). The need for climate change action in sectors such as education, training and public awareness was established by Article 6 of the United Nations Framework Convention on Climate Change (UNFCCC 1992). The implication is that climate change education becomes critical to firming up climate change awareness (Filho 2021). The essence of CCE is to encourage social transformation by reforming educational pedagogies and giving individuals the tools they need to gain the information, skills, values, and attitudes necessary for sustainable development (Bangay & Blum, 2010). This further emphasizes the significance of including sustainable development concepts in curriculums (UNESCO 2020).

Climate Change Education in Europe

When compared with efforts at entrenching climate change education with other parts of the world, the European continent is doing quite well with ensuring that countries in the continent embrace climate change and implement measures that do not only bring awareness but moderate behavior of individuals, corporate entities and governments in the region (Carter, 2019). This renewed initiative is premised on the believe that climate change is an international problem, hence, it requires international solutions (Riddersborg et al., 2022). With this realization, governments in about 27 European countries appear ready to take concrete actions towards this end. So understanding each European country and appreciation of the obstacles that may hinder the objectives becomes very imperative to achieve this and enable them to inspire each other with green success stories in the various European countries. This aspiration represents the root of European Climate Change Curriculum (ECCC).

The essence of creating a ECCC curriculum was to enable European students in high schools to identify and take action on different perspectives of climate change including, political, economic and technological (Riddersborg et al., 2022). This signifies that concern on educating people about climate change is now of top political agenda in many European countries that can no longer be limited. This explains why week in week out thousands of European youths are organized to voice out challenges they encounter with lack of real action aimed at meeting targets stated in the Paris Accord. European countries have ensured that European students learn about climate change and possible solutions to issues surrounding climate change (Riddersborg et al., 2022). However, in doing so, the emphasis is on national frameworks that emphasize on each country's unique challenges and its resolutions. It is important to state that the objective is to empower young European citizens to work in cooperation and advance cross-country innovative ways to combat climate change. Knowledge of the ideas and challenges in the different countries is imperative for the accomplishment of ECCC objectives. Towards this end, a document called the European e-book has been developed to guide European countries (Riddersborg et al., 2022).

Climate Change Education in North America

Climate change in North America is viewed from ideological standpoints with countries such as the United States of America sharply divided on partisan lines (Kamanetz, 2019). In spite this, Canada is doing much better with regards to Climate Change Education due to the fact that there is overwhelming consensus on the dangers inherent in it (Kamanetz, 2019). In Mexico, there is relatively less awareness compared to its northern neighbours but not so much with the ideological divide witnessed in its immediate northern neighbour (Kamanetz, 2019).

Climate Change Education in Asia

The Asian continent apart from being the largest in the world is also diverse in terms of population and geography with issues relating to climate change in the continent and its sub-regions towing same path. Countries of South Asia which include Bhutan, Bangladesh, India, Afghanistan Sri Lanka, the Maldives, Nepal, and Pakistan, have a population of 1.85 billion and the landscapes are diverse with mountains in the Himalayas, incredibly rich soil, forests, and coastal plains. The cities and hinterlands of South Asia have many distinctive characteristics, yet the region's geographic location makes it susceptible to natural disasters or obstacles brought on by the environment (Mbah et al., 2022). According to the Intergovernmental Panel on Climate Change (IPCC) assessment, South Asian nations will likely have to address problems including melting glaciers, rising temperatures, and floods in their tropical zones.

In fact, South Asia is already experiencing the effects of climate change, making it crucial to promote climate education in the region. (Mbah et al., 2022). In the case of Bhutan, the ministry of education has developed in addition to including knowledge of climate change issues at various levels of the educational systems, a curriculum on the environment and climate change should be included (Mbah et al., 2022). Furthermore, for the Maldives, efforts centre at engaging the youths through programmes that entrench information about coping with climate change at secondary and tertiary institutions of learning as well as in the field of vocational education (UNISDR 2021). The emphasis for Nepal is centered on the implementation of adaptation programs in educational institutions by incorporating climate knowledge into the formal and non-formal education sectors.

Climate Change Education in Latin America

The biodiversity of Latin America, poses concerns of threats by climate change to the huge indigenous population of the continent if nothing is done to advance deforestation of the rainforest thus fueling global warming. But the interesting thing about Latin America is that there are many higher institutions dedicated to research on climate change including; Universidad de los Andes in Colombia, the Universidad de Chile and the Center for Climate and Resilience Research (Siemen Stiftung, 2018). Academics at the aforementioned institutions take a keen interest in indigenous communities' knowledge of nature, which are in turn introduced in schools in the continent by educationalists. From the supranational platform, Latin American governments have also proclaimed their commitment to incorporate climate change education in national curricula across different level of their school system (Siemen Stiftung, 2018).

Climate Change Education in Africa

According to research by Afro-barometer, Africa lags behind in climate change education when compared with other parts of the world. In that research, climate change literacy, rate is 37% while that of Europe and North America, is over 80% (Simpson et al., 2021). Though climate change awareness varies considerably across African countries, and even within each country, examining it on country-specific basis; Mauritius has 66% climate change literacy rate, while in Uganda it is 62%. However, in Mozambique and Tunisia, it is 25% and 23% respectively.

The Afro barometer research in which 394 sub-national regions were surveyed shows that 8% (37 regions in 16 countries) scored lower than 20% on climate change literacy, whereas only 2% (8 regions) performed up to 80%. It also shows that there is a striking difference within countries as well. For instance, in Nigeria, the rates range from 71% in Kwara to 5% in Kano. In Botswana, the Lobatse region is 69% while Kweneng West is 6%. The difference in climate change literacy rates between the top and lowest subnational units is on average 33% (Simpson et al., 2022).

From the above narrative is is clear that West Africa can draw positive lessons from Europe. Hence,the European relative success in climate change education has become a template for developing countries in the West African sub-region to advance climate change education. The absence of a continental institution as is the case with Europe is certainly not the way to go for countries of West African though it has not hindered North American countries from making progress with climate change education. The reason being that North American countries have strong national institutions that allows individual countries to prioritize climate change issues including climate change education. Again, West African countries have lessons to draw from the experience of Asian countries which share so many in common in terms of state capacities and economic capabilities.

It is obviously clear that the best indicator of knowledge about climate change is education and West Africa has to embrace climate change education in its entirety. A comparison of people with no formal education and people who obtained their high school diploma in a study carried out by (Simpson et al., 2021), revealed that the former is likely to be 19% more climate change literate than the later. Those who completed a university education in the same vein are 36% likely to be knowledgeable about climate change (Simpson et al., 2021). The study found that wealthier and more mobile Africans, that are urban based are more climate change literate, while poverty ridden Africans are not climate change literate. Even though a robust climate change education has not been developed in the continent, Olubebe's study revealed that education will be effective in increasing climate change literacy in the continent (Olubebe, 2015).There is no doubt that climate literacy is imperative for all global citizens including citizens of West African states because climate change education produces climate change conscious citizens which reflects in the life style that is green friendly. The examples we cited in the cross -continent analysis presents bright prospect for West Africa that it is doable. A handy illustration of an African country that has borrowed a leaf from other continents is Zimbabwe which has created and implemented numerous programs to inform its population about climate change and provide them with the tools they require to respond to the global environmental catastrophe on a personal level. Zimbabwe is creating a comprehensive national climate law to promote public education campaigns and programs for coping with climate change (Simpson et al., 2021).Towing the same path is Kenya which due to her experience of climate emergencies such as flood, drought and locust invasion utilized music to raise public awareness on climate crisis (Simpson et al., 2021). This response mechanism apart from serving as a medium of education depicts ingenuity and pro-activeness in the midst of rising uncertainties.

Assessing Climate Change Education in West Africa

Climate change is a global challenge that affects the global international system with a plethora of untoward consequences that require urgent solutions of which climate change education represents one. It is important to state that the rampaging and ravaging incidences of climate change in the African continent continue to expose the need for governmental support of climate focused research and education(Magadza, 2000).This is so as human-induced climate change has heightened in West Africa with the resultant consequence of excessive rainfall a clear indicator of the paucity of climate change education (Tandon, 2022).It is disturbing to state that one of the sectors most negatively impacted by climate change in West Africa is the agricultural sector with threats of rampaging food security in the offing if left unchecked(Sorgho et al., 2020a, 2020b).

Conceptually, climate change education can be classified as an adaptation measure as it equips individuals with the requisite skills to reduce the effects of climate change or help them develop coping mechanisms. Policy experts and environmental scientist have recognized it as a necessity in the contemporary world system especially with the challenges of climatic dysfunctions staring her in the face (UNESCO & UNFCCC, 2016).

In recent years, there has been an increasing interest in climate change education (Monroe et al., 2019). Although climate change education is a new concept, it seems to be gaining heightened relevance and significance among international organizations and policy makers (Laessoe & Mochizuki, 2015).This surge in the study of climate change education might not be unconnected with the devastation that accompanies climatic dysfunctionalities. Although we agree that there is heightened interest in climate change education, we are not oblivious of the legion of problems it contends with especially in the African continent (Monroe et al., 2017). For instance it is challenged by issues such as paucity of climate skilled instructors, low receptability, and weak or non-existent government policies on climate change education among several others.

To state that there are impediments to the implementation of climate change education is to be parsimonious with words as CCE in West Africa is bedeviled by an avalanche of encumbrances. For instance, as catastrophic as the effects of climate change seem, some individuals dismiss it with a wave of the finger (Pruneau et al., 2010). A clear indicator of nonchalance with propensity to lead to acts laden with ineptitude. Such actions contribute significantly to impeding this valuable field of education which represents the panacea to the sub-region's woes with respect to climatic dysfunctionalities.

In the midst of these impediments plaguing climate change education, a legion of roles has been attributed to it. Key among them is the influence on people's actions from a carefree one to a climate sensitive culture. This reinforces Mochizuki & Bryan's position who opined that education has the capacity to play a critical role in improving climate literacy (Mochizuki & Bryan, 2015) and is in sync with the findings of McCright et al., who posited that the opportunities in climate change education are boundless as it is laden with transformative capacities (Mc Cright et al., 2013).It is fallout of the aforementioned literature that we assert that engaging in climate change education is mandatory for any society desirous of sustainability. The Chinese government for instance made climate change education compulsory for her nationals since the 1990's (Yi & Wu, 2009).We are therefore of the opinion that such actions be implemented across West Africa as a means of stemming the tide of climate dysfunctionalities.

It is equally pertinent to reiterate that climate change education requires a comprehensive approach as it encompasses a broad range of disciplines (Alexandru et al., 2013). It will therefore be apt to state that no discipline is excluded from this venture. Hence, it will be misleading and a grave error of unquantifiable magnitude to consign climate change education to the scientific domain as currently experienced. This erroneous classification apart from hampering the success of the venture has the potential of leading to a great deal of resistance from non-scientist (Cherry, 2011; Tomasevic, 2013) with the possibility of creating a sense of aloofness. According to (Walter Leal Filho et al., 2018) across the globe, universities have served as climate change research centers (Leal Filho et al., 2018) a clear indicator of this being the establishment of research hubs and clusters specifically addressing issues of climate change. However, in spite of the centripetal role educational institutions play especially in developed climes, a fundamental lacuna subsists which is the failure of school curricula in some countries to address issues of climate change hence, producing climate uneducated graduates (Eilam, 2021). While West Africa cannot shy away from the dangers inherent in churning out climate illiterate graduates and must ensure she nips this in the bud, it is equally not cheering news to observe that despite the number of schools and training institutes in the sub-region, they have not been utilized as platforms for the dissemination of climate change education. This reinforces a gap prior researches have brought to the fore hinged on the insufficient adoption of educational structures and frameworks to tackle the global climate change conundrum (Mochizuki and Bryan, 2015).

Taking a cue from Nigeria which is West Africa's most populous state (Statista, 2023), granted that there have been calls for full-scale implementation of climate change education in all strata of her educational institutions, it is however worrisome to state that one major factor militating against this seems to be the use of an outmoded curriculum which has heightened calls for a curriculum review (Amanchukwu, 2015). That CCE is present in school curricula in Nigeria can be regarded as cheering news as in some countries, climate change education has not been introduced into the school curricula which explains why at different levels, there have been calls for its inclusion. For Eilam, apart from arguing for the inclusion of climate change in school curricula he opines that it should be a stand-alone discipline (Eilam, 2021). His position is anchored on the fact that when the discipline is disentangled from others, it will be more effective and functional.

Another key challenge of climate change education in West Africa is the weak capacity of the teachers who seem not to be fully schooled in the rudiments of environmental education hence the inhibition on their ability to educate their students. This reinforces Nebechi's and Okoro's position who alluded to the fact that successful teaching of climate change can only happen when the teachers are adequately equipped with requisite skills (Nebechi and Okoro, 2016).This is so true because, climate change education is a specialized form of education that requires expertise or sufficient knowledge of the subject matter. Hence, we make bold to state that the need for train the trainer programs in West Africa cannot be over-emphasized as they obviously cannot give what they do not have.

Studies have shown that climate change education when administered in formal settings has the capacity to equip individuals with the necessary skills to respond to climatic dysfunctionalities' (Vaughter, 2016). However, we must sound a caveat that CCE can take place both in formal and informal settings. Hence, when engaging in discourses bothering on climate change education, it is important to be explicit and distinguish between formal and informal education as CCE is multipurpose and multi-dimensional in nature.

Frameworks for Enhancing Climate Change Education in West Africa

A general consensus has been reached that there is a strong allusion among climate experts that human influence is causing global warming. (Malla et al., 2022), and despite the fact that most nations have a clearly defined vision for long - term sustainability founded on the Sustainable Development Goals, certain initiatives taken to realize that vision, particularly those that fall under the purview of academic curricula are still not fully realized in the continent. This is so, because many West African countries are facing difficulty posed by climate change, yet it is obvious that they lack the means of advancing mitigation and adaptation strategies to checkmate this challenge. This shortcoming calls for the strengthening of capacities in these countries which can only be achieved through knowledge. This makes it imperative to include climate change education in colleges and other higher educational institutions while fostering cooperation among various actors.

The framework for enhancing Climate Change education centers on the UNESCO Climate Change Education for Sustainable Development (CCESD) which is anchored on the fact that education offers opportunity to combat climate change. Under this framework, education will be used to promote principles and practice of sustainable development. Though UNESCO has a role to play in the implementation of this programme at the regional and national levels,The overall objective of National CCESD Programme is to support the capacities of education policy makers and teacher training institutions in Member States to strengthen their educational responses to mitigate and adapt to climate change. The specific objective of CCESD country programme is to support the development and implementation of CCESD programmes in cooperation with national responsible authorities and change agents to ensure that education systems can respond to the needs of climate change mitigation and adaptation. Under the framework, education authorities at the national levels with limited technical capacity to address DRR and Education in Emergencies (EiE), receives such support through CCESD project. The information base would help to assess the loss and damage of the educational institutions induced from disaster and climate change impacts; to know about the degree of DRR/CCA/CCE interventions taken from different corners for protecting schools/colleges from disaster and climate change impacts that would eventually percolate to working places and homes as more and more students gain this knowledge.

Environmental education, education on climate change, and education about sustainable development are separately taught but these three are now integrated into a single curriculum under CCESD brings to achieve the objective of mitigation and adaption to climate disasters in the continent. So far, both teachers and non-governmental organizations have been responsible for implementing climate education. In order to build integrated climate change knowledge in scientific research, daily life, the workplace, and academia, it is vital to incorporate climate change education into daily communication as well as upper secondary level and higher education. This has become expedient as education about climate change aids decision-makers in comprehending the significance and urgency of the implementation of enhanced strategies to combat climate change nationally. Communities equally gain knowledge about how climate change will impact them, how to decrease their own carbon footprint and how to protect themselves from its adverse repercussions. The transmission of information on climate change, in particular, aids in strengthening the resilience with already vulnerable communities, which are the ones most likely to suffer negative effects from it.

Climate change is a major and disruptive global phenomenon with wide-ranging effects, including repercussions on the health of people and ecosystems, sea level rise, drought and severe weather (IPCC 2014; Moraci et al., 2020). It poses difficulties for students, science educators and researchers due to its complexity and scope (Brickhouse et al., 2017). Students will have a sound knowledge of what the issues of climate change and sustainable development are if they acquire and accept different facets of sustainable education in the right and proper method and gradually, this will aid in raising their awareness of the necessity to practice a more environment-oriented lifestyle having been imparted systematically. This aspiration can only be attained through proper utilization of CCE curriculum content and the adoption of courses relevant to climate change which have great divergences across the countries.

Conclusion

The relevance of climate change education to environmental sustainability implies that climate sensitization and education can take place anywhere although engaging in this activity from institutions of learning seem to come with a great deal of advantages. It is now a necessity and highly important that climate change education in West Africa be localized and environmentally specific as the causal factors for climate change have more often than not proven to differ from one location to the other.

In addition to this is the need for more commitment from various West African governments to issues of climate change education by providing funding for the training of climate change educators. These educations when trained can train others which can help reduce the risk of climate change in the sub-region. It is our firm belief that if the necessary gaps are filled and loopholes addressed, the sub-region will surmount the problems of CCE and attain bright prospects.

References

Alexandru, A., Ianculescu, M., Tudora, E., & Bica, O. (2013). ICT challenges and issues in climate change education. *Studies in Informatics and Control, 22*(4), 349–358.

Amanchukwu, R. N., Amadi-Ali, T. G., & Ololube, N. P. (2015). Climate change education in Nigeria: The role of curriculum review. *Education, 5*(3), 71–79.

Amanchukwu, R. N. (2015). Climate change education in Nigeria: The role of curriculum review. *Education, 5*(3), 71–79.

Apollo, A., & Mbah, M. F. (2021). Challenges and opportunities for climate change education (Cce) in East Africa: A critical review. *Climate, 9*(6), 93.

Armstrong, A. K., & Krasny, M. E. (2020). Tracing paths from research to practice in climate change education. *Sustainability, 12*(11), 4779.

Bafana, B. (2022). Africa's effort to improve climate change literacy as its effects worsen across the continent. https://allianceforscience.org/blog/2022/04/africa-struggles-to-improve-literacy-about-climate-change-as-its-effects-worsen-across-the-continent/

Bangay, C., & Blum, N. (2010). Education responses to climate change and quality: Two parts of the same Agenda? *International Journal of Educational Development, 30*(4), 359–450. https://doi.org/10.1016/j.ijedudev.2009.11.011

Bodzin, A. M., Anastasio, D., Sahagian, D., Peffer, T., Dempsey, C., & Steelman, R. (2014). Investigating climate change understandings of urban middle-level students. *Journal of Geoscience Education, 62*(3), 417–430.

Boon, H. J. (2010). Climate change? Who knows? A comparison of secondary students and pre-service teachers. *Australian Journal of Teacher Education, 35*(1), 104–120.

Brannon, L., Gold, L., Magee, J., & Walton, G. (2022). The potential of interactivity and gamification within immersive journalism & interactive documentary (i-docs) to explore climate change literacy and inoculate against misinformation. *Journalism Practice, 16*(2–3), 334–364.

Brickhouse, N., McGinnis, J. R., Shea, N., Drewes, A., Hestness, H., & Breslyn, W. (2017). Core idea ESS3: Earth and human activity. In *Disciplinary Core Ideas: Reshaping Teaching and Learning. Arlington, VA: NSTA* (pp. 223–240).

Busby, J. (2018). Warming world: Why climate change matters more than anything else. *Foreign Affairs, 97*, 49.

Carter, K. (2019). Climate change goes G local. NCSE Blog (blog). Accessed on 15 Aug 2019. https://ncse.com/blog/2019/07/climate-change-goes-glocal-0018925

Charles, A. O., Anetor, F. O., & Akpabio, E. S. (2023). The effect of climate change on the agricultural sub-sectors in Nigeria: An autoregressive distributed lag (Ardl) approach. *The Journal of Developing Areas, 57*(2), 17–29.

Cherry, L. (2011). Young voices on climate change: The Paul F-Brandwein 2010 NSTA lecture. *Journal of Science Education and Technology, 20*, 208–213.

Cordero, E. C., Centeno, D., & Todd, A. M. (2020). The role of climate change education on individual lifetime carbon emissions. *PLoS ONE, 15*(2), e0206266.

Csoknyai, T., Legardeur, J., Abi Akle, A., & Horváth, M. (2019). Analysis of energy consumption profiles in residential buildings and impact assessment of a serious game on occupants' behavior. *Energy and Buildings, 196*, 1–20. https://doi.org/10.1016/j.enbuild.2019.05.009

Dhal, P.K. (2019). *Education for climate change, environmental sustainability and world peace.* Paper presented at the Building the World Parliament International Conference with the theme 'Climate Change and the Earth Constitution. O.P. Jindal Global University, Sonipat, India.

Douglas, B. D., & Brauer, M. (2021). Gamification to prevent climate change: A review of games and apps for sustainability. *Current Opinion in Psychology, 42*, 89–94.

Duruji, M. M., & Urenma, D. M. F. (2017). The environmentalism and politics of climate change: A study of the process of global convergence through UNFCCC conferences. In *Natural Resources Management: Concepts, Methodologies, Tools, and Applications* (pp. 77–108). IGI Global.

Duruji, M. M., Olanrewaju, F. O., & Duruji-Moses, F. U. (2018). From Kyoto to Paris: An analysis of the politics of multilateralism on climate change. In *Promoting Global Environmental Sustainability and Cooperation* (pp. 31–56). IGI Global.

Duruji, M. M., & Duruji-Moses, F. U. (2017). Legislative powers and constituency project In Nigeria's fourth republic. *South East Journal of Political Science, 3*(1).

Eilam, E. (2021). Climate change education: The problem with walking away from discipline. *Studies in Science Education, 58*(2), 231–264.

Fernandez, G., Thi, T. T. M., & Shaw, R. (2014). Climate change education: recent trends and future prospects. *Education for sustainable development and disaster risk reduction*, 53–74.

Field, E et al. (2020). How should climate change be taught in schools. *Sustainability.* Retrieved December 16, 2022 from https://www.edcan.ca

Leal Filho, W. (2021). *Handling climate change education at universities: an overview Springer Open.* http://creativecommons.org/licenses/by/4.0/

Flood, S., Cradock-Henry, N. A., Blackett, P., & Edwards, P. (2018). Adaptive and interactive climate futures: Systematic review of 'serious games' for engagement and decision-making. *Environmental Research Letters, 13*(6), 063005.

Hapgood, S., & Palincsar, A. S. (2006). Where literacy and science intersect. *Educational Leadership, 64*(4), 56.

Hestness, E., McDonald, R. C., Breslyn, W., McGinnis, J. R., & Mouza, C. (2014). Science teacher professional development in climate change education informed by the next generation science standards. *Journal of Geoscience Education, 62*(3), 319–329.

Hsu, C. L. (2022). Applying cognitive evaluation theory to analyze the impact of gamification mechanics on user engagement in resource recycling. *Information & Management, 59*(2), 103602. https://www.researchgate.net/publication/283081778

IPCC. (2007). Climate Change 2007: Impacts, adaptation and vulnerability. In M.L. Parry, O.F. Canziani, J.P. Palutikof, P.J. Van der Linden and C.E. Hanson (Eds.), *Contribution of Working Group II to the Fourth Assessment Report of the Intergovernmental Panel on Climate Change* (p. 976). UK: Cambridge University Press.

IPCC. (2014). *Climate Change 2014: Synthesis Report. Contribution of Working Groups I, II and III to the Fifth Assessment Report of the Intergovernmental Panel on Climate Change* [Core Writing Team, R.K. Pachauri and L.A. Meyer (Eds.)] (pp.151). IPCC, Geneva, Switzerland.

Kamenetz, A. (2019). *8 Ways to Teach Climate Change in Almost Any Classroom.* NPR. 2019. Accessed on 15 August 2019. https://www.npr.org/2019/04/25/716359470/eight-ways-to-teach-climate-change-in-almost-anyclassroom

Kamenetz, A. (2019) *Most Teachers Don't Teach Climate Change; 4 in 5 Parents Wish They Did.* NPR. 2019. Accessed on 15 August 2019. https://www.npr.org/2019/04/22/714262267/most-teachers-dont-teach-climate-change-4-in-5-par

Koivisto, J., & Hamari, J. (2019). The rise of motivational information systems: A review of gamification research. *International Journal of Information Management, 45*, 191–210.

Kolenatý, M., Kroufek, R., & Činčera, J. (2022). What triggers climate action: The impact of a climate change education program on students' climate literacy and their willingness to act. *Sustainability, 14*(16), 10365.

Krajcik, J. S., & Sutherland, L. M. (2010). Supporting students in developing literacy in science. *Science, 328*(5977), 456–459.

Læssøe, J., & Mochizuki, Y. (2015). Recent trends in national policy on education for sustainable development and climate change education. *Journal of Education for Sustainable Development, 9*(1), 27–43.

Leal Filho, W., Morgan, E. A., Godoy, E. S., Azeiteiro, U. M., Bacelar-Nicolau, P., Ávila, L. V., & Hugé, J. (2018). Implementing climate change research at universities: Barriers, potential and actions. *Journal of Cleaner Production, 170*, 269–277.

Magadza, C. Climate change impacts and human settlements in Africa: Prospects for adaptation. *Environmental Monitoring and Assessment* 61, 193–205.

Mazur-Stommen, S., & Farley, K. (2016). Games for grownups: The role of gamification in climate change and sustainability. *S. Mazur Stommen, & K. Farley, Taxonomy of games*, 28–39.

Mahato, A. (2014). Climate change and its impact on agriculture. *International Journal of Scientific and Research Publications, 4*(4), 1–6.

Malla, F. A., Mushtaq, A., Bandh, S. A., Qayoom, I., & Hoang, A. T. (2022). Understanding climate change: Scientific opinion and public perspective. *Climate Change: The Social and Scientific Construct* (pp. 1–20). Springer International Publishing.

Mbah, M. F., Shingruf, A., & Molthan-Hill, P. (2022). Policies and practices of climate change education in South Asia: Towards a support framework for an impactful climate change adaptation. *Climate Action, 1*(1), 1–18.

McCright, A., O'Shea, B., Sweeder, R., et al. (2013). Promoting interdisciplinarity through climate change education. *Nature Clim Change, 3*, 713–716.

McGinnis, J. R., McDonald, C., Hestness, E., & Breslyn, W. (2016). An investigation of science educators' view of roles and responsibilities for climate change education. *Science Education International, 27*(2), 179–193.

Mochizuki, Y., & Bryan, A. (2015a). Climate change education in the context of education for sustainable development: Rationale and principles. *Journal of Education for Sustainable Development, 9*(1), 4–26.

Molthan-Hill, P., Worsfold, N., Nagy, G. J., Leal Filho, W., & Mifsud, M. (2019). Climate change education for universities: A conceptual framework from an international study. *Journal of Cleaner Production, 226*, 1092–1101.

Monroe, M. C., Plate, R. R., Oxarart, A., Bowers, A., & Chaves, W. A. (2019a). Identifying effective climate change education strategies: A systematic review of the research. *Environmental Education Research., 25*(6), 791–812.

Monroe, M. C., Plate, R. R., Oxarart, A., Bowers, A., & Chaves, W. A. (2019b). Identifying effective climate change education strategies: A systematic review of the research. *Environmental Education Research, 25*(6), 791–812.

Monroe, C. M., et al. (2017). Identifying effective climate change education strategies: A systematic review of the research. *Environmental Education Research, 25*(6), 791–812.

Moraci, F., Errigo, M. F., Fazia, C., Campisi, T., & Castelli, F. (2020). Cities under pressure: strategies and tools to face climate change and pandemic. *Sustainability, 12*(18), 7743.

Nebechi, A. A., & Okoro, C. O. (2016). The teacher and the teaching of climate change: a case study of Obio-Akpor local government area of rivers state Nigeria. *Scientific Research Journal (SCIRJ), 4*(1).

Noor, M., Rehman, N. U., Jalil, A., Fahad, S., Adnan, M., Wahid, F., & Hassan, S. (2020). Climate change and costal plant lives. *Environment, Climate, Plant and Vegetation Growth*, 93–108.

Olubebe, N. (2015). P 2015, Climate change education in Nigeria: The role of curriculum review. *Education, 5*(3), 71–79. https://doi.org/10.5923/j.edu.20150503.01

Peter, R. B., Kasper, T. S., Amanda, H. (2022). *European climate change curriculum national challenges and opportunities in a European context*. https://climateperspectives.eu/wp-content/uploads/2022/10/ECCC-e-book-final-version.pdf

Pielke, R. A., Jr. (2004). What is climate change? *Energy & Environment, 15*(3), 515–520.

Pruneau, D., et al. (2010). Challenges and possibilities in climate change education. *US-China Education Review, 7*(9), 15–24.
Rajanen, D., & Rajanen, M. (2019). Climate change gamification: A literature review. *GamiFIN*, 253–264.
Senevirathne, M., Amaratunga, D., Haigh, R., Kumer, D., & Kaklauskas, A. (2022). A common framework for MOOC curricular development in climate change education-findings and adaptations under the BECK project for higher education institutions in Europe and Asia. *Progress in Disaster Science, 14*, 100222.
Seow, T., & Ho, L. C. (2016). Singapore teachers' beliefs about the purpose of climate change education and student readiness to handle controversy. *International Research in Geographical and Environmental Education, 25*(4), 358–371.
Sharma, A. (2012). Global climate change: what has science education got to do with it? *Science & Education, 21*, 33–53.
Sibanda, A., & Manik, S. (2022). Reflecting on climate change education (CCE) initiatives for mitigation and adaptation in South Africa. *Environmental Education Research*, 1–18.
Siemen, S. (2018). *Climate change education in Latin America.* https://www.siemens-stiftung.org/wpcontent/uploads/medien/publikationen/publicationclimatechangeeducationinlatinamerica-siemensstiftung.pdf
Simpson, N. P., Trisos, C., Krönke, M., & Andrews, T. M. (2021). Africa's first continent-wide survey of climate change literacy finds education is key. https://reliefweb.int/report/world/africa-s-first-continent-wide-survey-climate-change-literacy-finds-education-key
Simpson, N. P., Mach, K. J., Constable, A., Hess, J., Hogarth, R., Howden, M., Lawrence, J., Lempert, R. J., Muccione, V., Mackey, B., New, M. G., & Trisos, C. H. (2021). A framework for complex climate change risk assessment. *One Earth, 4*(4), 489–501.
Sorgho, R et.al. (2020). Climate change policies in 16 West African countries: a systematic review of adaptation with a focus on agriculture, food security, and nutrition. *International Journal of Environmental Research and Public Health, 17*(23).
Sorgho, R et al. (2020). Climate change policies in 16 West African countries: a systematic review of adaptation with a focus on agriculture, food security, and nutrition. *International journal of environmental research and public health, 17*(23), 8897.
Statista (2023). African countries with the largest population as of 2020. Accessed on 16 February 2023. https://www.statista.com.
Sylla, M. B et.al (2016). Climate change over West Africa: Recent trends and future projections, Chapter 3, In Yaro, J. A & Hesselberg, J (Eds.), *Adapatation to climate change and variability in rural West Africa.* Springer Cham Switzerland.
Szczepankiewicz, E. I., Fazlagić, J., & Loopesko, W. (2021). A conceptual model for developing climate education in sustainability management education system. *Sustainability, 13*(3), 1241.
Tandon, A. (2022). West Africa's deadly rainfall in 2022 made '80 times more likely' by climate change, Carbon Brief. Accessed from https://www.weforum.org
Tomasevic, G. (2013). *Climate change now included in US curriculum.* Accessed on 23 January, 2023. http://rt.com/usa/climate-change-curriculum-school-653/
UNESCO & UNFCCC. (2016). *Action for Climate Empowerment: Guidelines for Accelerating Solutions Through Education, Training and Public.* Paris: UNESCO and UNFCCC. Accessed on 21 January, 2023. https://unfccc.int/sites/default/files/action_for_climate_empowerment_guidelines.pdf
UNESCO. (2020). Global education report, UNECSO Global education monitoring report, 2020: Inclusion and education: All means all, easy to read version, key messages, recommendations—UNESCO Digital Library.
UNFCCC. (1992). U.N. Doc. A/AC.237/18, reprinted in 31 I.L.M. 849 United Nations Framework Convention on Climate Change.
UNFCCC. (2020). Annual report 2020. UNFCCC.

UN. (2023). With climate crisis generating growing threats to global peace, security council must ramp up efforts, lessen risk of conflicts, speakers stress in open debate, meetings coverage and press releases. Accessed on April 8, 2023. https://press.un.org

UNEP. (2023). Climate change challenges for Africa: Evidence from selected EU-funded research projects, Report. Accessed on April 8, 2023. https://www.unep.org/

UNICEF. (2023). UNICEF's Children climate risk index places children from WCA as the most at risk. Accessed on April 8, 2024. https://www.unicef.org

UNISDR. (2021). *United Nations Office for Disaster Reduction 2021 Annual report.*

Vaughter, P. (2016). *Climate change education: from critical thinking to critical action.* Policy Brief No 4, Institute for the advanced study of sustainability, United Nations University.

Wee, S. C., & Choong, W. W. (2019). Gamification: predicting the effectiveness of variety game design elements to intrinsically motivate users' energy conservation behaviour. *Journal of Environmental Management, 233*, 97–106. https://doi.org/10.1016/j.jenvman.2018.11.127

Wade, J. A. (1999). Students as environmental change agents. *International Journal of Contemporary Hospitality Management, 11*(5), 251–255.

Wouters, P., Nimwegen Van, C., Oostendorp Van, H., and Spek Vander, E D. (2013). A meta-analysis of the cognitive and motivational effects of serious games. *Journal of Educational Psychology, 105*(2), 249.

Yi, J., & Wu, P. (2009). *Climate change and sustainable development: The response from education.* Beijing Normal University.

Chapter 17
Climate Change Education in African Higher Education Institutions: Insights into Current Practices and Future Directions

Perez L. Kemeni Kambiet and Marcellus Forh Mbah

Abstract Climate change education (CCE) is a vital tool for addressing Africa's vulnerability to climate change and advancing sustainable development. Its impact is even more significant when integrated into higher educational systems. Notably, Higher Education serves as a catalyst for innovation and nation-building and, hence, an essential medium for driving sustainability efforts across present and future generations. However, evidence suggests that the adoption of CCE in African Higher Education Institutions (HEIs) remains low. Few HEIs have specialized courses on the subject, while others have an integrated approach within related subjects like geography and environmental sciences. This underscores the necessity to identify the factors behind the slow uptake of CCE across the continent. Given this premise, our study seeks to explore the nature of the barriers as well as existing and future opportunities for scaling up climate change education in African HEIs. Among other factors, we identified resource limitations, institutional barriers, and socio-political challenges as the primary constraints to CCE uptake in Africa. On the other hand, the increasing climate commitment and the development of regional climate institutions equally provide a unique opportunity for capacity building, knowledge exchange and the spread of CCE across African HEI. Moreover, the rapid development of affordable digital communication and computing has the potential to increase networking and collaborative efforts between climate change learners and scientists across the region. These prospects signal a need for regional governments to multiply their efforts toward providing an environment for the development and uptake of suitable CCE programs.

Keywords Climate change education · Higher education · Sustainable development · Universities

P. L. Kemeni Kambiet (✉)
International Water Research Institute (IWRI), Mohammed VI Polytechnic University (UM6P), Ben-Guerir, Morocco
e-mail: perez.kambiet@um6p.ma

M. F. Mbah
Manchester Institute of Education, University of Manchester, Manchester, UK

M. F. Mbah et al. (eds.), *Practices, Perceptions and Prospects for Climate Change Education in Africa*, https://doi.org/10.1007/978-3-031-84081-4_17

Introduction

Climate change education (CCE) is crucial for the fight against climate change and its impact is far-reaching (Mochizuki and Bryan 2015). Fundamentally, CCE increases awareness of the existing threat which is a necessary precondition for effective adoption of new mitigation and adaptation technologies (Apollo & Mbah, 2021; Sato & Kitamura, 2023). Moreover, it provides a flexible ecosystem for the generation of new knowledge that may enable mitigation and adaptation overtime (Kagawa & Selby, 2012). Its dynamic and progressive merits make it the most malleable and adaptable to diverse situations regardless of spatiotemporal differences. Besides, CCE also presents communities with new knowledge and perspectives which enhances engagement with the phenomenon of climate change (Reid, 2019). Additionally, when channeled through higher educational systems, CCE can potentially make sustainability ambassadors out of those who go through, hence guaranteeing a long-term commitment to making the world safer (Mbah et al., 2022).

According to Aleixo et al. (2016), Higher Education Institutions (HEI) provide an ideal medium for developing a new generation of environmental stewards (Goyal et al., 2023; Kapitulčinová et al., 2017). In concept, the prospects of CCE in HEI are enormous partly because of the notion that Higher Education is charged with equipping the national and international labour force for all industries (Kapitulčinová et al., 2017; Maiya & Aithal, 2023; Ritcher and De Sousa, 2019). The wide range of courses and learning environment in HEIs provides an ideal avenue for the acquisition and transmission of specialized knowledge. When acquired, this knowledge enable individuals (students) to strive in various economic, social, and political settings. In context, CCE courses usually stimulate climate awareness and environmental consciousness while providing knowledge of feasible adaptative measures. This relieves future mitigation and adaptation anxiety while accelerating the adoption of relevant technologies in climate-hit communities. Thus, integrating CCE into higher educational systems could imply an effective approach to fighting short, mid, and long-term climate change. Secondly, Higher Education systems are a "youth assembly" (largely composed of youths) and an integration of CCE could provide youths with critical reasoning skills for the development of future mitigation and adaptation measures (Tilbury, 1995). Higher Education Institutions are also perfect grounds for sustainable development theory experimentation. Finally, education equally plays a crucial role in shaping dynamic climate action and environmental conservation behaviour among youths (Leal Filho et al., 2018a). Higher Education Institutions like universities are also potential incubators and testing units for countless ideas and operations which are tailored towards enhancing sustainability while reducing the carbon footprint of similar organizations.

Yet, despite the highlighted benefits of CCE, its adoption in African Higher Education systems is slow (Damoah, 2023; Leal Filho et al., 2018b; Molthan-Hill and Mbah, 2022). UNESCO 2021 notes that only 53% of African states have integrated climate change education into their curriculum. Among other challenges, they highlight that over 70% of instructors and educators had a poor understanding of the climate change concept. This begs two questions: why is CCE engagement low? And what opportunities exist for improving CCE engagement in African higher institutions? Literature suggests that such sluggishness is associated with peculiarities within the African Higher Education system. In many cases, authors fail to provide information about the nature (their primary aim and structure) of available CCE programs. Equally, there is little information on the effect of eLearning and other digital communication tools on the integration of CCE on HEI in Africa. Meanwhile, to drive more effective CCE programs, policy makers require extensive information on the nature and short falls of current CCE programs. Based on the above considerations, this work seeks to explore the nature, challenges and opportunities associated with implementing CCE programs in HEIs in Africa. By this, we contribute to the fast-growing literature on CCE while guiding the development and implementation of future climate education approaches for Africa.

Overview of Climate Change Education Efforts in African Higher Education Institutions

Arguably, the integration of climate-related programs in African HEIs has been a progression of two stages. The initial stage included the integration of environmental education into HEIs while the latter included more climate-specific programs. Fundamentally, these stages are either continuous progressions (in situations where a course on the environment transforms into a climate-specific course) or discrete progressions (when both programs develop independently of each other) depending on the context (country). Many countries have progressed in the discrete CCE pathway. Where, environmental programs and climate programs were developed independently. Accordingly, African countries started integrating environmental and sustainable development in their educational programs as early as the 1992 UNESCO education for sustainable development agenda (Agbedahin, 2019; Manteaw, 2012). These initial programs highlighted the relationship between human activities, the environment and (to a limited extent) climate change. Among other benefits, these courses served as a baseline for the development of an environmental consciousness among youths. However, they failed to emphasize the critical implications of climate change and the vulnerabilities of various communities in Africa. Of course, it may be logical especially since the period of their implementation coincides with the periods of high climate scepticism (and denial) globally. To a wider extent, some aspects of climate change education can be found sandwiched into other environmental-related courses and programs in universities and other higher institutes of learning all around Africa (Ritcher and De Sousa, 2019; Tilbury, 2011). Notably, most African universities offer some semblance of climate education within courses in environmental sciences, natural resource economics, geography, and climatology. On the other hand, the increased awareness and global acceptance of climate change served as a propeller for the development of more climate-specific programs.

Table 17.1 summarizes the outcome of our search of web-accessible CCE programs in African regional HEIs. The catalogue of HEIs (Table 17.1) provides a list of the currently operational mainstream (Bachelor, Masters and Doctoral) and short course programs in Africa. The catalogue equally highlights the country, specific HEI offering the courses, the nature of the courses offered and an accessible link. From the catalogue, we observe forty-one (thirty-five mainstream and six short) climate change programs. These programs are hosted by twenty-seven (out of over a thousand recognized) African Universities. Most notably, we note eighteen Master's programs and fifteen Doctoral programs as well as Bachelor programs are offered in eighteen of the fifty-four African nations. To provide more perspective into CCE in African HEIs, we further associated each course with their first cohort enrolment data and reclassified them under subregions (Fig 1b).

From Fig. 17.1a, we observe that 2013 (which coincides with the year of the recognition as a major challenge under Agenda 2063) records the largest number of implemented CCE programs in Africa. Beyond this, it is also observed that the implementation of HEI climate-specific programs has witnessed both high and low periods since 2012 with three intermediate peaks in 2015, 2019 and 2022. Given that Fig. 17.1b highlights a high number of Masters and Doctoral programs, it is possible that these intermediate peaks are associated with years of implementation of new programs whereas the troughs correspond to years of evaluation. In addition to the most recurrent Masters and Doctoral programs, we observed a few Bachelor's programs and short courses. Precisely we note only two Bachelor's programs with one in South Africa at Durbin University and another at Mesano University in Kenya (Fig. 17.1b and Table 17.1 in the appendix). We also note six short courses including one in central Africa (Rwanda), two in East Africa (Kenya and Tanzania) and three in south Africa (Zimbabwe and Tanzania). Visibly, some regions fare better than others in the implementation of CCE programs. West Africa records 44% followed by East (27%) and South Africa (23%) with the least regions being the central (2%), north Africa (2%) as well as some islands like Cabo Verde (2%) (Fig. 17.1b). Typically, regardless of the country or subregion, mainstream programs characteristically span two (for Masters) to four years (for Doctoral and Bachelor's programs) and are designed to include initial course work and some form of dissertation or thesis. This approach ensures that learners are imparted with a wide range of course materials that inform a better perspective of the climate problem, but the requirements are somewhat strict for many youths. Besides some of the most affected groups of youths are found in areas with little access to education that will enable them to integrate such university programs. In perspective, quality CCE programs (CCE programs that increase the sense of security among learners) carry four aspects including a cognitive, socio-emotional, action-oriented and a justice aspect. The UNESCO Global Education Monitoring Report (UNESCO, 2023) notes that most African universities and institutions have made progress in enhancing cognitive aspects of climate change with little consideration of other aspects related to socio-emotional, action-oriented and justice. Meanwhile, these aspects are the main propellers of sustainable attitudes in many communities.

Table 17.1 Catalogue of climate change programs in African higher education institutions

S/n	Country	University	Program	Source (Accessed: March 10, 2024)
I. Mainstream [Bachelors, Master snad Doctoral] Programs				
1.	Algeria	The Pan African University Institute of Water and Energy Science (PAUWES) in Abou Bakr Belkaïd University of Tlemcen	Masters in Climate Change (MCC) [Training for better communication, knowledge management and visualisation of complex datasets, scenarios, and models]	Pan African University Institute for Water and Energy Sciences (incl. climate change) - Institute for Environment and Human Security (unu.edu)
2.	Benin	WASCAL, BMBF and The University of Abomey-Calavi	Doctoral Program Climate Change and Water Resources [Training in Climate change and Water resource management]	Doctoral Programme Climate Change and Water Resources – WASCAL
3.	Burkina Faso	WASCAL, German Ministry of Education and Research (BMBF) and University Ouaga 1 Pr. Joseph Ki-Zerbo, Ouagadougou	Master Research Program in Informatics for Climate Change [Training on adequate scientific computation and climate data management]	Master Research Programme on Informatics for Climate Change – WASCAL
4.	Cabo Verde	WASCAL, BMBF and The Atlantic Technical University	Master Research Program on Climate Change & Marine Sciences [Training in management of climate change for marine environments]	Master Programme Climate Change and Marine Sciences – WASCAL

(continued)

Table 17.1 (continued)

S/n	Country	University	Program	Source (Accessed: March 10, 2024)
5.	Cote d'Ivoire	WASCAL, BMBF, Université Felix Houphouet Boigny	Doctoral Program Climate Change and Biodiversity (CCB) [Understanding and protecting species richness, genetic diversity, ecosystems and ecosystem services]	Doctoral Programme Climate Change and Biodiversity – WASCAL
6.	Ethiopia	Ethiopian Civil Service University (ECSU)	Master of Sciences (MSc) Program in Environment and Climate Change	l_www.ecsu.edu.et
7.	Ethiopia	Haramaya University	ACE Climate SABC Masters in Climate Smart Agriculture and Biodiversity Conservation PhD in Climate Smart Agriculture and Biodiversity Conservation	Climate SABC – Climate SABC (haramaya.edu.et)
8.	Gambia	West African Science Service Centre on Climate Change and Adapted Land Use (WASCAL) and the University of the Gambia	Doctoral Research Program in Climate Change and Education Master Program on Climate Change and Education	Doctoral Programme Climate Change and Education – WASCAL WASCAL Master Programme on Climate Change and Education - Institute for Environment and Human Security (unu.edu)

(continued)

Table 17.1 (continued)

S/n	Country	University	Program	Source (Accessed: March 10, 2024)
9.	Ghana	University of Ghana	PhD In Climate Change And Sustainable Development Master of Science in Climate Change and Sustainable Development Master of Philosophy in Climate Change and Sustainable Development	https://www.ug.edu.gh/announcements/call-applicants-phd-climate-change-and-sustainable-development Master of Science (Climate Change and Sustainable Development) | University of Ghana Business School (ug.edu.gh)
10.	Ghana	WASCAL, BMBF and Kwame Nkrumah University of Science and Technology (KNUST)	Doctoral Program Climate Change and Land Use [Training in Managing climate change with Remote data management and Geographical Information tools]	Doctoral Programme Climate Change and Land Use – WASCAL
11.	Kenya	University of Nairobi	Doctor of Philosophy in Climate Change and Adaptation Masters in Climate Change and Adaptation	DOCTOR OF PHILOSOPHY IN CLIMATE CHANGE AND ADAPTATION | Department of Earth & Climate Science (uonbi.ac.ke) MASTER OF CLIMATE CHANGE ADAPTATION (MCCA) | Department of Earth & Climate Science (uonbi.ac.ke)
12.	Kenya	Kenyatta University	Master of Environmental Studies (Climate Change and Sustainability)	MASTER OF SCIENCE (CLIMATE CHANGE AND ENVIRONMENTAL SUSTAINABILITY) (ku.ac.ke)

(continued)

Table 17.1 (continued)

S/n	Country	University	Program	Source (Accessed: March 10, 2024)
13.	Kenya	CGAIR and Taita Taveta University	Master of Science in Climate-Smart Agriculture [Training to enhance productivity, resilience and adaptive capacity]	Master of Science in Climate-Smart Agriculture Curriculum I AICCRA (cgiar.org)
14.	Kenya	Maseno University	Bachelor of Science in Climate Change and Development with IT	Bachelor of Science in Climate Change and Development with IT I Maseno university - School of education
15.	Kenya and Cote d'Ivoire	DAAD, University of Nairobi (UoN) And Université Félix Houphouët-Boigny (UFHB)	The African Climate and Environment Center – Future African Savannas (AFAS) Masters program in Nature Based Solutions regarding Climate Change Adaptation and Biodiversity Conservation in African Savannas	AFAS Graduate Programme - AFAS
16.	Mali	WASCAL, BMBF, University of Bamako and the University of Cape Coast (UCC) Ghana	Doctoral Program Climate Change and Agriculture	Doctoral Programme Climate Change and Agriculture – WASCAL
17.	Namibia	International University of Management	Master of Science in Climate Change Mitigation and Adaptation	Master of Science in Climate Change Mitigation and Adaptation - IUM (edu.na)
18.	Niger	WASCAL, BMBF and Université Abdou Moumouni- Niger	Doctoral Research Program on Climate Change and Energy	Doctoral Programme Climate Change and Energy – WASCAL

(continued)

Table 17.1 (continued)

S/n	Country	University	Program	Source (Accessed: March 10, 2024)
19.	Nigeria	University of Nigeria	Doctor of Philosophy (Ph.D) Degree Program in Climate Change Studies Masters of Science (M.Sc) Degree Program in Climate Change Studies	INSTITUTE-OF-CLIMATE-CHANGE.pdf (unn.edu.ng)
20.	Nigeria	WASCAL, BMBF and The Federal University of Technology, Akure (FUTA)	Doctoral Program West African Climate Systems [Training in Meteorology and Climatology]	Doctoral Pogramme West African Climate Systems – WASCAL
21.	Nigeria	WASCAL, BMBF and Federal University of Technology, Minna (FUT Minna)	Doctoral Program Climate Change and Human Habitat	Doctoral Programme Climate Change and Human Habitat – WASCAL
22.	Senegal	Cheikh Anta Diop University of Dakar (UCAD)	Doctoral Program Climate Change Economics (CCEcon) [Training on Effective Climate Change Adaptation and Mitigation]	Doctoral Programme Climate Change and Economics – WASCAL
23.	South Africa	Durblin City University	MSc in Climate Change: Policy, Media and Society BA. Climate and Environmental Sustainability	MSc in Climate Change: Policy, Media and Society I DCU Climate and Environmental Sustainability I School of History and Geography I Dublin City University (dcu.ie)

(continued)

Table 17.1 (continued)

S/n	Country	University	Program	Source (Accessed: March 10, 2024)
24.	South Africa	University of Capetown/ African Climate and Development Initiative (ACDI)	The ACDI Masters in Climate Change and Sustainable Development MSc/MPhil specializing in Climate Change and Sustainable Development The ACDI PhD Scholarships in Climate risk, Resilience, and Sustainable Development	ACDI Masters in Climate Change and Development ǀ University of Cape Town (uct.ac.za) Overview: Masters in Climate Change & Sustainable Development ǀ University of Cape Town (uct.ac.za) PhD Scholarships in Climate risk, Resilience, and Sustainable Development ǀ University of Cape Town (uct.ac.za)
25.	Tanzania	University Of Dar Es Salaam Centre For Climate Change Studies (Cccs)	MSc. Climate Change and Sustainable Development PhD in Climate Change and Sustainable Development	20240226_120151_Advertisement for UDSM Postgraduate Programmes 2024–2025 Academic Year.pdf
26.	Togo	WASCAL, BMBF and The University of Lomé (UL)	Doctoral Research Program on Climate Change and Disaster Risks Management [Training in Climate Change and Disaster Risk Management]	Doctoral Programme Climate Change and Disaster Risks Management – WASCAL
27.	Uganda	Makerera University	MSc climate change	Climate Change ǀ Makerere University Courses

(continued)

Table 17.1 (continued)

S/n	Country	University	Program	Source (Accessed: March 10, 2024)
II. Short-course programs				
1.	Kenya	UN Climate Change and Nairobi work programme (NWP)	UN Climate Change and Universities Partnership Program. Provides opportunity for graduate students to work closely with local, national, and regional partners to undertake a research project as a part of producing their master's thesis	UN Climate Change and Universities Partnership Programme (unfccc.int)
2.	Rwanda	United Nation University	Climate Academy aimed to enhance understanding of climate change and foster the science-policy-action interface	Climate Academy 2022 – Digitalization, Energy Transition and Climate Action - Institute for Environment and Human Security (unu.edu)
3.	South Africa	DAAD/ SASSCAL (Southern African Science Service Centre for Climate Change and Adaptive Land Management) and WASCAL	The DAAD climapAfrica program - Climate change research Alumni and Postdocs in Africa for future leaders in the field of climate research and protection	DAAD climapAfrica - Climate change research in Africa - DAAD
4.	South Africa	Dublin City University	INTRA internship program which gives Undergraduates a chance to work on real climate and environmental challenges	INTRA Internships I Dublin City University I INTRA (dcu.ie)

(continued)

Table 17.1 (continued)

S/n	Country	University	Program	Source (Accessed: March 10, 2024)
5.	Tanzania	University Of Dar Es Salaam Centre For Climate Change Studies (Cccs)	Short Courses Programs Climate Change and Sustainable Development	UDSM Admission
6.	Zimbabwe	Africa Initiative on Climate Change (AYICC)	Climate change Virtual School for Youths [Uses telecommunication technology to empower African youths to better understand climate change and environmental management approaches]	Climate Change Virtual School for Youths in Africa – Zimbabwe | UNFCCC

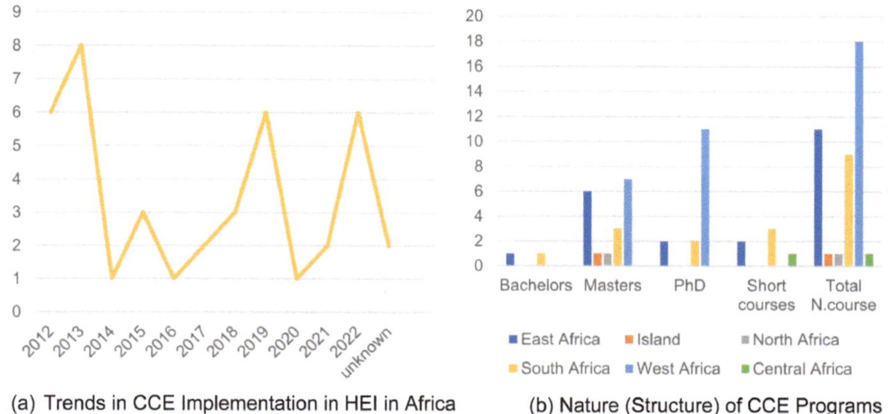

Fig. 17.1 Distribution of CCE programs in African HEIs **a** trends in CCE implementation in HEI in Africa **b** nature (Structure) of CCE programs. *Source* Computed from the outcome of Authors' search of web-accessible (implemented programs that have application websites) CCE programs in HEIs of each subregion in Africa, 2024

Stevenson and Peterson (2016) note that quality CCE education is gauged from its ability to help learners overcome feelings of grief, anxiety, denial, and apathy. This is intuitively linked to the socio-emotional, action-oriented and justice-focused aspects of climate change education. As observed from Table 17.1, African CCE programs are highly class-based. Hence, they are less desirable even to the vulnerable climate-anxious youths (Hickman et al., 2021). Similarly, the master's programs are research-based which restricts the action scope of learners to a sort of one-size-fits-all view of climate problems. These lapses in African HEI represent the lost opportunities and benefits of CCE. Many African HEIs incur higher costs over benefits of CCE. This is especially visible when training is targeted towards older age groups (Master's or Doctoral levels). To complement the lack of group specific climate change courses, shorter courses have been developed. Shorter courses require fewer preliminary conditions, but unfortunately their availability is still limited. Other HEIs in Africa have rather committed their efforts in research focused climate change programs.

Some African universities have established centres for environmental monitoring and control (Leal Filho et al., 2018a). For example, the University of Nigeria has set up a Centre for Environmental Management and Control (CEMAC) and a Centre for African Climate Change Adaptation Initiative (ACCAI). In South Africa, Rhodes University has set up the Environmental Learning Research Center (ELRC). Other South Africa universities like the University of South Africa, Pretoria, Witwatersrand, and Cape town have established departments for environmental studies and policy which research various aspects of climate change. Similar departments of environmental sciences are present in other African sub-regions. These research centres play critical roles in the (re)design of CCE programs. Firstly, they interact with communities (through surveys, workshop and local consultations) to produce research with policy relevance. Secondly, research institutions attract financial resources that are necessary to enhance the quality of research in mainstream study programs. Lastly, the centres also serve as conducive mediums through which new teaching approaches and innovations can be prototyped and tested. (Ritcher and De Sousa, 2019), highlights multiple research projects at the North-Western University which seek to investigate the possibilities of integrating environmental education for sustainability in primary and secondary schools in south Africa. Similar research on climate change can be seen in Cameroonian and Nigerian universities.

Interuniversity networks equally play a critical role in subsidizing the training cost and fostering collaborations between climate scientists in Africa. Frameworks such as the RUFORUM, the East African Inter-university Council, the African Research University Alliance and the Tertiary Education and Research Network of South Africa usually finance small and medium-sized research projects which permit researchers to live local realities from other regions through increased research collaborations (Sehai and Lemma, 2008). Ideally, interuniversity platforms permit researchers to interact and learn from shared research experiences. This accelerates collaboration and development of high impact policy driven research across Africa. Equally, there exist many inter-African scholarships and sandwich programs on environmental, economics and earth sciences which are tailored to enhance knowledge sharing. Other institutional development alliances also play similar roles. The South African Development Commission has played a key role in increasing the possibility of Indigenizing knowledge for climate change. Indigenizing climate change knowledge involves the development of climate-based solutions that suit the local context. Placing Indigenous knowledge at the centre of human development, South Africa's 2004 Indigenous knowledge policy has placed South Africa at the forefront of developing the protocol for Indigenous knowledge research (Kaya, 2013). Though currently not totally settling, many authors accept that Indigenous knowledge systems may play a critical role in designing more efficient mitigation and adaptation that could respond to the location specificities of the impacts of climate change. This is partly based on some preliminary notions which include the idea that people usually adopt strategies that are identifiable to their tribe and communities. Plus, many African cultures have emerged from a long progressive adaptation to various climatic factors. As such, these communities may hold some unexplored knowledge that can assist in the development of more efficient mitigation and adaptation strategies.

Challenges that Underpin Climate Change Education in African Higher Education Institutions

Among others, experts agree that a key challenge for climate change education in most African HEIs is its perceived overlap with established science disciplines like geography, meteorology, physics, biology, and environmental sciences.

This perceived lack of differentiation further translates into three interconnected sub-challenges: institutional barriers, resource constraints, and regional limitations.

(a) Resource Constraints

In theory, CCE programs are designed to respond to the impacts of climate change whereas, climate change is an evolving global phenomenon with local impacts (time and context varying concept) that cuts across social, economic, and political sectors of all nations. This peculiar characteristic (of climate change) has elicited the need for CCE program to constantly modify their content to suit the evolving climate narrative. In other words, the evolving and context-specific nature of CCE requires instructors to constantly research and develop new course material to meet the evolution of climate change (Alber, 2018). Plus, non-CCE programs are less context and time-specific. Particularly, courses like geography, physics, chemistry, climatology, mathematics, and other exact sciences have fundamentally maintained the same frameworks over a long period hence making the content development process easier. For example, trigonometry in mathematics and the types of winds in geography have remained the same over the years which makes it possible for researchers to easily develop course content with minimum effort.

Additionally, authors such as Tilbury (2011), Reid (2019) and Mubanga et al. (2022) highlight a general misconception of climate change among HEI climate instructors. In fact, only about 25% of educators of climate change in Zimbabwe attested to have some worth training in the science (Dzvimbo et al., 2022; Mubanga et al., 2022). Such trends are also seen in other African subregions. Moreover, the intersectoral implications (multi-sectoral impact) of climate change raise an imperative need for cross-disciplinary collaboration (Kaya, 2013; Ritcher and De Sousa, 2019; Reid, 2019). Meanwhile, prior to the current climate challenge, many instructors were trained to deliver courses that lay less emphasis on multidisciplinary research. As such, many HEI instructors face difficulties in embracing the new need for multidisciplinary research.

Beyond the complexities around CCE programs, HEIs in many African countries are inherently highly understaffed (Mubanga et al., 2022). In concept, understaffed institutions may have well-trained professionals who are enthusiastic about developing CCE programs but are overburdened by the workload of their current main course. In many understaffed universities, it is possible to see an instructor who is responsible for more than three courses with tens of thesis and dissertation supervision students (Dzvimbo et al.,2022). This makes it difficult for the available staff to allocate time for teaching or integrating new materials into their current course content. Oftentimes, in such cases, instructors usually turn to shy away from handling climate change courses because it requires enormous efforts to develop new course content (Reid, 2019; Tilbury, 2011).

Furthermore, because African HEIs are mostly government-owned they turn out to be underfinanced whereas the relative novelty of the science demands enormous research to drive decisions. Additionally, only 0.47% of the public investment budget of in African countries is allocated to research endeavors. As such many African universities lack adequate funding and resources to set up proper laboratories, acquire equipment, and employ experienced technologists. Tilbury (2011) and Reid (2019), argue that HEI responds to underfinancing by readjusting priorities which might sometimes imply cutting research resources while prioritizing staff remuneration. To some extent, this approach seeks to maintain the commitment of staff to educating learners. Institutions which turn to international sources of funding also complain of an inherent constraint in research agenda where most funding agencies determine the nature of research. Such funding generally defeats the initial reasons for the funding of interest while driving diverging research agendas.

(b) Institutional Barriers

Regardless of the right personnel, institutionalizing the subject still requires a consensus on the nature of a HEI climate change course program in Africa. At present, there is little consensus on the basis for such a program thus making institutionalization more difficult for universities (Reid, 2019; Tilbury, 2011). Also, there is little understanding of the most effective way to deliver climate change education materials (Mubanga et al., 2022). Studies (Ritcher and De Sousa, 2019; Rousell, Cutter-Mackenzie-Knowles, 2020) in developed nations indicate higher engagement from alternate education including games, interactive learning as well as excursion and field visits of museums and cultural heritages whereas in Africa, climate programs are widely teacher-based which might limit engagement and learning. Interactive games and practical experimentation on data build a sense of proximity to the climate problem that is not possible using traditional teaching approaches. Unfortunately, the use of such interactive games is highly dependent on available soft and hard infrastructure such as access to the internet, meteorological stations, databanks (trail datasets) and laboratories (Dzvimbo et al., 2022; Mubanga et al., 2022). Of course, combining current mainstream and alternative teaching seems promising even for Africa but it lacks the critical infrastructures that are needed to guarantee a low-cost CCE program. Without such infrastructure, even presumably "easy" solutions like excursions and field visits usually require learners to make some form of financial commitment. These are therefore deemed costly and undesirable, especially in the absence of subsidies for the less financially viable.

Conversely, university education is already subsidized (to accommodate the less financially privileged) in many African countries (World Bank, 2010). In the same light, the cost of stable internet access in many parts of Africa makes it difficult for most students to take advantage of the availability of interactive educational software. Many interactive games are simulated on the internet or require downloading and purchase of license keys. All of which are financially demanding to educational institutions and learners who are already financially constrained. Also, data-related practical programs require location-specific data from meteorological stations and data centres, but such centres are still rare. In cases where meteorological stations exist, they are either too sparsely distributed to collect representative data or are poorly maintained such that the data generated are arguably unreliable. This absence of meteorological data makes it extremely difficult to implement practical exercises with learners. Moreso, these data challenges have made it difficult for climate scientists to develop or experiment with African climate realities (Conway and Schipper, 2011). Some authors have resulted to depend on freely available satellite data but again the cost of stable internet may interfere with learner's desire to study. Partly because satellite data are stored in cloud storages which require internet access and a large offline storage unit for download. These are again not readily available for researchers in many African universities hence authors are forced to pay using (limited) personal resources. Such conditions equally reduce both learners' and instructors' interest in (studying) incorporating more practical courses into current CCE programs.

Furthermore, most African HEIs are government institutions and administrative positions are obtained through government appointment. This makes universities to be at the centre of political machinations (Dzvimbo et al., 2022; Mubanga et al., 2022). In some cases, appointment is not based on merit but rather based on political affiliations thus making plans and strategies unattainable. Besides, most appointments are short-lived. University and general HEI bureaucracy sometimes cause delays in the approval of new courses or programs into the general curriculum. Such political machines usually cause delays and even stiffening of some strategic visions and goals. Finally, Mubanga et al., (2022) remark that HEIs including universities fail to practice what they preach. For example, higher institutions might lecture on the ills of combustible fuels and fuel wood burning and yet their entire institution is run on dirty energy. Such laxity in the implementation of sustainable behaviour reduces learners' commitment to learning. It also makes learners minimize the potential climate threat and its implications for development.

(c) Socio-Political Hurdles

A significant number of African states and communities continue to grapple with an array of complex sociopolitical challenges and inter-tribal tensions. In the last decade, a minimum of 3 in 10 countries in every region has suffered at least one form of sociopolitical unrest or tribal conflict (Alobo et al., 2018; Leonid, 2013). Sudan underwent a major separation, dividing the country into two distinct nations. Meanwhile, Niger and Gabon have witnessed internal political struggles and transitions of power, often marked by tension and instability (Alobo et al., 2018). These crises and conflicts often force schools to shut down, requiring major adjustments to education systems. In the most extreme cases, youth lives are lost, resulting in the complete loss of educational opportunities for entire communities. Additionally, the prevalence of war-related crimes like kidnapping and armed robbery in these environments makes research inherently risky and deters international funding bodies from engaging in these areas (Arieff and Johnson, 2012; Muhammad, 2019; Osawe, 2015). At worst, HEI activities are stopped for prolonged periods partly due to either combined or independent student and staff strikes (Kaya, 2013). This was the case with the recent ASUU strike in Nigeria where lectures were interrupted for an entire academic calendar, resulting in an imminent break in the learning process.

Dzvimbo et al., (2022) and Mubanga et al., (2022) equally highlight a major challenge faced by most African universities, which is the low engagement from the surrounding rural communities. These communities often find the developed approaches recommended by the universities to be irrelevant to their realities. The situation highlights a possible disconnect between research and society. On the one hand, some authors have attributed this disconnect to the aforementioned data scarcity challenges in Africa. Following this premise, the unavailability of data that reflect community realities has led to the development of less desirable adaptation practices. Whereas other authors suggest that such disconnects are tied to the perceived Eurocentricity of African climate programs. Proponents of this view argue that climate research and CCE programs in Africa are conceived using a top–bottom approach which fails to capture local realities. In fact, their premise suggests that every funded research is based on a theme that fits the main vision of the (European) funding body (most of which might not fit the African context). Moreover, empiricism (the royal road to scientific research) also turns to disregarding other peculiarities in African cultures while promoting more empirically explained concepts. Meanwhile, these peculiarities of cultural factors have historically played key roles in the adaptation and survival of rural communities. Interestingly, Universities exist within communities with well-defined constructs which guide communal activities and livelihoods for many generations. Hence integrating new concepts (especially those which are perceived as "far off" from the local practices) could spark a loss of interest in the adoption of the designed mitigation and adaptation solutions. These sparks generally reduce the penetration of new knowledge in many African communities. And even though the study of Indigenous Knowledge Systems (IKS-local practices inherent to African community life) is fast gaining steam, its rate of development remains constraint by a lack of a consensus among IKS researchers and a lack of funding to mainstream the approaches learned.

By the later assertions and inferring from Tables 17.1 and 17.2, it is difficult to ignore the implication of the huge European funding that has permitted the establishment of the current CCE. Over 40% of the offered mainstream courses are funded by the German federal ministry of economic cooperation and development or the German academic exchange service. Additionally, all the short course programs and even associated free E-resources (in Table 17.2) are funded by a European agency. While some authors may argue that such funding has permitted African institutions to increase climate change knowledge and research despite the challenges of the African Higher Education ecosystem, others still argue that the constraints attached to this funding account in part for the reduced diffusion rates of CCE in Africa. According to this group, the fact that students are trained in a purely academic style usually constricts their understanding of the climate change problem whilst CC is ever-evolving. Not excluding the fact that purely academic programs exclude many brilliant youths who cannot either afford (the academic and financial requirements for) entry into the university. Such youths are also affected by climate change and their participation equally matters but this cannot be possible without a commensurate education.

Consequently, progress in Africa is observed to be less than that observed in other regions which have similar development realities. By intuition, it is fair to say that the development of more effective CCE programs may not be achieved without the support of external funding agencies but a purely Eurocentric design

Table 17.2 Catalogue of freely accessible United Nations and other African HEI accessible eLearning platforms

S/n	Course	Course type	Source (Accessed: March 10, 2024)
1	EO AFRICA Space academy	Practical training in the use and management of Earth Observation data to infer climate change-related occurrences	Space Academy – EO AFRICA (eoafrica-rd.org)
2	UNIVERSITY MOHAMMED VI POLYTECHNIC (UM6P) Morocco African Urban economic development in the context of climate change	Introductory Lecture series to urbanization and climate change	African Urban Economic Development in the Context of Climate Change I African Cities Lab
3	Implementation of national voluntary carbon footprint programmes (NVCFPs)	Theoretical course to guide participants in the different stages of design and implementation, costs, financing mechanisms and benefits of a voluntary national program	Courses – Learning for Nature
4	UNEP MOOC on Nature-based solutions for disaster and climate resilience	Theoretical guide to understanding and applying nature-based solutions that restore or protect natural or modified ecosystems and biodiversity	Massive Open Online Courses I UNEP - UN Environnent Programme
5	UNEP Disasters and ecosystems: resilience in a changing climate MOOC 2	Theoretical guide to interlinkages between ecosystems, disaster risk reduction and climate change adaptation	Disasters and Ecosystems: Resilience in a Changing Climate MOOC 2 I UNEP - UN Environment Programme
6	UNCC: Learn online course catalogue	Theoretical guides to various aspects of climate change, adaptation, mitigation and sustainable development	Home I One UN Climate Change Learning Partnership (uncclearn.org)
7	Climate change, peace and security	Guide to the interlinkages between climate, peace and security	adelphi and UN organisations develop self-paced course on climate change, peace and security I Climate-Diplomacy
8	United Nations food and agricultural organizations (FAO) e-leaning catalogue	Wide range of e-learning courses ranging from institutional, social and economic climate change-related modules	Learning I Integrating Agriculture in National Adaptation Plans (NAPs) I Food and Agriculture Organization of the United Nations (fao.org)

(continued)

Table 17.2 (continued)

S/n	Course	Course type	Source (Accessed: March 10, 2024)
9	Global research alliance e-learning catalogue	Wide range of e-learning courses ranging from institutional, social and economic climate change-related modules	E-Learning I Global Research Alliance
10	International Institution for Biosaline Agriculture	Training on climate change and agriculture for the Middle East, North Africa and other Arid regions of the world	E-learning I International Center for Biosaline Agriculture
11	The Moroccan digital academy [UM6P] and coursera data science and analysis in python and R	Training in data curation, management and big data inferencing with case studies on climate change	Data Science I Coursera
12	Data science and analysis in python and R	Training in data curation, management and big data inferencing with case studies on climate change	Get Started I DataCamp

Source Authors' computation from web-accessible African HEI CCE program websites and United Nation-related eLearning sites (UNFCC, UNEP, UNESCO, FAO), 2024

may not suit African nations the case for Africa especially given its unique socio-economic and political constructs. Rather CCE may become better if climate scientist manages to integrate IKS (the Indigenous perspective: technics, views, and strategies) into the current teaching-based and short-course approaches. The reason is that CCE programs are fundamentally designed to increase climate consciousness and action (this type of knowledge is available in the mainstream and short course programs). More precisely, the learner is trained to be apt at solving local climate problems whereas rural communities are peculiar and face difficulties in accepting new approaches. This bias in rural communities could be easily addressed through an understanding and design of mitigation and adaptation strategies from the Indigenous perspective. This is only possible through an understanding of IKS and a constant interaction of learners with the local communities.

Prospects for an Impactful Climate Change Education in African Higher Education Institutions

The CCE concept is a new concept which is directly and indirectly linked to many other sustainable development initiatives (Molthan-Hill et al., 2022). In this light, it will turn to share benefits and support from previously established initiatives and agendas but most importantly it will help in concretizing the sustainability agenda. While there are many obstacles to its dissemination across Africa, it appears there exist some semblance of opportunities within the current HEI ecosystem. These advantages are broadly categorized under the emergence of digital technologies (Digitalization, communication and computation which is linked to emerging eLearning opportunities) and Institutional improvements.

A. Emergence of Digital Technologies

   I. Digitalization, Communication and Computation

   Though initially a little unexpected, it is highly likely that the explosion of digital telecommunication technologies may open huge windows of opportunity for the development of future CCE. Ideally, the expansion of digital communication devices, services and tools provides three benefits to African institutions (Nuutinen and Leal Filho, 2018; Hassall, 2011). Firstly, affordable digital communication tools and devices increase the chances for the development of institutional digital infrastructures. Such infrastructure includes institutional servers, and data centres which have the capacity to achieve regional-specific research for future collaborations and explorations. Besides such centres may reduce the cost of acquisition of publicly available data while providing access to computational tools and automated software. These aspects will not only increase data availability but could potentially enhance precision in climate measurement. The combination of these aspects will generate renewed interest in the development of more hands-on approaches for CCE. Rousell and Cutter-Mackenzie-Knowles (2019) asserts that the introduction of novel technologies may usher in a universe where youths can engage with climate change in ways that are regionally significant.

   Apart from the ease of data acquisition and computation, the emergence of more affordable digital tools will also permit learners in African HEIs to access many available online courses. Bearing in mind that many basic concepts of climate change are currently available online, it holds that access to these online courses may reduce the burden associated with developing new course material. Consequently, many instructors may have more time to carryout research in areas that may help in the development of more sustainable mitigation and adaptation strategies for the future. This may involve intensifying research in promising areas like IKS. Perhaps this could also accelerate the rate of development and institutionalization of such concepts. The prospects of satellite communication also provide the possibility for research scientists in Africa to adopt existing experimented interactive teaching frameworks. Fransman et al., (2018) suggest the possibility of marked improvements in learners' skills when exposed to interactive software and games. In effect, Rooney-Varga et al., (2018), highlight that access to affordable internet and digital communication devices could permit African youths access to achieve an 81% higher learning effectiveness especially when compared to the current class-based methods. Finally, digital communications software (WhatsApp, Facebook, twitter(X),

Research Gate and LinkedIn) may increase the development of extended research networks. Such networks could have the potential of fertilizing interdisciplinarity across Africa while enabling the clustering of likeminded researchers in climate change and other fields. Social media platforms have also served as interesting media to encourage climate interest and engagement among youths. In any case, it is evident that the more internet services and digital communication tools become affordable, the more Africa HEIs and communities will potentially increase their commitment and interest in CCE initiatives.

II. Emerging E-Learning and E-Libraries

As highlighted above, the available of affordable digital communication tools and services has driven the development of many online courses. Most of these classes are funded by international institutions and cover topics which previously posed challenges to instructors in African higher institutions. Table 17.2 highlights an inexhaustible catalogue of freely available online courses that touch on aspects of climate change. In the first column, we highlight the institutions and associated courses. Since many of these courses are not widely known, we decided to provide a highlight of the main course objectives in the second column and finally in the third column, we attached a hyperlink to enable further research in its regards. From the table, observe that many international institutions have consistently developed open courses to enable an understanding of climate and its implications for regional realities (Haslett & Wallen, 2011). These courses cover regular subjects on the environment and sustainability but also extend to novel fields like data science, data analysis, remote sensing, and machine learning in association with climatic data. Thus, permitting the modelling and internalization of the climate change challenge on an individual basis. Most of these courses are available for free on platforms. In context, from the table observe that the FAO [8], United Nations Environmental program [UNEP 4 and 5], UN climate change [6], the European space agency [1] and other University institutions like UM6P [2 and 11] each have eLearning platforms or an e Library that is filled with interesting context specific materials that can be easily accessed with a click on the internet (Hassall, 2011; Nuutinen and Leal Filho, 2018; Okoye and Okoye, 2016).

En-ROADS is another freely available, transparent, fully documented, embedded in the latest climate science, and calculates a scenario in less than a second which makes it appropriate for use in classrooms (Rayhan, 2022; Ryder et al., 2023). This permits the study of climate energy dynamics through workshops or interactive games. The framework of these eLearning materials could serve as a solid framework for the development of a CCE curriculum that suits the African narrative. In situations of delays in the development of curriculum, these lessons could be used as teaching aids.

We used a keyword search on associated United Nation related environment and climate change eLearning websites such as UNFCC, UNEP, FAO and UNESCO.

These results were complemented by the outcome of our initial search of web-accessible CCE programs in HEIs in Africa to produce a list of freely available digital learning platforms as highlighted in Table 17.2. From our point of view, the highlighted courses provide a wide range of practical and theoretical courses which may aid learners in familiarizing themselves with the climate narrative. Of course, even though internet accessibility is still a challenge in some parts of Africa, it is visible from Table 17.2 that the internet is bringing interesting learning alternatives that might reduce the burden that is currently carried by instructors while permitting them the time required to carry out policy-driven research in the area of climate change. Moreover, some platforms provide freely accessible trial datasets which span all the African continents. Hence actualizing the learning-by-doing approach. In addition, it is necessary to note that most online courses (including those highlighted in the table) are self-paced and readily available to all learners regardless of their nationality or academic standing. These modules are equally adapted to suit the learners' realities and the learners can take them at any time of the year. This additional flexibility makes them suitable for all age groups and for reinforcing other associated materials. Quite unfortunately, this course cannot be easily afforded by African youths partly because of the high cost of internet and communication devices. We however anticipate that the rapid development of less costly digital tools and digital communication infrastructure in HEIs may enable many African youths to take advantage of the wide range of online courses.

B. Institutional Improvements

While it is true that the African Higher Education systems are plagued by many challenges, it remains equally true that some aspects could be improved. For example, the CCE mainstream approach (Bachelor, Masters and Doctoral programs) has excluded many youths who have the potential to contribute positively to developing more sustainable mitigation and adaptation solutions for their countries. To increase opportunities for all irrespective of socio-economic class and educational status, many African nations have resolved to develop innovation hubs for climate change (Molthan-Hill et al., 2022). These hubs are spaces that are designed to promote free thinking and framing of solutions and products that could aid climate change mitigation and adaptation. Innovation hubs are generally either independent within communities or associated with universities and other related institutions that promote critical thinking and problem-solving. In principle, the combination of innovation hubs with universities generally yields the biggest returns. This is because hubs within universities provide a learning and exchange platform where the local stakeholders can interact and co-create sustainable solutions for the community. A recent example occurred when the government of Zimbabwe launched several innovation hubs in Zimbabwean universities aimed at producing more adoptable climate-resilient technologies (Dzvimbo et al., 2022; Mubanga et al., 2022).

Beyond the potential that the innovative hub could have for the development of more CCE programs, other factors have emerged which signal the potential for increased implementation of CCE programs in African HEIs. Notably, Togo & Lotz-SisKita (2013) highlights an increased awareness of the importance of sustainability across African communities and University institutions. This increased awareness may perhaps increase the interest of HEIs to seek CCE courses for learners to address the sustainability needs of their communities. Other HEIs may increase the financial allocation for research in areas of sustainable developed as well as related topics like climate change mitigation and adaptation. An example is UM6P in Morocco where the sustainability culture is reflected in its buildings, courses, and research. The university is highly research-driven with 70% of its research projects incorporating aspects of climate change and environmental sustainability and thus setting the pace for a new Higher Education perspective in Africa. Another good example is Uganda's Martyrs University which has taken significant steps to improve the livelihood of the surrounding communities and has contributed to improving income, food security, water conservation, and sustainable living. Tilbury, (2011) highlights that this initiative has had a huge positive impact on the relationship between the university and neighbouring communities. In some cases, it is likely that the success of such projects usually restores the trust of the local community in HEI solutions, hence, setting the stage for the dissemination of new technics and concepts like CCE.

At a macro level, the increased awareness of climate vulnerability in many African countries has initiated a new set of agreements and regional partnerships. These initiatives are set to assist African states in revamping the research and climate adaptive capacity (the ability of a community to cope despite climate extreme events) of developing communities. For example, the United Nations University (UNU) accredited over 63 Regional Centers of Expertise (RCE) across Africa, Australia, the Asia–Pacific region, Europe, the Middle East, South America, the Caribbean, and North and Central America (Tilbury, 2011). These institutions are dedicated to promoting regional collaboration, knowledge dissemination, and practical action for sustainability. Apparently, the increased interest and engagement from participants in the Mainstreaming Environment and Sustainability in African Universities (MESA) Partnership could largely be attributed to these RCEs (Ogbuigwe, 2009; Togo & Lotz-Sisitka, 2013). Among others, the operationalization of RCEs permits universities to combine scarce resources to address the sustainability imperative. Apart from the RCEs, there exist various agreements which bind states to fulfil their ratified regional commitments. Other special cases like the successful collaboration of WWF and BUSE in Zimbabwe equally assert the benefits of collaborations between academic institutions and non-governmental institutions (Mugambe et al., 2023). However, there are many research centres, initiatives, and programs that have diverse interests including environmental policy, energy, global change, environmental health, and environment (Dzvimbo et al., 2022; Mubanga et al., 2022; Rousell and Cutter-Mackenzie-Knowles, 2019).

Beyond the resource mobilization aspects, there are two benefits of most climate-related agreements. Firstly, they establish regional and subregional baselines from which progress can be measured in the future. Secondly, agreements constrain countries and regions to contribute quotas or meet various standards under strict time frames thereby limiting bureaucracy. By extension, the establishment of climate agreements may intrinsically suggest the prospects of probable improvements in the area. We build on data from the UNESCO, 2023 Global Education Monitoring

Report, Apollo and Mbah (2021), web-accessible HEI databases, and extended web search on various United nation environment and climate-specific initiatives (UNFCC, UNEP, UNITAR, UNESCO) to curate a catalogue.We consider the potential of regional and subregional agreement and the relative absence of reference materials on aspects of CCE (Table 17.3). The catalogue (Table 17.3) highlights a list of CCE-related agreements which are likely to increase future regional implementation of CCE programs. To increase the general understanding of these programs, Table 17.3 is broken into four sections including the precise name of each initiative, and the country or subregions which are directly involved (in columns one and two). In the third column, we highlight the objectives of these programs and in the fourth, we provide a reference link to guide further research.

This table adds to the work of Apollo and Mbah (2021) which had previously highlighted a list of national institutional engagements in West African states. The table covers agreements that cut across countries in other subregions (North, East, South and Central) and Africa. From Table 17.3, we count 12 initiatives which have been advanced to enhance access to CCE education in African states. These agreements are grouped into three categories including regional financial commitments, regulatory (regional consulting) agreements and subregional strategies that are tailored to fight climate change. Financial commitments include agreements (highlighted in italic on the table [10, 11, 12, 14, 15, 17]) that will most likely be created to increase regional access to climate finance. The presence of these finances may increase government investments in HEIs research and perhaps also research in critical aspects like climate change. Regulatory agreements on the other hand refer to strategic commitments that foster regional investments (highlighted in bold [1, 3]), strategic initiatives which foster capacity building (still in bold, [4, 5, 7, 19, 20]) and subregional collaboration among climate scientists [2, 8, 9].

Notably, we highlight the Annual Action for Climate Change Empowerment Dialogue and the Local Communities and Indigenous People Platform (LCIPP) of the United Nations. The initiative was strategically kick-started to accelerate the creation of relevant Afrocentric knowledge for CCE. The LCIPP is the first-ever platform which collects local experiences across the region to create shared values which can be harnessed for a tailored and effective CCE in Africa. Such platforms may renew interest in IKSs while contributing to the acceleration of research and consensus on an acceptable African protocol for integrating IKS. This consensus could imply a first step towards developing CCE programs which reflect local realities and generate applicable mitigation and adaptation solutions. Subregional strategies (highlighted in white on the table) include strategic plans and commitments to increasing adaptive capacities. Sub-regional strategies vary from national adaptation plans (as is the case for Namibia [16], Côte d'Ivoire [18] and Ethiopia [13]) to regional adaptation plans (as is the case with the subregion of South Africa [6]).

Regional strategies play key roles in driving the implementation of new concepts. Arguably, the SADC-ESD regional strategic framework has permitted South Africa to implement an innovative approach to climate change education. South Africa is visibly the only region that has successfully integrated Indigenous knowledge across its academic CCE programs, including specialized masters, undergraduate

Table 17.3 Catalogue of regional CCE initiatives in Africa

s/n	Initiative	Country	Institutions objective	Source (Accessed: March 10, 2024)
1	Regional declaration on climate education	West Africa	This highlights the commitment of some West African nations, including Burkina Faso, Cote d'Ivoire, Senegal, Togo, and Guinea to increase funding for comprehensive climate literacy efforts in the region	West African nations signed a Regional Declaration on Climate Change Education I UNITAR
2	Mission du Conseil Economique, Social et Environnemental	20 African states	Set up to assist African nations in their bid for climate financing from other international institutions. They also strengthen research cooperation between climate scientists across Africa	Conseil Economique Social et Environnemental du Royaume du Maroc (cese.ma)
3	Comité inter-état de lutte contre la sècheresse dans le sahel	13 African Nations mostly located in the Sahel	Promoting food security and ecological balance in the Sahel involves strategic formulation, scientific cooperation, information dissemination, capacity-building, and program implementation	Bienvenue sur le Portail Web du CILSS

(continued)

Table 17.3 (continued)

s/n	Initiative	Country	Institutions objective	Source (Accessed: March 10, 2024)
4	Regional universities forum for capacity building in agriculture (RUFORUM) "Transforming African agricultural universities to meaningfully contribute to Africa's growth and development" (TAGDev) Program	Africa	The TAGDev Program, implemented by RUFORUM in collaboration with The MasterCard Foundation, aims to transform African agricultural universities and their graduates. This transformation involves enhancing the application of science, technology, business, and innovation to better address developmental challenges and promote rural agricultural transformation in Africa	TAGDev Flagship I RUFORUM
5	UNESCO greening Education partnership	Global	The Greening Education Partnership is a global initiative that prepares learners for climate action by emphasizing the critical role of education	Greening Education Partnership I UNESCO
6	South African Development Education for Sustainable Development program [(SADC I ESD) Regional Strategic Framework]	South Africa Region	The goal of this program is to create and implement programs that enhance human capabilities across various domains, including education, health, poverty eradication, employment, food security, environmental management, and gender equality	SADC Education For Sustainable Development Strategic Framework I SADC

(continued)

Table 17.3 (continued)

s/n	Initiative	Country	Institutions objective	Source (Accessed: March 10, 2024)
7	UN action for climate change empowerment (ACE)	Global	According to UNFCC, ACE aims to empower everyone for climate action through education, awareness, training, participation, information access, and international cooperation	Action for Climate Empowerment I UNFCCC
8	UN local communities and indigenous people platform (LCIPP)	Global	LCIPP is an open and inclusive space that brings together people and their knowledge systems to build a climate-resilient world for all	Homepage I Local Communities and Indigenous Peoples Platform (unfccc.int)
9	UN Annual (ACE) empowerment dialogue	Global	The platform provides a regular forum for Parties and other stakeholders to share their experiences, and exchange ideas, good practices and lessons learned regarding the ACE framework	ACE Dialogue 2023 - Day 1 I UNFCCC
10	African Development Bank (AfDB) and the Global Center for Adaptation (GCA)- Youth Adapt Challenge	Africa	The challenge seeks to boost sustainable job creation through support for entrepreneurship and youth-led innovation in climate change adaptation and resilience across Africa	Africa Adaptation Acceleration Program I African Development Bank Group (afdb.org)

(continued)

Table 17.3 (continued)

s/n	Initiative	Country	Institutions objective	Source (Accessed: March 10, 2024)
11	African Development Bank Climate Action Window	African	The AfDB set up this project to assist financial mobilization of climate funds by African nations. The resources mobilized under this dedicated Window will be allocated through three sub-windows: 75% will finance Adaptation action in line with Africa's most urgent priorities 15% will support Mitigation, and 10% will be used to deliver Technical Assistance	Climate Action Window I African Development Bank Group (afdb.org)
12	Africa Climate Fund	Africa	Since 2014, this AfDB program has supported over 26 African countries in developing capacities to access international climate finance and in the implementation of small-scale adaptation projects to enhance their resilience to the impacts of climate change	Africa Climate Change Fund I African Development Bank Group (afdb.org)
13	The Climate Change Education Strategy (2017–2030)	Ethiopia	US$2 million to fund the development of climate change materials for primary schools, refresher training for schoolteachers, and monitoring and evaluation of the strategy	Ethiopia Launches the Implementation of its National Climate Change Education Strategy and Priority Actions – Knowledge Sharing Platform (uncclearn.org)

(continued)

Table 17.3 (continued)

s/n	Initiative	Country	Institutions objective	Source (Accessed: March 10, 2024)
14	Funds for CC awareness in general education	Zambia	Allocating (US$200,000) for climate change awareness in general education and (US$200,000) for Higher Education from 2021 to 2024	Education in Zambia | Global Partnership for Education
15	A climate change education plan	Cabo Verde	Implementing a climate education plan from 2022 to 2028, estimated at approximately US$1,500,000 with the national adaptation plan	NAP_Cabo Verde_EN.pdf (unfccc.int)
16	The Namibian Communication, Education & Public Awareness Strategy (2019–2030)	Namibia	The strategy aims to ensure 75% of key groups understand climate change adaptation and mitigation by 2030	CEPA Strategy MET - final - 13 05 2019.pdf (cbd.int)
17	The national climate change learning strategy (2020)	Zimbabwe	The National Climate Change Learning Strategy (2020) has allocated an estimated budget of US$120,000 for this initiative. Above others, the educational system in Zimbabwe evaluates the level of understanding of climate-related study modules	Zimbabwe Has Launched its National Climate Change Learning Strategy, Cementing the Country's Commitment to Advancing Climate Literacy. – Knowledge Sharing Platform (uncclearn.org)
18	National climate change learning strategy	Côte d'Ivoire	Côte d'Ivoire, the Ministry of Environment and Sustainable Development, through the National Climate Change Program, organized a public validation workshop of the National Learning Strategy on Climate Change 2022–2026	afdb_cote_divoire_final_2018_english.pdf

(continued)

Table 17.3 (continued)

s/n	Initiative	Country	Institutions objective	Source (Accessed: March 10, 2024)
19	The Moroccan Eco-Schools program	Morocco	Mohammed VI Foundation and the Ministry of Education launched this project with the aim of improving education for environmental conservation and sustainable development	Eco-Schools – La Fondation Mohammed VI pour la Protection de l'Environnement (fm6e.org)
20	African Association for Green Cities (AVIVE) initiated the Environmental Education project,	Cameroon	The project promotes the development of school and community botanical gardens, and environmental education concepts	Environmental Education in Cameroon I Climate Chance (climate-chance.org)

Source Computed from UNESCO Global Education Monitoring Report 2023 and Authors' search of web-accessible (implemented programs that have application websites) CCE programs in HEIs of each subregion in Africa, 2024

courses, and doctoral dissertations. This is suggestive of the power of institutional collaboration. It equally suggests the huge possibilities that exist for the dissemination of CCE in African HEIs especially with the right institutional support. In hindsight, intensifying the adoption of CCE in African HEI needs a mix of strategies including the creation of relatable knowledge which is complemented by diverse forms of intrinsic learning motivations. Such motivations may vary with socio-demographics and local context, but the objectives remain the same.

Conclusion

This study examines the current challenges and future opportunities that could contribute to the development and integration of climate change education (CCE) into African Higher Education institutions (HEIs). With a keen focus on current challenges and the implication of future opportunities, the study identified three main constraints to CCE implementation in HEIs including resource limitations, institutional barriers, and socio-political constraints. These factors either indirectly affect the allocation of critical resources or directly affect the daily operations of HEIs. In many instances, the result usually leads to reduced interest of instructors to commit to implementing CCE programs. Conversely, we highlight two major areas for enhancing CCE dissemination: affordable digitalization communication and computation and institutional support mechanisms. These factors generally seek to reduce the burden on instructors while permitting them the time to carry out research that could enrich the development of more suitable CCE programs. Based on these findings, we conclude that the rapid implementation of CCE programs requires a mix of strong subregional institutions, human capital development, increased investment in climate change-related research and access to affordable digital communication tools and devices. Such frameworks could aid African climate scientists in building a consensus and developing culturally relevant approaches. Additionally, strengthening regional research networks can foster collaboration and knowledge sharing. These efforts could pave the way for an "African model" of climate change education in universities, tailored to address local realities. African HEIs can further leverage existing resources like climate data libraries to enhance education for both staff and students.

Acknowledgements The authors are grateful to Christian Shey Ndogmi Yoniwo and Francis Ebai Ndip for their valuable comments which helped in improving this work.

References

Agbedahin, A. V. (2019). Sustainable development, education for sustainable development, and the 2030 Agenda for sustainable development: emergence, efficacy, eminence, and future. *Sustainable Development, 27*(4), 669–680. https://doi.org/10.1002/sd.1931

Alber, B. (2018). *A Lesson in Climate Change Education: Examining how Climate Change is taught inthe Nova Scotia Public School Curriculum* (Doctoral dissertation).

Aleixo AM, Leal S, Azeiteiro UM. (2016). Conceptualization of sustainable higher education institutions, roles, barriers, and challenges for sustainability: An exploratory study in Portugal. *Journal of Cleaner Production.* https://doi.org/10.1016/j.jclepro.2016.11.010

Alobo, E. E., Niebebu, M., & Sampson, E. (2018). *Uncovering the Bond between Colonialism and Conflict: Perspective of the Causes, Cases and Consequences of Territorial Disputes in Africa.* Globeedu Group.

Apollo, A., Mbah, M.F. (2021). Challenges and opportunities for climate change education (CCE) in East Africa: A critical review. *Climate, 9*, 93. https://doi.org/10.3390/cli9060093

Arieff, A., & Johnson, K. (2012). Crisis in Mali. Congressional Research Service. Retrieved from https://www.refworld.org/pdfid/506c05282.pdf

Arieff, A., & Johnson, K. (2019). Crisis in Mali.

Conway, D., & Schipper, E. L. F. (2011). Adaptation to climate change in Africa: Challenges and opportunities identified from Ethiopia. *Global Environmental Change, 21*(1), 227–237. https://doi.org/10.1016/j.gloenvcha.2010.07.013

Damoah, B. (2023). Reinvigorating climate change education in universities a social transformative Agenda. *Environmental Science & Sustainable Development*, 19–26. https://doi.org/10.21625/essd.v8i4.1013

Dzvimbo, M. A., Mashizha, T., Zhanda, K., & Mawonde, A. (2022). Promoting sustainable development goals: Role of higher education institutions in climate and disaster management in Zimbabwe. *Jàmbá: Journal of Disaster Risk Studies*, *14*(1), a1206. https://doi.org/10.4102/jamba.v14i1.1206

Filho, W. L., Azeiteiro, U., Alves, F., Pace, P., Mifsud, M., Brandli, L., Caeiro, S. S., & Disterheft, A. (2018). Reinvigorating the sustainable development research agenda: The role of the sustainable development goals (SDG). *International Journal of Sustainable Development & World Ecology, 25*(2), 131–142. https://doi.org/10.1080/13504509.2017.1342103

Fransman, A., Richter, B., & Raath, S. (2018). An interactive computer program for South African urban primary school children to learn about traffic signs and rules. *African safety promotion, 16*(1), 57–67. https://hdl.handle.net/10520/EJC-173a3342bd

Goyal, N., Tripathy, M., Singh, V. Sharma G. P. (2023). Transformative potential of higher education institutions in fostering sustainable development in India. *Anthropocene Science 2*, 112–122. https://doi.org/10.1007/s44177-023-00061-5

Haslett, S. K., & Wallen, J. (2011). A component-based approach to open educational resources in climate change education. *Planet, 24*(1), 89–92. https://doi.org/10.11120/plan.2011.00240089

Hassall, G. (2011). Proof of concept: using search technologies to enhance teaching public policy issues facing small developing states.

Hickman, C., Marks, E., Pihkala, P., Clayton, S., Lewandowski, R. E., Mayall, E. E., Wray, B., Mellor, C., & van Susteren, L. (2021). Climate anxiety in children and young people and their beliefs about government responses to climate change: A global survey. *The Lancet Planetary Health, 5*(12), e863–e873.

Kagawa, F., & Selby, D. (2012). Ready for the storm: Education for disaster risk reduction and climate change adaptation and mitigation1. *Journal of Education for Sustainable Development, 6*(2), 207–217. https://doi.org/10.1177/0973408212475200

Kapitulčinová, D., AtKisson, A., Perdue, J., Will, M. (2017). Towards integrated sustainability in higher education—Mapping the use of the accelerator toolset in all dimensions of university practice. *Journal of Cleaner Production*. https://doi.org/10.1016/j.jclepro.2017.05.050

Kaya, H. O. (2013). Integration of African indigenous knowledge systems into higher education in South Africa: Prospects and challenges. *Alternation, 20*(1), 135–153.

Leal Filho, W., Pallant, E., Enete, A., Richter, B., & Brandli, L. L. (2018a). Planning and implementing sustainability in Higher Education institutions: An overview of the difficulties and potentials. *International Journal of Sustainable Development & World Ecology, 25*(8), 713–721. https://doi.org/10.1080/13504509.2018.1461707

Leal Filho, W., Raath, S., Lazzarini, B., Vargas, V. R., de Souza, L., Anholon, R., & Orlovic, V. L. (2018b). The role of transformation in learning and education for sustainability. *Journal of Cleaner Production, 199*, 286–295. https://doi.org/10.1016/j.jclepro.2018.07.017

Leonid, G. (2013). State and socio-political crises in the process of modernization. *Social Evolution & History, 12*(2), 35–76.

Maiya, A. K., & Aithal, P. S., (2023). A review based research topic identification on how to improve the quality services of higher education institutions in academic, administrative, and research areas. *International Journal of Management, Technology, and Social Sciences (IJMTS), 8*(3), 103–153. Available at SSRN: https://ssrn.com/abstract=4575687

Manteaw, O. O. (2012). Education for sustainable development in Africa: The search for pedagogical logic. *International Journal of Educational Development, 32*(3), 376–383. https://doi.org/10.1016/j.ijedudev.2011.08.005

Mbah, M. F., Shingruf, A., & Molthan-Hill, P. (2022). Policies and practices of climate change education in South Asia: Towards a support framework for an impactful climate change adaptation. *Climate Action, 1*(1), 1–18. https://doi.org/10.1007/s44168-022-00028-z

Molthan-Hill, P., Blaj-Ward, L., Mbah, M. F., & Ledley, T. S. (2022). Climate change education at universities: Relevance and strategies for every discipline. *In Handbook of Climate Change Mitigation and Adaptation* (pp. 3395–3457). Cham: Springer International Publishing. https://doi.org/10.1007/978-3-030-72579-2_153

Mubanga, K. H., Mazyopa, K., Chirwa, B., Musonda-Mubanga, A., & Kayumba, R. (2022). Trained climate change educators: Are secondary school pupils getting quality climate change education? Views from teachers and pupils in Lusaka, Zambia. *European Journal of Development Studies, 2*(4), 14–23. https://doi.org/10.24018/ejdevelop.2022.2.4.127

Mugambe, R. K., Ssekamatte, T., Isunju, J. B., et al. (2023). Facilitators and barriers to the utilisation of sanitation-related decision-making support tools among environmental health practitioners in Uganda. *J Public Health (Berl.)*. https://doi.org/10.1007/s10389-023-02087-w

Muhammad, A. A. (2019). Terrorism as a threat for economic development in Nigeria. *IJUS| International Journal of Umranic Studies, 2*(1), 55–66.

Masahisa Sato, Yuto Kitamura (2023). Current status of climate change education and suggestions for its integrative development in Japan, IATSS Research, Volume 47, Issue 2, 2023, Pages 263-269, ISSN 0386-1112. https://doi.org/10.1016/j.iatssr.2023.04.002

Nuutinen, M., Filho, W. L. (2018). Online communities of practice empowering members to realize climate-smart agriculture in developing countries. In Azeiteiro, U., Leal Filho, W., Aires, L. (Eds.), *Climate Literacy and innovations in climate change education. Climate change management*. Cham: Springer. https://doi.org/10.1007/978-3-319-70199-8_5

Ogbuigwe, A. (2009). The possibility generation: Empowering students in the mainstreaming environment and sustainability in African Universities Partnership Programme. Young people, education, and sustainable development: Exploring principles, perspectives, and praxis, 143. https://doi.org/10.3920/978-90-8686-691-5

Okoye, N. J. C., & Okoye, M. O. (2016). Information delivery in developing countries: current status and observance of database agreements in academic libraries.

Osawe, C. O. (2015). Increase wave of violent crime and insecurity: A threat to socio-economic development in Nigeria. *Journal of Humanities and Social Science, 20*(1), 123–133.

Rayhan, M. (2022). *Climate change and sustainable development–Public awareness measure with climate simulator En-ROADS*. Master's thesis, University of South-Eastern Norway. https://hdl.handle.net/11250/3004775

Reid, A. (2019). Climate change education and research: Possibilities and potentials versus problems and perils? *Environmental Education Research, 25*(6), 767–790. https://doi.org/10.1080/13504622.2019.1664075

Richter, B. W., & De Sousa, L. O. (2019). The implementation of environmental education to promote sustainability: An overview of the processes and challenges. *International Journal of Sustainable Development & World Ecology, 26*(8), 721–731. https://doi.org/10.1080/13504509.2019.1672220

Rooney-Varga, J. N., Sterman, J. D., Fracassi, E., Franck, T., Kapmeier, F., Kurker, V., & Rath, K. (2018). Combining role-play with interactive simulation to motivate informed climate action: Evidence from the world climate simulation. *PLoS ONE, 13*(8), e0202877. https://doi.org/10.1371/journal.pone.0202877

Rousell, D., & Cutter-Mackenzie-Knowles, A. (2020). A systematic review of climate change education: giving children and young people a 'voice'and a 'hand'in redressing climate change. *Children's Geographies, 18*(2), 191–208. https://doi.org/10.1080/14733285.2019.1614532

Ryder, M., Evro, S., Brown, C., & Tomomewo, O. S. (2023). Multi-model approach of global energy model validation: Times and EN-ROADS models. *American Journal of Energy Research, 11*(2), 63–81.

Sehai, E., & Lemma, T. (2008). Forging partnership to enhance the relevance of Ethiopian graduate schools research in agriculture: Report on stakeholder workshop, Hawassa University, Ethiopia, 23–24.

Stevenson, K., and Peterson, N. (2016). Motivating action through fostering climate change hope and concern and avoiding despair among adolescents. *Sustainability, 8*(1), 6. https://www.mdpi.com/2071-1050/8/1/6

Tilbury, D. (1995). Environmental education for sustainability: Defining the new focus of environmental education in the 1990s. *Environmental Education Research, 1*(2), 195–212. https://doi.org/10.1080/1350462950010206

Tilbury, D. (2011). Higher Education for sustainability: A global overview of commitment and progress. *Higher Education in the World, 4*(1), 18–28.

Togo, M., Lotz-Sisitka, H. (2013). The unit-based sustainability assessment tool and its use in the UNEP mainstreaming environment and sustainability in African Universities Partnership. In Caeiro, S., Filho, W., Jabbour, C., Azeiteiro, U. (Eds.), *Sustainability assessment tools in higher education institutions*. Cham: Springer. https://doi.org/10.1007/978-3-319-02375-5_15

UNESCO (2021). Global Education Monitoring Report 2021/2: Non-state Actors in Education: Who Chooses? Who Loses?. UNESCO. Retrieved from https://gem-report-2021.unesco.org/
UNESCO (2023). Global Education Monitoring Report 2023: Climate change communication and education country profiles: Approaches to greening education around the world. https://doi.org/10.54676/XBVG6945
World Bank. (2010). Financing higher education in Africa. The World Bank.

Correction to: Practices, Perceptions and Prospects for Climate Change Education in Africa

Marcellus Forh Mbah, Petra Molthan-Hill, and Ernest L. Molua

Correction to:
M. F. Mbah et al. (eds.), *Practices, Perceptions and Prospects for Climate Change Education in Africa*, https://doi.org/10.1007/978-3-031-84081-4

The original version of the book was inadvertently published with minor errors in the Foreword, Chapters 4, 9, 11, 12, 14 and 17 which have now been corrected. The book and the chapters have been updated with the changes.

The updated version of this book can be found at
https://doi.org/10.1007/978-3-031-84081-4

M. F. Mbah et al. (eds.), *Practices, Perceptions and Prospects for Climate Change Education in Africa*, https://doi.org/10.1007/978-3-031-84081-4_18